Gene Regulation

Fifth Edition

D0084290

Gene Regulation

A eukaryotic perspective

Fifth Edition

David S Latchman

Master of Birkbeck, University of London
Professor of Genetics, Birkbeck
and Dean, Institute of Child Health,
University College London

Taylor & Francis Group

Published by:
Taylor & Francis Group

In US: 270 Madison Avenue
          New York, NY 10016

In UK: 4 Park Square, Milton Park
          Abingdon, OX14 4RN

© 2005 by Taylor & Francis Group

First published 1990; Fifth edition published 2005

ISBN: 0-4153-6510-4

This book contains information obtained from authentic and highly regarded sources. Reprinted material is quoted with permission, and sources are indicated. A wide variety of references are listed. Reasonable efforts have been made to publish reliable data and information, but the author and the publisher cannot assume responsibility for the validity of all materials or for the consequences of their use.

All rights reserved. No part of this book may be reprinted, reproduced, transmitted, or utilized in any form by any electronic, mechanical, or other means, now known or hereafter invented, including photocopying, microfilming, and recording, or in any information storage or retrieval system, without written permission from the publishers.

A catalog record for this book is available from the British Library.

Library of Congress Cataloging-in-Publication Data

Latchman, David S.
  Gene regulation : a eukaryotic perspective / Davis S. Latchman.-- 5th ed.
    p. cm.
  Includes bibliographical references and index.
  ISBN 0-415-36510-4
  1. Genetic regulation. 2. Gene expression. 3. Eukaryotic cells. 4. Genetic transcription. I. Title.
  QH450.2.L37 2006
  572.8'65--dc22
                                                        2005009966

Editor: Elizabeth Owen
Editorial Assistant: Chris Dixon
Production Editor: Karin Henderson
Typeset by: Phoenix Photosetting, Chatham, Kent, UK
Printed by: Cromwell Press, Trowbridge, Wiltshire, UK

Printed on acid-free paper

10 9 8 7 6 5 4 3 2 1

Taylor & Francis Group
is the Academic Division of T&F Informa plc.   Visit our web site at http://www.garlandscience.com

To my mother and in memory of my father

CONTENTS

Abbreviations

AEV	avian erythroblastosis virus
ALV	avian leukosis virus
AP	activator protein
APC	adenomatous polyposis coli
bp	base pairs
CBP	CREB binding protein
CGRP	calcitonin-gene related peptide
ChIP	chromatin immunoprecipitation
CPSF	cleavage and polyadenylation specificity factor
CstF	cleavage stimulation factor
CTD	C-terminal domain
eEF	eukaryotic elongation factor
eIF	eukaryotic initiation factor
GRE	glucocorticoid response element
HAT	histone acetyl transferase
HIV	human immunodeficiency virus
hnRNP	heterogeneous nuclear RNP
HSE	heat-shock element
HSF	heat-shock transcription factor
HSV	herpes simplex virus
ICR	imprinting control region
IGF	insulin-like growth factor
IRE	iron response element
IRES	internal ribosome entry site
LAP	liver activator protein
LCR	locus control region
LIP	liver inhibitor protein
LTR	long terminal repeat
MALDI	matrix-assisted laser desorption/ionization
MAR	matrix attachment regions
NES	nuclear export signal
NURF	nucleosome remodeling factor
PML	promyelocytic leukemia
PPARγ	peroxisome proliferator-activated receptor gamma
RAR	retinoic aid receptor
Rb	retinoblastoma
RNP	ribonucleoprotein
RSV	Rous Sarcoma virus
SDS	sodium dodecyl sulfate
snRNP	small nuclear ribonucleoprotein
SUMO	small ubiquitin-related modifier
TAF	TBP-associated factors
TBP	TATA binding protein
tRNA	transfer RNA
U2AF	U2 accessory factor
UBF	upstream binding factor

Preface to the fifth edition

As in previous editions, the aim of this work is to provide an up-to-date and comprehensive account of the processes involved in gene expression and its regulation, as well as an introduction to the methods by which these processes are studied. To assist in fulfilling these aims, two major changes in format have been introduced for this edition. Firstly, each chapter now begins with a summary of the key points of the chapter. This is intended to allow students (as well as other more senior readers!) to obtain a rapid overview of the main points to which they can refer during their reading of the chapter. Second, the major methods used to study gene regulation are now described in separate methods boxes inserted at the appropriate point in the book. This has the benefit of allowing the reader to refer to these boxes at any point in the book where the method is mentioned and also aids the flow of the main part of the text by removing detailed descriptions of methodology.

As well as these stylistic changes, the process of updating the book to reflect extensive new advances has continued in this edition, as in previous editions. Apart from general updating, the sections dealing with aspects of gene regulation, where particularly extensive progress has been made, have been extensively revised. These include the discussion of histone modifications in the control of chromatin structure where each of the individual modifications now merits a separate sub-section of Chapter 6. Similarly, the interaction between transcription and post-transcriptional processes, which was added as a new section in the Fourth Edition, now merits considerably more extensive discussion (see particularly Chapters 3 and 4) to reflect our understanding of its increasing importance. In addition, a new section (Section 5.7) has been added discussing the exciting and novel work on the inhibition of gene expression by small RNA molecules.

As well as these changes in the account of basic processes in gene regulation, Chapter 9 on gene regulation and human disease has been extensively modified. Although this chapter still stresses the important role of abnormalities in gene regulation which occur in human cancer, there is now a much more extensive discussion of the role of aberrant gene regulation in other diseases, reflecting our increasing knowledge of this topic. Similarly, as our understanding of this area increases, it has become increasingly clear that it offers therapeutic possibilities for the improved treatment of human disease using methods which modulate gene expression. This topic is now discussed therefore in a new section of Chapter 9 (Section 9.7).

Overall therefore, it is hoped that these changes in the content of the book and in its layout will allow it to continue to fulfil its aim of providing an account of gene expression and its regulation which is accurate, up-to-date and easily understandable to students, scientists and clinicians interested in this topic which is so central to our understanding of human development, health and disease.

Finally, I would like to thank Maruschka Malacos for continuing to deal so efficiently with the many textural and stylistic changes to the word-processed text which have been required for this new edition. I would also like to thank the staff of Garland/BIOS Science Publishers and, particularly, Dr Nigel Farrar, for commissioning a new edition of this book and for their efficiency in producing it.

David S. Latchman

Preface to the first edition

An understanding of how genes are regulated in humans and higher eukaryotes is essential for the understanding of normal development and disease. For these reasons this process is discussed extensively in many current texts. Unfortunately, however, many of these works devote most of their space to a consideration of gene regulation in bacteria and then discuss the complexities of the eukaryotic situation more briefly. Although such an approach was the only viable one when only the simpler systems were understood, it is clear that sufficient information is now available to discuss eukaryotic gene regulation as a subject in its own right.

This work aims to provide a text which focuses on how cellular genes are regulated in eukaryotes. For this reason bacterial examples are included only to illustrate basic control processes, while examples from the eukaryotic viruses are used only when no comparable cellular gene example of a type of regulation is available.

It is hoped that this approach will appeal to a wide variety of students and others interested in cellular gene regulation. Thus, the first four chapters provide an introduction to the nature of transcriptional regulation as well as a discussion of the other levels at which genes can be regulated, which is suitable for second-year undergraduates in biology or medicine. A more extensive discussion of the details of transcriptional control, at a level suitable for final-year undergraduates, is contained in Chapters 5–7. The entire work should also provide an introduction to the topic for starting researchers, or those moving into this area from other fields and wishing to know how the gene which they are studying might be regulated. Clinicians moving into molecular research should find the concentration on the mechanisms of human cellular gene regulation of interest, and for these readers in particular, Chapter 8 focuses on the malregulation of gene expression which occurs in human diseases and, specifically, in cancer.

I would like to thank Professor Martin Raff and Dr Paul Brickell for their critical reading of the text, and Dr Joan Heaysman who suggested that I should write this book. I am also extremely grateful to Mrs Rose Lang for typing the text and tolerating my continual changes, and to Mrs Jane Templeman who prepared the illustrations.

David S. Latchman

Preface to the second edition

In the four years since the first edition of *Gene Regulation* was published, considerable advances have been made in understanding the details of gene regulation. Nonetheless, the basic principles outlined in the first edition have remained unchanged and this second edition therefore adopts the same general approach as its predecessor. The opportunity has been taken, however, of updating the discussion of most topics to reflect recent advances. This has required the complete rewriting of some sections where progress has been particularly rapid such as those dealing with alternative splicing factors (Chapter 4) or the mechanism of transcriptional activation (Chapter 7). In addition, several new sections have been added reflecting topics which are now known to be of sufficient importance in gene regulation to merit inclusion as well as others suggested by comments on the first edition. These include regulation of transcription by polymerase pausing (Chapter 3), X chromosome inactivation and genomic imprinting (Chapter 5), locus control regions (Chapter 6), negative regulation of transcription (Chapter 7) and anti-oncogenes (Chapter 8). It is hoped that these changes and additions will allow the new edition of *Gene Regulation* to build on the success of the first edition and provide a comprehensive overview of our current understanding in this area.

Finally, I would like to thank Mrs Rose Lang who has coped most efficiently with the need to add, delete or amend large sections of the original word processed text of the first edition. I am also most grateful to Rachel Young and the staff at Chapman & Hall for deciding to commission a new edition of this work and their efficiency in producing it.

David S. Latchman

Preface to the third edition

As before, considerable progress has been made in the study of gene regulation in the three years since the second edition of this book was published. Once again, this has necessitated the re-writing of many sections of the book including, as previously, those on alternative splicing factors (Chapter 5) and transcriptional activation (Chapter 8) where progress has again been rapid. In addition, several new topics are discussed in detail for the first time to reflect their increasing importance, these include regulation of RNA transport (Chapter 4) the SWI-SNF complex (Chapter 6), the effect of specific transcription factors on histone acetylation (Chapters 6 and 8) as well as the role of co-repressors/co-activators and the mediator complex by which the inhibitory or activating effects of transcription factors are often mediated (Chapters 8 and 9).

The most significant change in this edition, however, is the addition of an extra chapter (Chapter 3) which describes the basic processes (transcription, capping, polyadenylation, splicing, transport and translation) by which DNA is converted into RNA and then into protein. Since it is this process which is regulated by the mechanisms described in the remainder of the book, this chapter renders the work complete in itself as a textbook dealing with gene expression and its regulation. In addition, this extra chapter allows the extensive discussion of the RNA polymerases themselves and their associated factors which carry out the basic process of transcription and are hence the ultimate target for the transcriptional regulatory processes which are the major regulatory mechanisms operating to control gene expression.

Lastly, I would like to thank Sarah Chinn who has taken over the difficult tasks of typing the new material and extensively amending the text of the previous edition, as well as Jane Templeman who has once again prepared the illustrations. I am also most grateful to Nigel Balmforth and the staff at Chapman & Hall for commissioning another edition and for their continued efficiency in its production.

David S. Latchman

Preface to the fourth edition

Once again, considerable information on the process of Gene Regulation has accumulated since the third edition of this book was published. Interestingly, much of this information involves the two topics of modulation of chromatin structure by histone modifications/remodelling complexes and the role of co-activators/co-repressors which were discussed in detail for the first time in the third edition. To reflect this, the topic of histone modifications now warrants a separate section (Section 6.6) and the discussion of chromatin remodelling complexes (Section 6.7) and of co-activators/co-repressors (Section 8.4) has been considerably expanded.

As well as general updating of the work, enhanced consideration has also been given to the methods used to analyse gene expression such as transcriptomic and proteomic techniques (Chapter 1) with a new section added on methods for examining DNA binding by transcription factors (Section 7.2.4). Similarly, new sections have been added on topics which have become of increasing importance such as the coupling of transcription with post-transcriptional processes (Section 3.3.4) and negatively acting sequence elements (Section 7.4). In addition, the more extensive discussion of the aberrations in gene regulation which result in diseases other than cancer in this new edition has led to the title of Chapter 9 being changed to "Gene regulation and human disease" (from "Gene regulation and cancer").

Overall therefore, the book provides an up-to-date and comprehensive account of the processes involved in gene expression and the mechanisms by which such expression is regulated, as well as an introduction to the methods used to study these processes.

Finally, I would like to thank Maruschka Malacos who has taken over the difficult task of extensively modifying the existing word processed text and inserting much new material. I would also like to thank the staff of Nelson Thornes Ltd, and particularly Catherine Shaw, for their enthusiasm to commission a new edition of this book and their efficiency in producing it.

David S. Latchman

Acknowledgments

I would like to thank all those colleagues who have given permission for material from their papers to be reproduced in this book and have provided prints suitable for reproduction. I am also especially grateful to those who have provided the colour plates.

Tissue-specific expression of proteins and messenger RNAs

SUMMARY

- The central dogma of molecular biology is that DNA produces RNA which in turn produces proteins.
- Different tissues and cell types of the same organism show differences in the different proteins which are present and in their relative abundances.
- Similarly, different tissues and cell types show differences in the RNAs which are present and in their relative abundances.
- Gene regulation must therefore operate to produce different amounts of different RNAs in different cell types which in turn produce different amounts of different proteins.

1.1 Introduction

The evidence that eukaryotic gene expression must be a highly regulated process is available to anyone visiting a butcher's shop. The various parts of the mammalian body on display differ dramatically in appearance, ranging from the muscular legs and hind quarters to the soft tissues, of the kidneys and liver. However, all these diverse types of tissues arose from a single cell, the fertilized egg or zygote, raising the question of how this diversity is achieved. It is the aim of this book to consider the processes regulating tissue-specific gene expression in mammals and the manner in which they produce these differences in the nature and function of different tissues.

1.2 Tissue-specific expression of proteins

The fundamental dogma of molecular biology is that DNA produces RNA which in turn produces proteins. Thus the genetic information in the DNA specifying particular functions is converted into an RNA copy, which is then translated into protein. The action of the protein then produces the phenotype, be it the presence of a functional globin protein transporting oxygen in the blood or the activity of a proteinaceous enzyme capable of producing the pigment causing the appearance of brown rather than blue eyes. Hence, if the differences in the appearance of mammalian tissues described above are indeed caused by differences in gene expression in different tissues, they should be produced by differences in the proteins present in these tissues. Such differences can be detected both by specific methods which study the expression of one particular protein and by general methods aimed at studying the expression of all proteins in a given tissue.

Specific methods for studying the protein composition of tissues

A simple means of determining whether differences in protein composition exist in different tissues involves investigating the expression of individual known proteins in such tissues. These methods might involve the isolation of widely differing amounts of an individual protein (or none of the protein at all) from different tissues, using an established purification procedure, or the detection of an enzymatic activity associated with the protein in extracts of only one particular tissue. We shall consider in detail, however, only methods where the expression of a specific protein is monitored by the use of a specific antibody to it. Normally, such an antibody is produced by injecting the protein into an animal such as a rabbit or mouse. The resulting immune response results in the presence in the animal's blood of antibodies which specifically recognize the protein and can be used to monitor its expression in particular tissues.

Such antibodies can be used in conjunction with the one-dimensional polyacrylamide gel electrophoresis technique to investigate the expression of a particular protein in different tissues. In this technique, proteins are denatured by treatment with the detergent sodium dodecyl sulfate (SDS) and then subjected to electrophoresis in a polyacrylamide gel which separates them according to their size (Laemmli, 1970). Subsequently, in a technique known as Western blotting (Gershoni and Palade, 1983), the gel-separated proteins are transferred to a nitrocellulose filter which is incubated with the antibody. The antibody reacts specifically with the protein against which it is directed and which will be present at a particular position on the filter, dependent on how far it moved in the electrophoresis step and hence on its size. The binding of the antibody is then visualized by a radioactive or enzymatic detection procedure. If a tissue contains the protein of interest, a band will be observed in the track containing total protein from that tissue and the intensity of the band observed will provide a measure of the amount of protein present in the tissue. If none of the particular protein is present in a given tissue, no band will form (Fig. 1.1). Hence this method allows the presence or absence of a specific protein in a particular tissue to be assessed using one-dimensional gel electrophoresis without the complicating effect of other unrelated proteins of similar size, since these will fail to bind the antibody. Similarly quantitative differences in the expression of a particular protein in different tissues can be detected on the basis of differences in the intensity of the band obtained when extracts prepared from the different tissues are used in this procedure.

The specific reaction of a protein with an antibody can also be used directly to investigate its expression within a particular tissue. In this method, known as immunofluorescence, thin sections of the tissue of interest are reacted with the antibody, which binds to those cell types expressing the protein. As before, the position of the antibody is visualized by an enzymatic detection procedure, or more usually, by the use of a fluorescent dye which can be seen in a microscope when appropriate filters are used (Fig. 1.2). Therefore this method can be used not only to provide information about the tissues expressing a particular protein but, since individual cells can be examined, it also allows detection of the specific cell types within the tissue which are expressing the protein.

The results of experiments using these specific methods to study the expression of particular proteins, indicate that while some proteins are

A   B

Figure 1.1

Western blot with an antibody to guinea-pig casein kinase showing the presence of the protein in lactating mammary gland (A) but not in liver (B). Photograph kindly provided by Dr A. Moore, from Moore *et al.*, *Eur. J. Biochem.*, **152**, 729 (1985), by permission of Springer-Verlag.

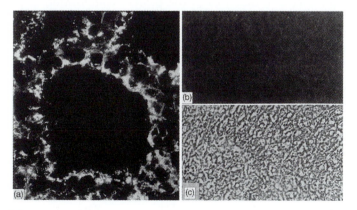

Figure 1.2

Use of the antibody to casein kinase to show the presence of the protein (bright areas) in frozen sections of lactating mammary gland (a) but not in liver (b). (c) A phase-contrast photomicrograph of the liver section, confirming that the lack of staining with the antibody in (b) is not due to the absence of liver cells in the sample. Photograph kindly provided by Dr A. Moore, from Moore *et al.*, *Eur. J. Biochem.*, **152**, 729 (1985), by permission of Springer-Verlag.

present at similar abundances in all tissues, others are present at widely different abundances in different tissues, and a large number are specific to one or a few tissues or cell types. Hence the differences between different tissues in appearance and function are correlated with qualitative and quantitative differences in protein composition, as assayed using antibodies to individual proteins.

General methods for studying the protein composition of tissues

As well as examining the expression of individual proteins in different tissues, it is also possible to use more general methods to compare the overall protein composition of different tissues. All the proteins present in a tissue can be subjected to one-dimensional gel electrophoresis (see above) and examined by using a stain which reacts with all proteins. Hence this method allows the visualization of all cellular proteins, separated according to their size on the gel rather than using Western blotting with a particular antibody to focus on a specific protein. However, this method is of relatively limited use to investigate variations in the overall protein content of different tissues. This is because of the very large number of proteins in the cell and the limited resolution of the technique which separates proteins solely on the basis of their size. Thus, for example, two entirely different proteins in two tissues may be scored as being the same protein simply on the basis of a similarity in size.

A more detailed investigation of the protein composition of different tissues can be achieved by two-dimensional gel electrophoresis (O'Farrell, 1975). In this procedure (Fig. 1.3), proteins are first separated on the basis of differences in their charge, in a technique known as isoelectric focusing, and the separated proteins, still in the first gel, are layered on top of an SDS-polyacrylamide gel. In the subsequent electrophoresis the proteins are separated by their size. Hence a protein moves to a position determined both by its size and its charge. The much greater resolution of this method allows a number of differences in the protein composition of particular tissues to be identified. Thus some spots or proteins are found in only one or a few tissues and not in many others, while others are found at dramatically different abundances in different tissues (Fig. 1.4).

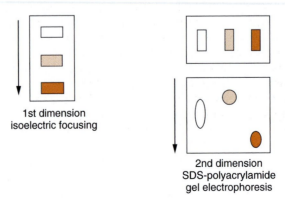

1st dimension
isoelectric focusing

2nd dimension
SDS-polyacrylamide
gel electrophoresis

Figure 1.3

Two-dimensional gel electrophoresis.

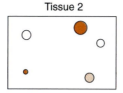

Tissue 1   Tissue 2

○ Proteins present at similar levels in both tissues

◔ Proteins present in only one tissue

● Proteins present in both tissues but at different levels

Figure 1.4

Schematic results of two-dimensional gel electrophoresis allowing the detection of proteins specific to one tissue or expressed at different levels in different tissues.

Hence, the different appearances of different tissues are indeed paralleled by both qualitative and quantitative differences in the proteins present in each tissue. It should be noted, however, that some proteins can be shown by two-dimensional gel electrophoresis to be present at similar levels in virtually all tissues. Presumably such so-called housekeeping proteins are involved in basic metabolic processes common to all cell types.

Initial experiments using two-dimensional gel electrophoresis simply involved examining the patterns of spots generated by different tissues and describing which spots were present or absent in specific tissues or showed different intensities in different tissues. More recently however, methods have been developed which allow the protein responsible for forming a spot of particular interest to be identified and characterized. These methods have allowed the development of a new field of study, known as proteomics, in which the power of two-dimensional gels to separate a wide range of proteins is combined with the ability to study specific proteins individually.

In these methods (for review see Blackstock and Weir, 1999; Pandey and Mann, 2000) individual spots of interest are excised from the two-dimensional gel and digested into their constituent peptides using the proteolytic enzyme, trypsin. The individual peptides from a single protein spot are then analyzed by mass spectrometry. This can involve both matrix-assisted laser desorption/ionization (MALDI), which allows the molecular weight of the peptide to be determined and nanoelectrospray mass spectrometry, which allows the amino acid sequence of a particular peptide to be obtained (Fig. 1.5). Often these two techniques are performed sequentially in a tandem mass spectrometer which first determines the molecular weight of a peptide produced by trypsin digestion and then fragments it further, allowing its sequence to be determined.

This peptide molecular weight and sequence information can then be used to search databases of known protein sequences which have been obtained either by direct protein sequencing or more frequently predicted from the ever-expanding amount of DNA sequence information. As

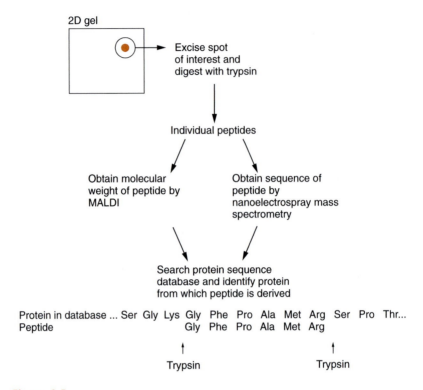

2D gel

Excise spot
of interest and
digest with trypsin

Individual peptides

Obtain molecular
weight of peptide by
MALDI

Obtain sequence of
peptide by
nanoelectrospray mass
spectrometry

Search protein sequence
database and identify protein
from which peptide is derived

Protein in database ... Ser Gly Lys Gly Phe Pro Ala Met Arg Ser Pro Thr...
Peptide   Gly Phe Pro Ala Met Arg

Trypsin   Trypsin

Figure 1.5

Use of mass spectrometry to determine the molecular weight and amino acid
sequence of a particular protein spot on a two-dimensional gel. The information
obtained in this way can be used to search databases of known protein
sequences and hence identify the protein producing the spot.

trypsin always cleaves proteins after a lysine or arginine residue, it is possible to align the peptide sequence against that of a known protein and
hence identify the protein responsible for the original spot (Fig. 1.5; color
plate 1) (for further discussion of the techniques of mass spectrometry see
Yates, 2000).

   This method can thus be used to identify an individual protein which
shows an interesting pattern of expression in different tissues or whose
expression is altered in a specific disease. Hence it combines the power of
two-dimensional gel electrophoresis to look at a wide range of proteins
with the ability to identify an individual protein. It is thus being widely
used, for example, to identify the proteins associated with individual cellular organelles when these are isolated and subjected to two-dimensional
gel electrophoresis. Similarly, it can be used to identify other proteins associated with a known protein in the cell. This can be achieved by carrying
out an immunoprecipitation with an antibody to the known protein and
then analyzing the immunoprecipitate by two-dimensional gel electrophoresis to identify the other proteins which have been isolated by
virtue of their association with the immunoprecipitated protein.

   In terms of gene regulation, proteomic methods involving two-dimensional gel electrophoresis and mass spectrometry allow the identification

of specific proteins whose expression varies between different tissues. It thus combines the methods for looking at specific proteins (e.g. with antibodies) or looking at the general population of proteins (using two-dimensional gel electrophoresis alone) which were described previously and reinforces the conclusion that the protein content of different tissues is both qualitatively and quantitatively different.

1.3 Tissue-specific expression of messenger RNAs

Proteins are produced by the translation of specific messenger RNA molecules on the ribosome (see Section 3.3). Hence, having established that quantitative and qualitative differences exist in the protein composition of different tissues, it is necessary to ask whether such differences are paralleled by tissue-specific differences in the abundance of their corresponding mRNAs. Thus, although it seems likely that differences in the mRNA populations of different tissues do indeed underlie the observed differences in proteins, it is possible that all tissues have the same mRNA species and that production of different proteins is controlled by regulating which of these are selected by the ribosome to be translated into protein.

As with the study of proteins, both specific and general techniques exist for studying the mRNAs expressed in a given tissue.

Specific methods for studying the mRNAs expressed in different tissues

A number of different specific methods exist which can be used to detect and quantify one specific mRNA using a cloned DNA probe derived from its corresponding gene. In the most commonly used of such methods, Northern blotting (Thomas, 1980), the RNA extracted from a particular tissue is electrophoresed on an agarose gel, transferred to a nitrocellulose filter, and hybridized to a radioactive probe derived from the gene encoding the mRNA of interest. The presence of the RNA in a particular tissue will result in binding of the radioactive probe and the visualization of a band on autoradiography, the intensity of the band being dependent on the amount of RNA present (Fig. 1.6). Although such experiments do detect the RNA encoding some proteins, such as actin or tubulin, in all tissues, very many others are found only in one particular tissue. Thus the RNA for the globin protein is found only in reticulocytes, that for myosin only in muscle, while (in the example shown in Fig. 1.6) the mRNA encoding the fetal protein, α-fetoprotein, is shown to be present only in the embryonic yolk sac and not in the adult liver. Similar results can also be obtained using other methods for detecting specific RNAs in material isolated from different tissues such as nuclease S1 analysis, RNAse protection and primer extension.

As with protein studies, methods studying expression in RNA isolated from particular tissues can be supplemented by methods allowing direct visualization of the RNA in particular cell types. In such a method, known as *in situ* hybridization, a radioactive probe specific for the RNA to be detected is hybridized to a section of the tissue of interest in which cellular morphology has been maintained. Visualization of the position at which the radioactive probe has bound (Fig. 1.7) not only allows an assessment of whether the particular tissue is expressing the RNA of interest but

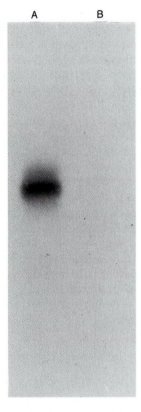

Figure 1.6

Northern blot hybridization using a probe specific for the α-fetoprotein mRNA. The RNA is detectable in the embryonic yolk sac sample (track A) but not in the adult liver sample (track B). Figure from Latchman *et al.*, *Biochim. Biophys. Acta* **783**, 130–136 (1984), by permission of Elsevier Science Publishers.

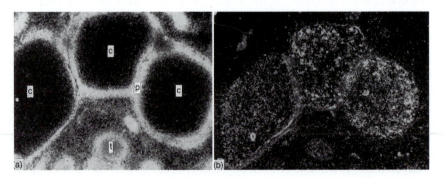

Figure 1.7

Localization of the RNA for type 1 collagen (A) and type II collagen (B) in the 10-day chick embryo leg by *in situ* hybridization. Note the different distributions of the bright areas produced by binding of each probe to its specific mRNA. c, cartilage; t, tendons; p, perichondrium. Photograph kindly provided by Dr P. Brickell, from Devlin *et al.*, *Development* **103**, 111–118 (1988), by permission of the Company of Biologists Ltd.

also of which individual cell types within the tissue are responsible for such expression. This technique has been used to show that the expression of some mRNAs is confined to only one cell type, paralleling the expression of the corresponding proteins by that cell type.

General methods for studying the mRNAs expressed in different tissues

As in the case of proteins, methods for studying individual RNAs can be supplemented by methods for studying RNA populations. Early experiments in this area made use of the *Rot* curve (Bishop *et al.*, 1974). In this method a complementary DNA (cDNA copy) of the RNA is first made, using radioactive precursors so that the cDNA is made radioactively labeled. After dissociating the cDNA and the RNA, the rate at which the radioactive cDNA reanneals or hybridizes to the RNA is followed. In this process, more molecules of an abundant RNA will be present than of an RNA which is less abundant. The cDNA from an abundant RNA will find a partner more rapidly than one derived from a rare one, simply because more potential partners are available. Hence, when used with the RNA of one tissue, the rate of hybridization or reannealing of any cDNA molecule provides a measure of the abundance of its corresponding RNA in that tissue. The higher the abundance of a particular RNA species, the more rapidly it anneals to its complementary cDNA (Fig. 1.8). Typical *Rot* curves, showing three predominant abundance classes of RNA in any given tissue, can be constructed readily (Fig. 1.9).

This method can be extended to provide a comparison of the RNA populations of two different tissues (Hastie and Bishop, 1976). In this method the RNA of one tissue is mixed with labeled cDNA prepared from the RNA of another tissue and the mixture annealed. Clearly, the rate of hybridization of the radiolabeled cDNA will be determined by the abundance of its corresponding RNA in the tissue from which the RNA is derived. If the RNA is absent altogether in this tissue (being specific to that from which the cDNA is derived), no reannealing will occur. If the corresponding RNA is present, the rate at which the cDNA hybridizes will be determined by its abundance in the tissue from which the RNA is derived. Hence, by carrying out an inter-tissue *Rot* curve, both qualitative and quantitative differences between the RNA populations of different tissues can be observed. Such experiments (see Fig. 1.10) lead to the conclusion that, although sharing of rare RNAs (presumably encoding housekeeping proteins) does occur, both qualitative and quantitative differences between the RNA populations of different tissues are observed, with RNA species that are abundant in one cell type being rare or absent in other tissues. Thus, in a comparison of kidney and liver messenger RNAs, Hastie and Bishop (1976) found that, while between 9500 and 10 500 of the 11 000 rare messenger RNA species found in kidney were also found in liver, the six highly abundant RNA species found in the kidney were either absent or present at very low levels in the liver (Fig. 1.10).

More recently, methods have been developed that can combine the ability to look at variation in the total RNA population of different tissues with the ability to look specifically at the variation in a specific mRNA. By analogy with the proteomic methods which achieve this at the protein level (see Section 1.2), these methods are often referred to as transcriptomic methods (for reviews see Lockhart and Winzeler, 2000; Young, 2000).

In these methods, so called "gene chips" are prepared in which DNA sequences derived from thousands of different mRNAs are laid out in a regular array at a density of over 10 000 different DNA sequences per

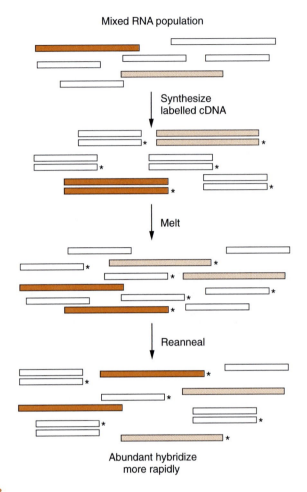

Figure 1.8

Hybridization of messenger RNA and its radioactive complementary DNA copy (indicated by the asterisks) in a *Rot* curve experiment. The abundant RNA finds its partner and hybridizes more rapidly.

square centimeter (Fig. 1.11). This is achieved either by spotting out cDNA clones derived from the different mRNAs at high density on a glass slide or by synthesizing short oligonucleotides derived from each mRNA *in situ* on the gene chip.

Once this has been achieved the gene chips can be hybridized with radioactive probes prepared from the total mRNA present in different tissues and the pattern of hybridization observed (Fig. 1.11). Thus, where a specific mRNA is present in a tissue, it will be present in the radioactive probe and a signal will be obtained on the gene chip, with the strength of the signal being proportional to the amount of the RNA which is present. Similarly, if the mRNA is absent then it will be absent from the probe and no signal will be obtained. As more and more DNA sequence information is obtained, it is possible to include increasing numbers of DNA sequences derived from individual mRNAs on the chips, taking advantage

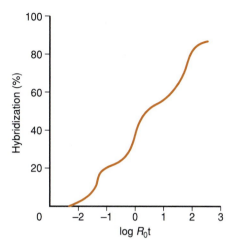

Figure 1.9

Typical *Rot* curve, produced using the mRNA of a single tissue. Note the three phases of the curve, produced by hybridization of the highly abundant, moderately abundant, and rare RNA species. Figure redrawn from Bishop *et al.*, *Nature* **250**, 199–204 (1974), by permission of Dr J. Bishop and Macmillan Magazines Ltd.

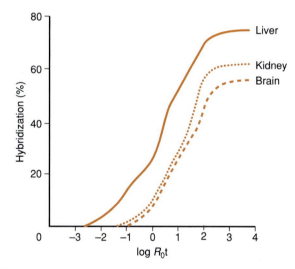

Figure 1.10

Inter-tissue *Rot* curve, in which mRNA from the liver has been hybridized to cDNA prepared from liver, kidney, or brain mRNA. Note the lower extent of hybridization, with the kidney and brain samples indicating the absence or lower abundance of some liver RNAs in these tissues. Figure redrawn from N.D. Hastie and J.O. Bishop, *Cell* **9**, 761–774 (1976), by permission of Dr J. Bishop and Cell Press.

of the ability to spot out many different DNA sequences onto a very small chip. Hence, the expression of virtually all the mRNAs in a cell can now be compared in different tissues and cell types (Fig. 1.11; color plate 2).

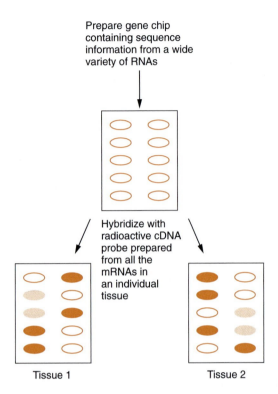

Figure 1.11

Gene chip analysis of mRNA expression patterns in different tissues. Radioactive cDNA is prepared from the total mRNA present in each tissue. Hybridization of this radioactive cDNA to the gene chip containing many different gene sequences will reveal the expression level of each gene in each tissue. The figure shows this for genes which are not expressed in a particular tissue (open symbol) genes which are expressed at low level (pale symbol) and genes which are expressed at high level (dark symbol).

This very powerful method thus allows the expression profile of individual tissues and cell types to be obtained and compared in a manner which combines the ability of Northern blotting or *in situ* hybridization to study individual mRNAs with the ability of *Rot* curve analysis to look generally at the entire mRNA population. The use of this method has provided further support for the conclusion that the mRNA populations of different tissues show both qualitative and quantitative differences. Because of its great power in combining both general and specific methods of looking at RNA populations, methods of this type have also been used to probe for subtle differences between closely related samples. Thus, for example, it has been used to characterize the gene expression differences between different types of breast cancer with different characteristics and different outcomes for the patient (Perou *et al.*, 2000). Similarly, this method has been used to examine the expression of all the genes in the *Drosophila* genome at different stages of the life cycle (Stolc *et al.*, 2004).

1.4   Conclusions

It is clear, therefore, that the qualitative and quantitative variation in protein composition between different tissues is paralleled by a similar variation in the nature of the mRNA species present in different tissues. Indeed, in a study of a wide range of yeast proteins and mRNAs the abundance of an individual protein as determined by two-dimensional gel electrophoresis generally correlated very well with the abundance of the corresponding mRNA as determined by a gene chip approach (Futcher *et al.*, 1999) and a similar correlation has also been observed in a survey of human liver mRNAs and proteins (Anderson and Seilhamer, 1997). Hence, the production of different proteins by different tissues is not regulated by selecting which of a common pool of RNA molecules is selected for translation by the ribosome, although some examples of such regulated translation exist (see Section 5.6). In order to understand the basis for such tissue-specific variation in protein and mRNA composition, it is therefore necessary to move one stage further back and examine the nature of the DNA encoding these mRNAs in different tissues.

References

Anderson, L. and Seilhamer, J. (1997). A comparison of selected mRNA and protein abundances in human liver. *Electrophoresis* 18, 533–537.

Bishop, J.O., Morton, J.G., Rosbash, M. and Richardson, M. (1974). Three abundance classes in HeLa cell messenger RNA. *Nature* 250, 199–204.

Blackstock, W.P. and Weir, M.P. (1999). Proteomics: quantitative and physical mapping of cellular proteins. *Trends in Biotechnology* 17, 121–127.

Futcher, B., Latter, G.I., Monardo, P., McLaughlin, C.S. and Garrels, J.I. (1999). A sampling of the yeast proteome. *Molecular and Cellular Biology* 19, 7357–7368.

Gershoni, J.M. and Palade, G.E. (1983). Protein blotting: principles and applications. *Analytical Biochemistry* 131, 1–15.

Hastie, N.B. and Bishop, J.O. (1976). The expression of three abundance classes of messenger RNA in mouse tissues. *Cell* 9, 761–774.

Laemmli, U.K. (1970). Cleavage of structural proteins during the assembly of the head of the bacteriophage T4. *Nature* 227, 680–685.

Lockhart, D.J. and Winzeler, E.A. (2000). Genomics, gene expression and DNA arrays. *Nature* 405, 827–836.

O'Farrell, P.H. (1975). High resolution two-dimensional electrophoresis of proteins. *Journal of Biological Chemistry* 250, 4007–4021.

Pandey, A. and Mann, M. (2000). Proteomics to study genes and genomes. *Nature* 405, 837–846.

Perou, C.M., Sørlie, T., Eisen, M.B., Van de Rijn, M., Jeffrey, S.S., Rees, C.A. *et al.* (2000). Molecular portraits of human breast tumours. *Nature* 406, 747–752.

Stolc, V., Gauhar, Z., Mason, C., Halasz, G., van Batenburg, M.F., Rifkin, S.A. *et al.* (2004). A gene expression map for the euchromatic genome of *Drosophila melanogaster*. *Science* 306, 655–660.

Thomas, P.S. (1980). Hybridization of denatured RNA and small DNA fragments transferred to nitrocellulose. *Proceedings of the National Academy of Sciences of the USA* 77, 5201–5205.

Yates, J.R. (2000). Mass spectrometry from genomics to proteomics. *Trends in Genetics* **16**, 5–8.

Young, R.A. (2000). Biomedical discovery with DNA arrays. *Cell* **102**, 9–15.

The DNA of different cell types is similar in both amount and type

2

SUMMARY

- Unlike the RNA and protein content, the DNA of different tissues and cell types is generally the same in a specific organism.
- Some cases of DNA loss, DNA amplification or DNA rearrangement do occur in specific cell types.
- However, these are exceptions caused by the special requirements of a particular situation.
- Gene regulation must therefore operate to produce different amounts of different RNAs in different cell types, from the same DNA.

2.1 Introduction

Having established that the RNA and protein content of different cell types shows considerable variation, the fundamental dogma of molecular biology, in which DNA makes RNA which in turn makes protein, directs us to consider whether such variation is caused by differences in the DNA present in each tissue. Thus, in theory, DNA which corresponded to an RNA required in one particular tissue only might be discarded in all other tissues. Alternatively, it might be activated in the tissue where RNA was required, by a selective increase in its copy number, in the genome via an amplification event or by some rearrangement of the DNA necessary for its activation. The possible use of each of these mechanisms (DNA loss, amplification, or rearrangement) will be discussed in turn (for a review see Brown, 1981).

2.2 DNA loss

DNA loss as a mechanism of gene regulation

In theory, a possible method of gene control would involve the genes for all proteins being present in the fertilized egg, followed by a selective loss of genes which were not required in particular tissues (Fig. 2.1). Thus only the genes for housekeeping proteins and the proteins characteristic of a particular cell type would be retained in that cell type. Hence the presence of the RNA for immunoglobulin only in the antibody-producing B cells would be paralleled by the presence of the corresponding gene only

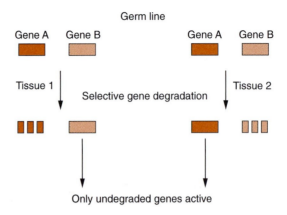

Figure 2.1

Model for gene regulation by the specific degradation of genes whose products are not required in a particular tissue.

in such cells, whereas the gene for a muscle-specific protein (such as myosin) would be absent in B-cell DNA but present in that of a muscle cell, where the immunoglobulin gene had been deleted. Such a mechanism would obviate the need for any other means of regulating gene activity, the tissue-specific RNA and protein populations described in Chapter 1 being produced by the constitutive activity of the genes retained in any particular cell type.

A variety of evidence suggests, however, that selective DNA loss is not a general mechanism of gene control. Such evidence is available from chromosomal, functional, and molecular studies, which will be considered in turn.

Chromosomal studies

The DNA-containing chromosomes, which can be visualized shortly before cell division, differ from one another in both size and the location of the centromere, by which they attach to the mitotic spindle prior to cell division. They can be further distinguished by a variety of staining techniques, which produce a specific banded appearance. The most widely used of these, known as G-banding, involves brief proteolytic or chemical treatment of the chromosome, followed by staining with Giemsa stain, and produces a characteristic pattern of bands (Fig. 2.2), allowing the cytogeneticist to identify individual chromosomes in humans and other species and to distinguish cells from different species.

In general, when such techniques are applied to chromosomes prepared from different differentiated cell types of the same species, however, the numbers and sizes of the chromosomes observed and their banding patterns are indistinguishable. Thus the patterns obtained from a liver cell or a brain cell are very similar and cannot be distinguished even by experienced cytogeneticists. Hence such karyotypic studies provide strong evidence against the loss of whole chromosomes or specific parts of chromosomes (visualized as G-bands) in particular differentiated cell types, and argue against selective DNA loss as a general mechanism for gene regulation.

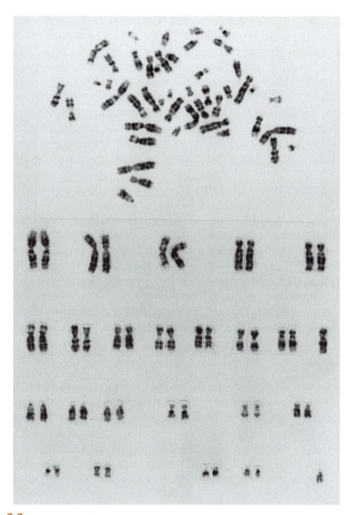

Figure 2.2

Human chromosomes stained with Giemsa stain. The upper panel shows a chromosome spread; in the lower panel the chromosomes have been arrayed in order of their chromosome number. Photograph kindly provided by Dr A.T. Sumner.

Nonetheless, a small number of cases of such a loss of DNA have been observed in particular cell types and these will be considered in turn.

Loss of the nucleus in mammalian red blood cells

The best known case occurs in the development of the mammalian red blood cell, or erythrocyte (for a review see Bessis and Brickal, 1952), which is a highly specialized cell containing large amounts of the blood pigment hemoglobin and functions in the transport of oxygen in the blood. Such cells are produced from progenitors, known as erythroblasts, which divide to produce two daughter cells, one of which maintains the stem cell population and later divides again, while the other undergoes differentiation. In this differentiation, the region of the cell containing the nucleus is pinched off, surrounded by a region of the cell membrane, and is eventually destroyed (Fig. 2.3). The resulting cell, known as the reticulocyte, is thus entirely anucleate and completely lacks DNA. However, it continues synthesis of large amounts of globin (the protein part of the hemoglobin molecule) as well as small amounts of other reticulocyte proteins, by repeated translation of messenger RNA molecules produced prior to the loss of the nucleus. Such RNA molecules must therefore be highly stable and resistant to degradation (see Section 5.5). Eventually, however, other cytoplasmic components (including ribosomes) are lost, protein synthesis ceases, and the cell assumes the characteristic structure of the erythrocyte, which is essentially a bag full of oxygen-transporting hemoglobin molecules.

Although this process offers us an example of DNA loss during differentiation, it is clearly a highly specialized case in which the loss of the nucleus is primarily intended to allow the cell to fill up with hemoglobin and to assume a shape facilitating oxygen uptake, rather than as a means of gene control. Thus the genes for globin and other reticulocyte proteins are not selectively retained but are lost with the rest of the nuclear DNA. Hence the tissue-specific pattern of protein synthesis observed in the reticulocyte is achieved not by selective gene loss but by the processes which, earlier in development, resulted in high-level transcription of the globin genes and the production of long-lived, stable messenger RNA molecules.

Loss of specific chromosomes in somatic cells

The reticulocyte case does not, therefore, provide us with an example of selective DNA loss and, indeed, no such case has as yet been reported in mammalian cells. However, a number of such cases in which specific chromosomes are lost during development have been described in other eukaryotes, including some nematodes, crustaceans and insects (Wilson, 1928). In these cases a clear distinction is made early in embryonic development between the cells which will produce the body of the organism (the somatic cells) and those which will eventually form its gonads and allow the organism to reproduce (the germ cells). Early on in the development of the embryo, during the cleavage stage, the somatic cells lose certain chromosomes, with only the germ cells, which will give rise to subsequent generations, retaining the entire genome. Such losses can be quite extensive, for example in the gall midge *Miastor* only 12 of the total

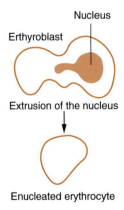

Nucleus

Erthyroblast

Extrusion of the nucleus

Enucleated erythrocyte

Figure 2.3

Extrusion of the nucleus from an erythroblast, resulting in an anucleate erythrocyte.

48 chromosomes are retained in the somatic cells. It is assumed that the lost DNA is only required for some aspects of germ cell development. It is known to be rich in the highly repeated (or satellite) type of DNA, which is thought to be involved in homologous chromosome pairing during the meiotic division that occurs in germ cell development. Hence it is more economical for the organism to dispense with such DNA in the development of somatic tissue, rather than to waste energy replicating it and passing it on to all somatic cells where it has no function.

Although the mechanism by which this selective loss of DNA occurs is not yet fully understood, there is evidence that it is mediated by cytoplasmic differences between regions of the fertilized egg or zygote. Thus, in the nematode worm *Parascaris* the distinction between somatic and germ cells is established by the first division of the zygote, the smaller of the two cells produced losing chromosomes and giving rise to the soma while the larger retains a full chromosome complement and gives rise to the germ line (Fig. 2.4). If, however, the zygote is centrifuged, the plane of the first division can be altered, resulting in the production of two similar-sized cells, neither of which loses chromosomes. Hence the centrifugation process has abolished some heterogeneity in the cytoplasm responsible for this effect. Interestingly, although both daughter cells retain a full complement of chromosomes, their inability to lose chromosomes prevents normal embryonic development from occurring.

Although cases of DNA loss observable by examination of the chromosomes have been observed in some organisms, it must be emphasized that in the vast majority of cases no such loss of DNA is observed by this means. However, because the amount of DNA in a Giemsa-stained band is very

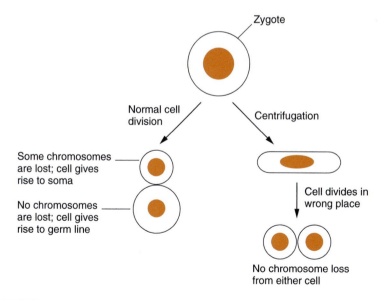

Figure 2.4

Early development in the nematode, and the consequences of changing the plane of the first cell division by centrifugation of the zygote. Redrawn from Cove, D.J., *Genetics* (1971), Cambridge University Press, by permission of Professor D.J. Cove and Cambridge University Press.

large (about 10 million base pairs), it is possible that losses of short regulatory regions necessary for gene expression, or losses of individual genes, could occur and not be observed by this technique. The large-scale losses seen in chromosome studies would then represent the tip of the iceberg in which varying degrees of gene loss occur in different situations. A consideration of this possibility requires the use of more sensitive methods to search for gene deletions. These involve both functional and molecular studies.

Functional studies

A model in which selective gene loss controls differentiation must view differentiation as essentially an irreversible process. Thus, for example, once the gene for myosin has been eliminated from an antibody-producing B-lymphocyte it will not be possible for such a cell, whatever the circumstances, to give rise to a myosin-containing muscle cell. Although the requirement for such a change from one differentiated cell type to another is not likely to occur in normal development, such a phenomenon, known as transdifferentiation, has been achieved experimentally in Amphibia (Yamada, 1967). Thus, if the lens is surgically removed from an eye of one of these organisms, some of the neighboring cells in the iris epithelium lose their differentiated phenotype and begin to proliferate. Eventually these cells differentiate into typical lens cells and produce large quantities of the lens-specific proteins, such as the crystallins, which are readily detectable with appropriate antibodies. The genes for the lens proteins must therefore be intact within the iris cells, even though in normal development such genes would never be required in these cells.

The case against selective loss of DNA is strengthened still further by the very dramatic experiments in which a whole new organism has been produced from a single differentiated cell. This was achieved a number of years ago in both plants (Steward, 1970) and Amphibia (Gurdon, 1968).

In plants such regeneration has been achieved in several species, including both the carrot and tobacco plants. In the carrot, for example, regeneration can occur from a single differentiated phloem cell, which forms part of the tubing system by which nutrients are transported in the plant. Thus, if a piece of root tissue is placed in culture (Fig. 2.5), single quiescent cells of the phloem can be stimulated to grow and divide, and an undifferentiated callus-type tissue forms. The disorganized cell mass can be maintained in culture indefinitely but, if the medium is suitably supplemented at various stages, embryonic development will occur and eventually result in a fully functional flowering plant, containing all the types of differentiated cells and tissues normally found. The plant that forms is fertile and cannot be distinguished from a plant produced by normal biological processes.

The ability to regenerate a functional plant from fully differentiated cells eliminates the possibility that genes required in other cell types are eliminated in the course of plant development. It is noteworthy, however, that in this case, as with lens regeneration, one differentiated cell type does not transmute directly into another, rather a transitional state of undifferentiated proliferating cells serves as an intermediate. Hence although differentiation does not apparently involve permanent irreversible changes in the DNA, it appears to be relatively stable and, although

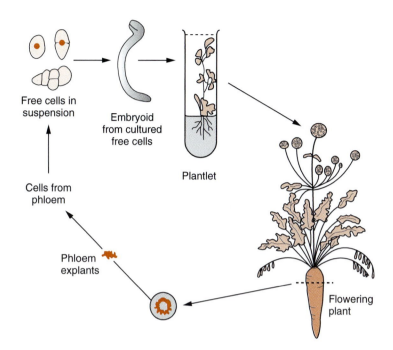

Figure 2.5

Scheme for the production of a fertile carrot plant from a single differentiated cell of the adult plant phloem by growth in culture as a free cell suspension which develops into an embryo and then into an adult plant. Redrawn from Steward *et al.*, *Science* **143**, 20–27 (1964), by permission of Professor F.C. Steward and the American Association for the Advancement of Science (AAAS).

reversible, requires an intermediate stage for changes to occur. This semi-stability of cellular differentiation will be discussed further in Chapter 6.

Although no complex animal has been regenerated by culturing a single differentiated cell in the manner used for plants, other techniques have been used to show that differentiated cell nuclei are capable of giving rise to very many different cell types (Gurdon, 1968). These experiments involve the use of nuclear transplantation (Fig. 2.6). In this technique the nucleus of an unfertilized frog egg is destroyed, either surgically or by irradiation with ultraviolet light, and a donor nucleus from a differentiated cell of a genetically distinguishable strain of frog is implanted. Development is then allowed to proceed in order to test whether, in the environment of the egg cytoplasm, the genetic information in the nucleus of the differentiated cell can produce an adult frog. When the donor nucleus is derived, for example, from differentiated intestinal epithelial cells of a tadpole, an adult frog is indeed produced in a small proportion of cases (about 2%) and development to at least a normal swimming tadpole occurs in about 20% of cases, with the organisms having the genetic characteristics of the donor nucleus. Such successful development supported by the nucleus of a differentiated frog cell is not unique to intestinal cells and has been achieved with the nuclei of other cell types, such as the skin cells of an adult frog. Hence, although these experiments are technically difficult and have a high failure rate, it is possible for the nuclei

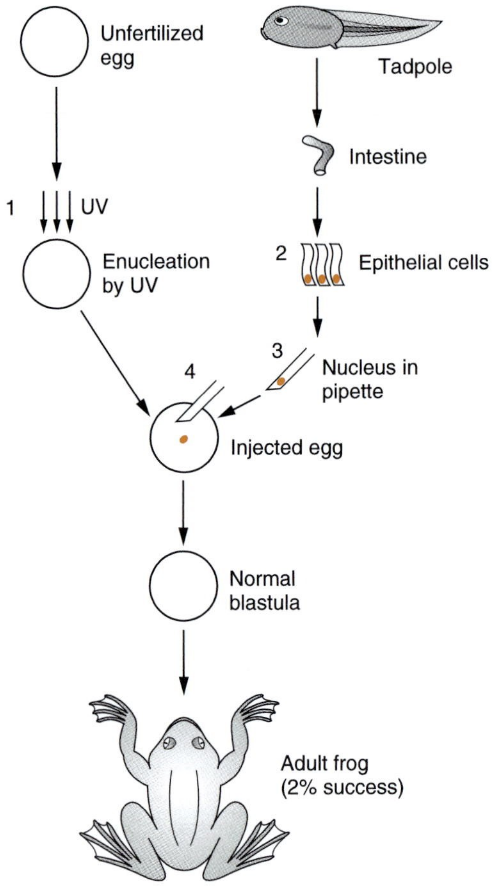

Figure 2.6

Nuclear transplantation in Amphibia. The introduction of a donor nucleus from a differentiated cell into a recipient egg whose nucleus has been destroyed by irradiation with ultraviolet light can result in an adult frog with the genetic char-acteristics of the donor nucleus. Redrawn from Gurdon, *Control of Gene Expression in Animal Development* (1974), Oxford University Press, Oxford, by permission of Professor J.B. Gurdon and Oxford University Press.

of differentiated cells specialized for the absorption of food to produce organisms containing a range of different cell types and which are perfectly normal and fertile.

Although these experiments were carried out in Amphibia over 30 years ago, it is only within the last few years that similar experiments have been successfully performed in mammals. In 1997 however, Wilmut and co-work-ers reported experiments in which a donor nucleus derived from a mammary gland of a 6-year-old sheep was introduced into an unfertilized egg whose own nucleus had been destroyed (Wilmut *et al.*, 1997; Stewart, 1997). In this manner, they produced an adult sheep (Dolly) which, as in Gurdon's experiments with frogs, had the genetic characteristic of the sheep from which the mammary gland cell nucleus had been derived. Similar "cloning" experiments have now been successfully carried out in several other mammalian species including mice (Wakayama *et al.*, 1998) and pigs (Polejaeva *et al.*, 2000). Although these studies have provoked considerable ethical discussion on the potential cloning of humans, their relevance for the area of gene regulation is that they indicate that in mammals, as in Amphibia, a nucleus from an adult differentiated cell is capable of supporting embryonic development and producing the full range of differentiated cell types present in an adult organism.

As with plants, it is therefore clear that, in animals, selective loss of DNA is not a general mechanism by which gene control is achieved.

Molecular studies

The conclusions of functional studies carried out in the 1960s and early 1970s were confirmed abundantly in the late 1970s and 1980s by molecular investigations involving the use of specific DNA probes derived from individual genes. Such probes can be used in Southern blotting experiments (Southern, 1975) to investigate the structure of a particular gene in DNA prepared from different tissues (see Methods Box 2.1).

As with the use of Northern blotting and Western blotting (see Chapter 1) to study, respectively, the RNA and protein content of different tissues, Southern blotting can be used to investigate the structure of the DNA in individual tissues. When this is done, no losses of particular genes are observed in specific tissues. Thus, the gene for β-globin is clearly present in the DNA of brain or spleen where it is never expressed, as well as in

---

METHODS BOX 2.1

Southern blotting (Fig. 2.7)

- Digest DNA with a restriction endonuclease which cuts it at specific sequences.
- Electrophorese on an agarose gel.
- Transfer DNA in gel to nitrocellulose filter.
- Hybridize filter with radiolabeled probe for gene of interest.
- Analyze structure of a specific gene based on patterns of hybridization obtained when DNA is cut with different restriction endonucleases.

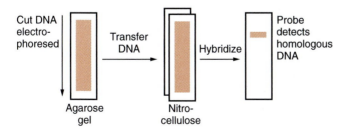

Figure 2.7

Procedure for Southern blot analysis, involving electrophoresis of DNA which has been cut with a restriction enzyme, its transfer to a nitrocellulose filter, and hybridization of the filter with a specific DNA probe for the gene of interest.

the DNA of erythroid tissues where expression occurs (Jeffreys and Flavell, 1977) and numerous other examples have been reported (Fig. 2.8).

These techniques, which could detect losses of less than 100 bases (as opposed to the millions of bases resolvable by chromosomal techniques), have now been supplemented by cases where different laboratories have isolated particular genes from the DNA of different tissues which either do or do not express the gene. When the DNA sequence of the gene was determined in each case, no differences were detected in the expressing or non-expressing tissues. Similarly, the full DNA sequence of the human genome (International Human Genome Sequencing Consortium, 2001; Venter *et al.*, 2001) did not reveal any deletions in the sequence of specific genes where the DNA had been obtained from a tissue where that gene was not expressed. Such techniques, which could detect the loss of even a single base of DNA, have therefore confirmed the conclusions of all other studies, indicating that selective DNA loss is not a general mechanism of gene regulation.

2.3   DNA amplification

DNA amplification as a mechanism of gene regulation

Having eliminated selective DNA loss as a general means of gene control, it is necessary to consider the possible role of gene amplification in the regulation of gene expression (for reviews see Stark and

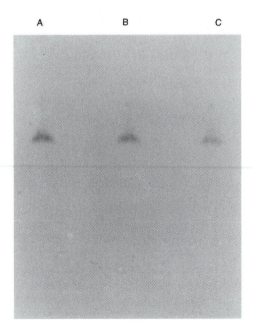

Figure 2.8

Southern blot hybridization of a radioactive probe specific for the α-fetoprotein gene to *Eco*RI-cut DNA prepared from fetal liver (track A), adult liver (track B) and adult brain (track C). Note the presence of an identically sized gene fragment in each tissue, although the gene is only expressed in the fetal liver.

Wahl, 1984; Kafatos *et al.*, 1985). A possible mechanism of gene control (Fig. 2.9) would involve the selective amplification of genes that were expressed at high levels in a particular tissue, the high expression of such genes simply resulting from normal rates of transcription of the multiple copies of the gene. Unlike gene-deletion models, this possibility is not incompatible with the ability of nuclei from differentiated cells, such as frog intestinal cells, to regenerate a new organism (see Section 2.2). Thus, although amplification of intestine-specific genes would have occurred in such cells, this would presumably not prevent the amplification of other tissue-specific genes in the individual cell types of the new organism. The evidence against this idea comes therefore from chromosomal and molecular rather than functional investigations.

Chromosomal studies

As with DNA loss, the identity of the chromosomal complement in individual cell types provides strong evidence against the occurrence of large-scale DNA amplification. Nonetheless, some cases of such amplification have been observed by this means and these will be discussed.

Chromosome polytenization in *Drosophila*

One such case, which has been extensively used in cytogenetic studies, occurs in the salivary gland cells of the fruit fly *Drosophila melanogaster*. In these cells the DNA replicates repeatedly and the daughter molecules do not separate but remain close together, forming a single giant, or polytene, chromosome (Fig. 2.10) which contains approximately 1000 DNA molecules. Although such giant chromosomes have proved very useful in experiments involving visualization of DNA transcription (see Section 4.2), they cannot really be considered an example of gene control since virtually the whole of the genome (with the exception of some repeated DNA) participates in this amplification and there is no evidence for selective amplification of the genes for proteins required in the salivary gland.

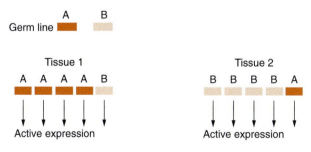

Figure 2.9

Model for gene regulation in which a gene that is expressed at high levels in a particular tissue is selectively amplified in that tissue. Gene A is amplified in tissue 1 and gene B in tissue 2.

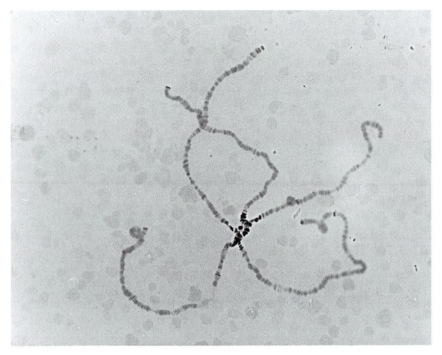

Figure 2.10

Polytene chromosomes of *Drosophila melanogaster*. Photograph kindly provided by Dr M. Ashburner.

Amplification of ribosomal DNA

Amplification of a specific part of the genome does occur, however, in some cases. Thus, amplification of the ribosomal DNA encoding the structural RNAs of the ribosomes is observed in the embryonic development of multicellular organisms, even though the germ line DNA of these organisms contains multiple copies of the ribosomal DNA prior to any amplification event. Thus, early in oogenesis the frog oocyte, like all other frog cells, contains approximately 900 chromosomal copies of the DNA encoding ribosomal RNA. During oogenesis, some of the copies are excised from the genome and replicate extensively to produce up to 2 million extra-chromosomal copies of the ribosomal DNA (Brown and Dawid, 1968).

Such ribosomal DNA amplification clearly represents a response to the intense demand for ribosomal RNA during embryonic development. This requirement has to be met by DNA amplification since the ribosomal RNA represents the final product of ribosomal DNA expression. By contrast, in the case of protein-coding genes, high-level expression can be achieved both by very efficient transcription of the DNA into RNA and also by similarly efficient translation of the RNA into protein. Hence the observed amplification of ribosomal RNA genes does not imply that this mechanism is used in higher organisms to amplify specific protein-coding genes. Indeed the general similarity of the chromosome complement in different tissues argues against large-scale amplification of such genes. As with gene deletion, however, the sensitivity of such techniques is such that small-scale amplification of a few genes would not be detected, and it is necessary to use molecular techniques to see whether such amplification occurs.

Molecular studies

In addition to its use in searching for gene loss (see Section 2.2) the technique of Southern blotting (see Methods Box 2.1; Fig. 2.7) can also be used to search for DNA amplification in tissues expressing a particular gene. Thus, either new DNA bands hybridizing to the specific probe will appear, or the same band, present in other tissues, will be observed but it will hybridize more intensely because more copies are present. Despite very many studies of this type only a very small number of cases of such amplification have been reported, the vast majority of tissue-specific genes being present in the same number of copies in all tissues. Thus, for example, the fibroin gene is detectable at similar copy numbers in the DNA of the posterior silk gland of the silk moth, where it is expressed, and in the DNA of the middle silk gland, where no expression is detectable (Fig. 2.11; Manning and Gage, 1978).

However, one exception to this general lack of amplification of protein-coding genes is particular noteworthy. This involves the genes encoding

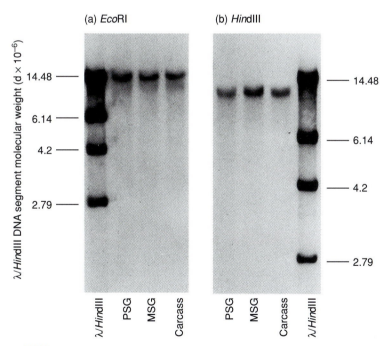

Figure 2.11

Southern blot of DNA prepared from the posterior silk gland (PSG), the middle silk gland (MSG) or the carcass of the silk moth, *Bombyx mori*, with a probe specific for the fibroin gene. Note the identical size and intensity of the band produced by *Eco*RI or *Hind*III digestion in each tissue, although the gene is only expressed in the posterior silk gland. Photograph kindly provided by Dr R. Manning, from R.F. Manning and L.P. Gage, *J. Biol. Chem.*, **253**, 2044–2052 (1978).

the eggshell, or chorion, proteins in *Drosophila melanogaster*. As first shown by Spradling and Mahowald (1980), the chorion genes are selectively amplified (up to 64 times) in the DNA of the cells which surround the egg follicle, allowing the synthesis in these cells of the large amounts of chorion mRNA and protein needed to construct the eggshell. Such amplification can be observed readily in a Southern blot experiment using a recombinant DNA probe derived from one of the chorion genes (Fig. 2.12). Unlike ribosomal DNA (see above), amplification occurs within the chromosome without excision of the template DNA or the newly replicated copies.

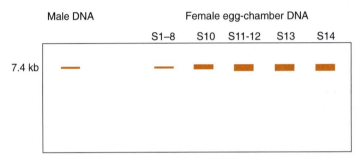

Figure 2.12

Amplification of the chorion genes in the ovarian follicle cells of *Drosophila melanogaster*. Note the dramatic increase in the intensity of hybridization of the chorion gene specific probe to the DNA samples prepared from Stage 10 to Stage 14 egg chambers compared with that seen in Stages 1–8 or in male DNA. Redrawn from A.C. Spradling and A.P. Mahowald, *PNAS* **77**, 1096–1100 (1980), by permission of Dr A. Spradling.

This amplification event is of particular interest because it occurs in normal cells as part of normal embryonic development, in contrast, for example, to the amplification of cellular oncogenes that occurs in some cancer cells (see Section 9.3) or the amplification of the gene encoding the enzyme dihydrofolate reductase (which occurs in mammalian cells in response to exposure to the drug methotrexate; Schimke, 1980). Nonetheless, the amplification of the chorion genes appears to be a response to a very specialized set of circumstances, requiring a novel means of gene regulation. Thus eggshell construction in *Drosophila* occurs over a very short period (about 5 h) and probably necessitates the synthesis of mRNA for the chorion proteins at very high rates, too large to be achieved even by high-level transcription of a single unique gene. It is this combination of high-level synthesis and a very short period to achieve it which necessitates the use of gene amplification. The restricted use that is made of this mechanism is well illustrated, however, by the fact that in silk moths, in which eggshell production occurs by a similar mechanism, multiple copies of the genes for the chorion proteins are present in the germ line. Thus they are present in all somatic cells, including those that do not express these genes, and are not further amplified in the ovarian follicle cells.

Such a finding serves to reinforce the conclusion of both molecular and chromosomal experiments that, as with DNA loss, DNA amplification is not a general method of gene regulation and is used only in isolated specialized cases.

2.4   DNA rearrangement

Having established that, generally, gene regulation does not occur by DNA deletion or amplification, it is necessary to consider the possibility that it might occur via specific rearrangements of genes in tissues where their protein products are required. Such activation could occur, for example, by translocation of the gene, removing it from a repressive effect of neighboring sequences, or by bringing together the part of the gene encoding the protein and a promoter element necessary for its transcription (Fig. 2.13).

As with DNA amplification, the occurrence of such rearrangements in one particular cell type might not preclude the subsequent rearrangement of genes required in other tissues. Hence the ability of a single differentiated cell type to give rise to a variety of other differentiated cells in functional studies (see Section 2.2) cannot be used to argue against the occurrence of such rearrangements. Similarly, these rearrangements might be on too small a scale to be detectable in chromosomal studies.

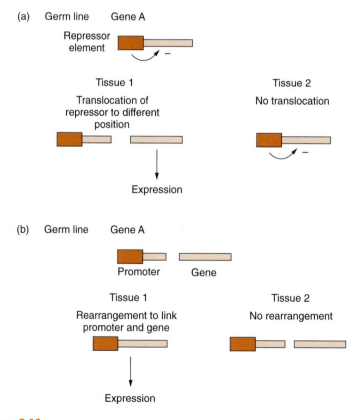

Figure 2.13

Model for activation of a gene by a DNA rearrangement involving either the removal of a repressor element (A) or the joining of the gene and its promoter (B).

The evidence from molecular studies, however, shows unequivocally that these types of rearrangement are not involved in the selective activation of the vast majority of genes in particular tissues. Thus such rearrangements would produce novel bands when the DNA from a tissue in which the gene was active was cut with restriction enzymes and used in Southern blot experiments (Fig. 2.14). In fact, no such novel bands are observed in the vast majority of cases when the DNA of a tissue in which the gene is active is compared with tissues where it is inactive (Figs 2.8 and 2.11).

The evidence against DNA rearrangement as a general means of gene control, obtained in such Southern blot experiments, has been abundantly confirmed by experiments in which the DNA sequence of a particular gene has been obtained both from a tissue where it is inactive and from one where it is active. Such experiments have failed to reveal even a single base difference in these two situations, and indicate that the gene is potentially fully functional in tissues where it is inactive. Despite these clear findings that, as with deletion and amplification, DNA rearrangement is not a general means of gene control, it is used in a few situations such as in the mammalian immunoglobulin genes.

Antibody production in mammalian B cells

When mammals are exposed to foreign bacteria or viruses, their immune systems respond by synthesizing specific antibodies directed against the

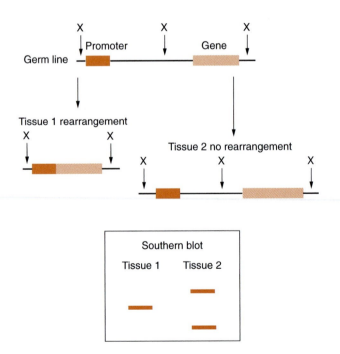

Figure 2.14

Generation of new restriction fragments detectable on a Southern blot following gene rearrangement. Sites for a particular restriction enzyme are indicated by an X and the consequences of digestion with this enzyme of unrearranged or rearranged DNA are indicated.

proteins of these organisms, with the aim of neutralizing the harmful effects of the infection. The potential requirement for the synthesis of diverse kinds of antibody by the mammalian immune system is obviously vast. Thus, the body must be able to defend itself, for example, against a bewildering variety of challenges from infectious organisms by the production of specific antibodies against their proteins (for recent reviews of the immune system, emphasizing the molecular aspects, see Abbas and Lichtman, 2003; Janeway *et al.*, 2005). These antibodies, or immunoglobulins, are produced by the covalent association of two identical heavy chains and two identical light chains to produce a functional molecule (Fig. 2.15). The combination of specific heavy and light chains produces the specificity of the antibody molecule. In particular, each chain contains, in addition to a constant region (which is relatively similar in different antibody molecules), a highly variable region which differs widely in amino-acid sequence in different antibodies. It is this variable region that actually interacts with the antigen and hence determines the specificity of the antibody molecule.

As any heavy chain can associate with any light chain, and as approximately 1 million types of antibody specificities can be produced, at least 1000 genes encoding different heavy chains and a similar number encoding the light chains would be required. Copy-number studies have shown, however, that the germ line does not contain anything like this number of intact immunoglobulin genes, and no specific amplification events have been detected in the DNA of antibody-producing B cells.

Rather, as first suggested by Dreyer and Bennett (1965), functional immunoglobulin genes are created by DNA rearrangements in the B-cell lineage. Thus the germ-line DNA contains a large number of tandemly repeated DNA segments encoding different variable regions, which are separated by over 100 kbp from a much smaller number of DNA segments encoding the constant region of the molecule (Fig. 2.16). This organization is maintained in most somatic cell types, but in each individual B cell a unique DNA rearrangement event occurs by which one specific variable

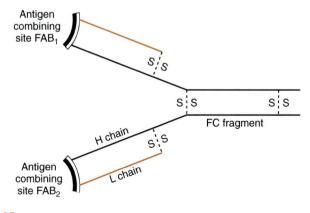

Figure 2.15

Association of two heavy (H) and two light (L) chains to form a functional antibody molecule. The disulfide bonds linking the chains and the region that binds antigen are indicated.

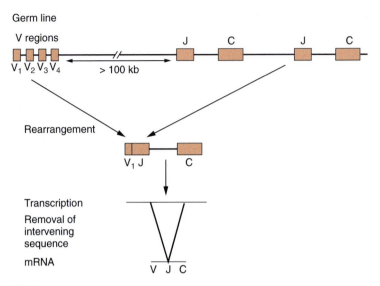

Figure 2.16

Rearrangement of the gene encoding the λ light chain of immunoglobulin results in the linkage of one specific variable region (V) to a specific joining region (J) with removal of the intervening DNA. The gene is then transcribed into RNA and the region between the joining (J) and constant regions (C) is removed by RNA splicing to create a functional mRNA.

region is brought together with one constant region by deletion of the intervening DNA. In this manner, a different functional gene, containing specific variable and constant regions, is produced in each individual B cell.

These rearrangements can be detected readily by the appearance of novel bands in Southern blot analyses of the structure of the immunoglobulin genes in the DNA of antibody-producing B cells. In the example shown in Fig. 2.17 (based on the data of Hozumi and Tonegawa, 1976) the digestion of DNA from non-B cells with the restriction enzyme *Bam*H1 produces fragments of 6.5 kb and 9.5 kb containing the variable and constant regions, respectively, whereas digestion of rearranged B-cell DNA generates a single 3.5 kb band containing both parts of the gene. This combination of different variable (V) and constant (C) regions in the production of a functional immunoglobulin gene obviously creates a wide range of different potential antibody specificities by the joining of partic-ular V and C segments.

Despite this, more and more complex rearrangements are used in differ-ent types of immunoglobulin gene to generate still more potential variety. Thus, in the simplest case that we have discussed so far, that of the λ light chain (Fig. 2.16), one of many V regions is brought into association with a constant region, consisting of a single J segment (encoding the junction between the variable and constant regions) which is separated by a short region from the DNA encoding the C segment itself. The rearranged gene is then transcribed and the intervening sequence between the J and C segments removed by RNA splicing to produce a functional messenger RNA (see Section 5.2). In the genes for another type of light chain, known as κ, the situation is more complex. Thus these genes contain multiple J

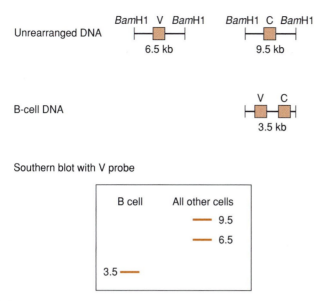

Figure 2.17

The rearrangement of the immunoglobulin gene locus creates a different *BamH*1 restriction fragment hybridizing to immunoglobulin DNA-specific probes in the DNA of B cells, compared with those that are present in other cell types where no rearrangement has occurred.

segments linked to individual constant segments and hence further diversity is created by the recombination of one of many V regions with one of several different J regions.

The case of the genes encoding the heavy-chain molecule is still more complex (Fig. 2.18) in that these genes also contain a D, or diversity, segment. A functional heavy-chain gene is thus created by two successive rearrangements. In the first of these, one of several D segments is brought together with one of the J segments and the associated C region. Subsequently, a V region segment is recombined with this DJC element exactly as occurs in the VJ joining found in the light-chain genes. These two combinatorial events allow the selection of one particular copy of each of four regions to produce the functional gene, generating many different possible heavy-chain molecules.

Interestingly, the generation of diversity by DNA rearrangement is supplemented by another DNA alteration, so far unique in mammalian

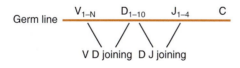

Figure 2.18

Rearrangement in the gene encoding the heavy chain of the immunoglobulin molecule involves the joining of a variable (V) and a diversity (D) region as well as the similar joining of a diversity and a junction (J) region, with removal of the intervening DNA in each case.

cells, to the genes of the immune system. It appears that immunoglobulin genes can undergo point mutations within particular B cells, resulting in changes in the amino acids they encode and hence in their specificity. Such somatic mutation (for a review see Tian and Alt, 2000) produces forms of immunoglobulin not encoded by the DNA present in other tissues or in the germ line, and represents another form of tissue-specific gene alteration occurring within the immunoglobulin genes.

It is clear that the primary role of the rearrangements and alterations which occur in the immunoglobulin genes in B cells is the generation of the diversity of antibodies required to cope with the vast number of different possible antigens, rather than to produce the high-level expression of the immunoglobulin genes which occurs in such cells. Nonetheless, high-level immunoglobulin-gene expression does occur as a consequence of such rearrangements. Thus the DNA encoding the variable regions of the antibody molecule is closely linked to the promoter elements that direct the transcription of the gene. Such elements show no activity in the unrearranged DNA of non-B cells, but become fully active only when the variable region is joined to the constant region. In the case of the heavy-chain genes, this is because the intervening sequence between the J and C regions of these genes contains an enhancer element (see Section 7.3) which, although not a promoter itself, greatly increases the activity of the promoter adjacent to the V-region element. Hence the rearrangement not only brings the promoter element into a position where it can produce a functional mRNA containing VDJ and C, regions but also facilitates the high-level transcription of this gene by juxtaposing promoter and enhancer elements that are separated by over 100 kb of DNA before the rearrangement event (Fig. 2.19).

The element of gene control in this process is, however, entirely secondary to the need to produce diversity by rearrangement. Thus, if the immunoglobulin enhancer is linked artificially close to the immunoglobulin promoter and introduced into a variety of cells, it will activate transcription from the promoter only in B cells and not in other cell types. Hence the enhancer is active only in B cells and would not activate immunoglobulin transcription even if functional immunoglobulin genes containing closely linked VDJ and C segments were present in all other tissues. The immunoglobulin genes are therefore regulated by tissue-

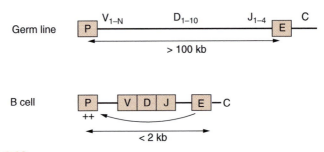

Figure 2.19

The rearrangement of the heavy-chain gene brings an enhancer (E) in the intervening sequence between J and C close to the promoter (P) adjacent to V and results in the activation of the promoter and gene transcription.

specific activator sequences in much the same way as other genes (see Chapter 7), but such activation has been made to depend on the occurrence of a DNA rearrangement whose primary role lies elsewhere.

Interestingly, a further link between gene regulation and gene rearrangement processes exists, apart from that provided by the juxtaposition of promoter and enhancer elements which is produced by the gene rearrangement of the immunoglobulin genes. Thus, enhancer elements in the unrearranged locus act to recruit histone acetylase enzymes which modify the chromatin structure to render it more accessible to the enzymes which will catalyze the DNA rearrangement (Fig. 2.20) (for reviews see Schlissel, 2000; Mostoslavsky *et al.*, 2003). This exactly parallels the role of histone acetylation in opening up the chromatin structure of genes which are not subject to rearrangement, allowing them to be transcribed (see Section 6.6).

Despite these relationships between gene rearrangement and gene regulation and the fact that similar DNA rearrangements are used in other situations where diversity must be generated, for example, in the genes encoding the T-cell receptor which is involved in the response to foreign antigen of T-lymphocytes, it is clear that they do not represent a general means of gene regulation applicable to other situations. Such a conclusion is entirely in accordance with the results of Southern blot and sequencing experiments which, as previously discussed, have failed to detect such rearrangements in the vast majority of genes active only in particular tissues.

2.5   Conclusions

Although we have discussed in this chapter a number of cases where DNA loss, amplification or rearrangement regulate gene expression, it is clear that these represent isolated special cases where the dictates of a particular situation have necessitated the use of such mechanisms. Thus the loss

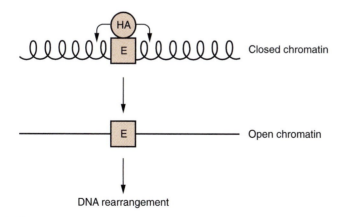

Figure 2.20

Recruitment of histone acetylase (HA) enzymes to an enhancer element results in a more open chromatin structure of immunoglobulin or T-cell receptor genes. This in turn allows access to the enzymes which catalyze the DNA rearrangement necessary for proper functioning of these genes.

of DNA in red blood cells is dictated by the need to fill the cell with hemoglobin, the amplification of the chorion genes in *Drosophila* by the requirement to produce the corresponding proteins in a very short time; and the rearrangement of the immunoglobulin genes by the need to generate a diverse array of antibodies. Hence these cases are not representative of a general mechanism of gene regulation by DNA alteration and, in general, the DNA of different cell types is quantitatively and qualitatively identical. Given that we have established previously that the RNA content of different cell types can vary dramatically, it is now necessary to investigate how such differences in RNA content can be produced from the similar DNA present in such cell types. As a background to this discussion we will first discuss, in the next chapter, the processes by which DNA is transcribed into RNA and the subsequent post-transcriptional events which convert that RNA into its final form.

References

Abbas, A.K. and Lichtman, A.H. (2003). *Cellular and Molecular Immunology* (5th edition). Saunders, Philadelphia PA.

Bessis, N. and Brickal, M. (1952). Aspect dynamique des cellules du sang. Revues *Hématologique* 7, 407–435.

Brown, D.D. (1981). Gene expression in eukaryotes. *Science* 211, 667–674.

Brown, D.D. and Dawid, B. (1968). Specific gene amplification in oocytes. *Science* 160, 272–280.

Dreyer, W.J. and Bennett, J.C. (1965). The molecular basis of antibody formation: a paradox. *Proceedings of the National Academy of Sciences of the USA* 54, 864–869.

Gurdon, J.B. (1968). Transplanted nuclei and cell differentiation. *Scientific American* 219 (Dec.), 24–35.

Hozumi, N. and Tonegawa, S. (1976). Evidence for somatic rearrangement of immunoglobulin genes coding for variable and constant regions. *Proceedings of the National Academy of Sciences of the USA* 73, 3628–3632.

International Human Genome Sequencing Consortium (2001). Initial sequencing and analysis of the human genome. *Nature* 409, 860–921.

Janeway, C.A., Travers, P., Walport, M. and Schlomchick, M.J. (2005). *Immunobiology* (6th edition). Garland Sciences, New York, NY.

Jefferys, A.J. and Flavell, R.A. (1977). The rabbit β-globin gene contains a large insert in the coding sequence. *Cell* 12, 1097–1108.

Kafatos, F.C., Orr, W. and Delidakis, C. (1985). Developmentally regulated gene amplification. *Trends in Genetics* 1, 301–306.

Manning, R.F. and Gage, L.P. (1978). Physical map of the *Bombyx mori* DNA containing the gene for silk fibroin. *Journal of Biological Chemistry* 253, 2044–2052.

Mostoslavsky, R., Alt, F.W., and Bassing, C.H. (2003). Chromatin dynamics and locus accessibility in the immune system. *Nature Immunology* 4, 603–606.

Polejaeva, I.A., Chen, S-H., Vaught, T.D., Page, R.L., Mullins, J., Ball, S., et al. (2000). Cloned pigs produced by nuclear transfer from adult somatic cells. *Nature* 407, 86–90.

Schimke, R.T. (1980). Gene amplification and drug resistance. *Scientific American* 243, 50–59.

Schlissel, M.S. (2000). A tail of histone acetylation and DNA recombination. *Science* 287, 438–440.

Southern, E.M. (1975). Detection of specific sequences among DNA fragments separated by gel electrophoresis. *Journal of Molecular Biology* 98, 503–17.

Spradling, A.C. and Mahowald, A.P. (1980). Amplification of genes for chorion proteins during oogenesis in *Drosophila melanogaster*. *Proceedings of the National Academy of Sciences of the USA* **77**, 1096–100.

Stark, G.R. and Wahl, G.M. (1984). Gene amplification. *Annual Review of Biochemistry* **53**, 447–491.

Steward, F.C. (1970). From cultured cells to whole plants: the induction and control of their growth and morphogenesis. *Proceedings of the Royal Society, Series B* **175**, 1–30.

Stewart, C. (1997). An udder way of making lambs. *Nature* **385**, 769–771.

Tian, M. and Alt, F.W. (2000). RNA editing meets DNA shuffling. *Nature* **407**, 31–33.

Venter, J.C., Adams, M.D., Myers, E.W. (2001). The sequence of the human genome. *Science* **291**, 1304–1351.

Wakayama, T., Perry, A.C.F., Zuccotti, M., Johnson, K.R. and Yanagimachi, R. (1998). Full-term development of mice from enucleated oocytes injected with cumulus cell nuclei. *Nature* **394**, 369–374.

Wilmut, I., Schnieke, A.E., McWhis, J., Kind, A.J. and Campbell, K.H.S. (1997). Viable offspring derived from fetal and adult mammalian cells. *Nature* **385**, 810–813.

Wilson, E. B. (1928). *The Cell in Development and Heredity*. Macmillan, New York, NY.

Yamada, T. (1967). Cellular and sub-cellular events in Wolffian lens regeneration. *Current Topics in Developmental Biology* **2**, 249–283.

Gene expression

3

SUMMARY

- The conclusion that gene regulation produces distinct mRNA populations from the same DNA focuses attention on the processes which produce mRNA from the DNA. These processes are:
- Transcription in which DNA is copied into the primary RNA transcript.
- Capping in which the 5′ end of the RNA transcript is modified.
- Splicing in which introns are removed from the RNA transcript.
- Polyadenylation in which the RNA transcript is cleaved at its 3′ end and a run of A residues is added.
- Transport in which the mature mRNA is transported from the nucleus to the cytoplasm and then translated into protein by the process of translation.

3.1 Levels of gene regulation

The observation that differences in the RNA and protein content of different tissues are not paralleled by significant differences in their DNA content indicates that the process whereby DNA produces mRNA must be the level at which gene expression is regulated in eukaryotes.

In bacteria this process involves only a single stage, that of transcription, in which an RNA copy of the DNA is produced by the enzyme RNA polymerase. Even while this process is still occurring, ribosomes attach to the nascent RNA chain and begin to translate it into protein. Hence cases of gene regulation in bacteria, such as the switching on of the synthesis of the enzyme β-galactosidase in response to the presence of lactose (its substrate), are mediated by increased transcription of the appropriate gene (Jacob and Monod, 1961). Clearly, a similar regulation of gene transcription in different tissues, or in response to substances such as steroid hormones which induce the synthesis of new proteins, represents an attractive method of gene regulation in eukaryotes.

In contrast to the situation in bacteria, however, a number of stages intervene between the initial synthesis of the primary RNA transcript and the eventual production of mRNA (Fig. 3.1). The initial transcript is modified at its 5′ end by the addition of a cap structure containing a modified guanosine residue and is subsequently cleaved near its 3′ end, followed by the addition of up to 200 adenosine residues in a process known as polyadenylation. Subsequently, intervening sequences or introns, which interrupt the protein-coding sequence in both the DNA and the primary transcript of many genes (Jeffreys and Flavell, 1977; Tilghman *et al.*, 1978) are removed by a process of RNA splicing (for reviews see Adams *et al.*, 1996; Staley and Guthrie, 1998). Although this produces a functional

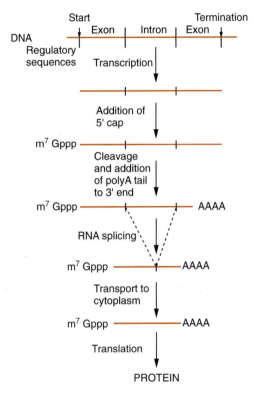

Figure 3.1

Stages in eukaryotic gene expression which could be regulated.

mRNA, the spliced molecule must then be transported from the nucleus, where these processes occur, to the cytoplasm where it can be translated into protein.

In theory any of these stages, all of which are essential for production of a functional mRNA, could be used to regulate the expression of specific genes in particular tissues. For example, it has been suggested that gene regulation could occur by transcribing all genes in all tissues and selecting which transcripts were appropriately spliced by correct removal of intervening sequences, thereby producing a functional mRNA (Davidson and Britten, 1979). The existence of such a plethora of possible regulatory stages has led, therefore, to much investigation as to whether all or any of these are used.

In order to provide the essential background for considering these studies, this chapter will consider the processes which occur at each of these stages between the DNA and the production of the functional mRNA. In addition, since some cases of regulation at the level of translation of the mRNA have been described, the final section of this chapter will consider the mechanism of translation. Once this background information has been provided, subsequent chapters will consider the evidence that the primary control of gene expression occurs at the level of transcription (Chapter 4) as well as a number of cases of post-transcriptional regulation (Chapter 5).

3.2   Transcription

Enzymes which are capable of copying the DNA so that a complementary RNA copy is produced by the polymerization of ribonucleotides are referred to as RNA polymerases. In prokaryotes a single RNA polymerase enzyme is responsible for the transcription of DNA into RNA. In eukaryotes, however, this is not the case, and three such enzymes exist. All three RNA polymerases are large multi-subunit enzymes and several subunits are held in common between the three enzymes. The three enzymes can be distinguished, however, by their relative sensitivity to the fungal toxin α-amanitin and each of them is active on a distinct set of genes (Table 3.1). Thus, whereas all genes capable of encoding a protein, as well as the genes for some small nuclear RNAs involved in RNA splicing (Section 3.3), are transcribed by RNA polymerase II, the genes encoding the 28S, 18S and 5.8S ribosomal RNAs are transcribed by RNA polymerase I, and those encoding the transfer RNAs and the 5S ribosomal RNA are transcribed by RNA polymerase III.

Table 3.1   Eukaryotic RNA polymerases

Genes transcribed	Sensitivity to α-amanitin
I Ribosomal RNA (45S precursor of 28S, 18S and 5.8S rRNA)	Insensitive
II All protein-coding genes, small nuclear RNAs U1, U2, U3, etc.	Very sensitive (inhibited 1 µg/ml)
III Transfer RNA, 5.S ribosomal RNA, small nuclear RNA U6, repeated DNA sequences: Alu, B1, B2, etc.; 7SK, 7SL RNA	Moderately sensitive (inhibited 10 µg/ml)

In considering the transcriptional regulatory processes that produce tissue-specific variation in specific mRNAs and proteins, our primary concern will therefore be with the regulation of transcription by RNA polymerase II. Transcription by RNA polymerases I and III is also subject to regulation, however. Moreover, the nature of the components involved in transcription by these polymerases is much simpler than those which are involved in transcription by RNA polymerase II. A prior understanding of the processes involved in basal transcription by RNA polymerases I and III therefore assists a subsequent understanding of the more complex processes involved in basal transcription by RNA polymerase II. Transcription by RNA polymerases I, II and III and their common processes is therefore considered in this section.

RNA polymerase I

RNA polymerase I is responsible for the transcription of the tandem arrays of genes encoding ribosomal RNA, such transcription constituting about one half of total cellular transcription. As with all RNA polymerases, RNA polymerase I itself does not recognize the DNA sequences around the start site of transcription. Rather, other protein factors recognize such sequences and then recruit the RNA polymerase by a protein–protein interaction. In the case of RNA polymerase I, the essential DNA sequences which are recognized are located within the 50 bases immediately upstream of the start site of transcription. As in all genes, the sequences adjacent to the transcriptional start site which control the expression of the gene are known as the gene promoter. Note that in the case of all the RNA polymerases the site at which transcription begins is denoted as +1 with bases within the transcribed region being denoted as +100, +200, etc., whilst bases upstream of the start site are denoted as –100, –200, etc., as one proceeds further and further upstream (see Fig. 3.2).

In the case of RNA polymerase I, sequences around –50 are recognized by a protein transcription factor known as UBF (upstream binding factor) (for reviews see Jacob, 1995; Paule and White, 2000; Grummt, 2003). Subsequently another regulatory protein known as SL1 (also known as TIF-IB) is recruited via protein–protein interaction with UBF and in turn SL1 recruits the RNA polymerase itself and its associated factors (Fig. 3.2). Hence the initiation of transcription by RNA polymerase I is relatively simple, with one essential transcription factor (SL1) being necessary to recruit the RNA polymerase itself. In turn the binding of this essential factor is facilitated by the prior binding to a specific DNA sequence of another transcription factor UBF (for review see Jacob, 1995; Paule and White, 2000; Grummt, 2003).

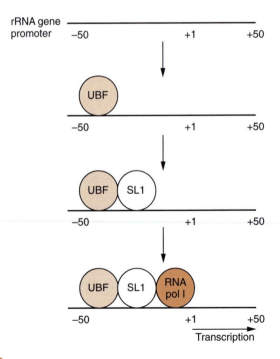

Figure 3.2

Transcription initiation at the ribosomal RNA gene promoter. Binding of UBF (upstream binding factor) is followed by the binding of the SL1 factor which in turn recruits RNA polymerase I via protein–protein interaction. Note that in this and all subsequent figures, +1 refers to the first base transcribed into RNA, with other + numbers denoting bases within the transcribed region whilst – signs denote bases upstream of the transcriptional start site.

RNA polymerase III

The involvement of a specific transcription factor which acts to recruit the RNA polymerase as well as of other factors which recruit the specific factor is also illustrated by RNA polymerase III (for reviews see Paule and White, 2000; Geiduschek and Kassavetis, 2001; Schramm and Hernandez, 2002).

The situation is complicated, however, by the fact that in different genes which are transcribed by RNA polymerase III, the essential promoter DNA sequences recognized by the transcription factors which recruit the polymerase can be located either upstream or downstream of the transcribed region. Thus genes transcribed by RNA polymerase III can have either an upstream promoter (as is the case for genes transcribed by RNA polymerase I and II) or a downstream promoter located within the transcribed region (Fig. 3.3).

This type of downstream promoter, which is unique to RNA polymerase III, was first identified by detailed studies which focused on the genes that encode the 5S RNA of the ribosome. In an attempt to identify the sequences important for the expression of this gene, sequences surrounding it were deleted and the effect on the transcription of the gene in a cell-free system investigated (Sakonju et al., 1980). Somewhat surprisingly, the entire upstream region of the gene could be deleted with no effect on gene expression (Fig. 3.4a,b). Indeed, deletions within the transcribed region of the 5S gene also had no effect on its expression until a boundary, 40 bases within the transcribed region, was crossed. By this means, an internal control region essential for the transcription of the 5S RNA gene was defined which is located entirely within the transcribed region.

This region of the 5S gene was shown subsequently to bind a transcription factor known as TFIIIA (for review see Pieler and Theunissen, 1993) (Fig. 3.5). Subsequently, another transcription factor TFIIIC binds to the DNA adjacent to TFIIIA and in turn TFIIIC functions to recruit a further transcription factor TFIIIB to form a stable transcription complex (Fig. 3.6). This complex, which is stable through many cell divisions, promotes the subsequent binding of the RNA polymerase III which is recruited via a protein–protein interaction with TFIIIB and binds at the transcriptional start site (Fig. 3.6). The binding of RNA polymerase III is dependent on the

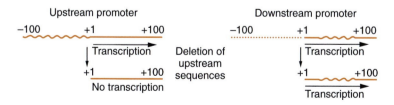

Figure 3.3

Consequences of deleting sequences upstream of the transcriptional start site in genes which have either an upstream promoter or a downstream promoter. Note that, when the promoter (wavy line) is located upstream of the transcriptional start site, deletion of upstream sequences results in an absence of transcription. In contrast, when the promoter is located within the transcribed region, deletion of upstream sequences has no effect on transcription.

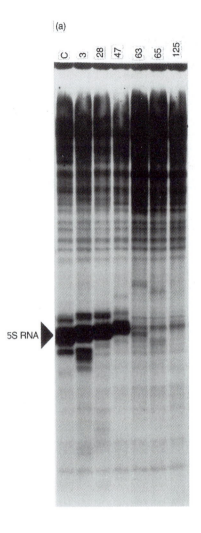

(a)

C 3 28 47 63 65 125

5S RNA ▶

Figure 3.4

Effect of deletions in the 5S rRNA gene on its expression. (a) Transcription assay in which the production of 5S rRNA (arrowed) by an intact control 5S gene (C) and various deleted 5S genes is assayed. The numbers indicate the end point of each deletion used, 47 indicating that the deletion extends from the upstream region to the 47th base within the transcribed region, etc. (b) Summary of the extent of the deletions used and their effects on transcription. The use of these deletions allows the identification of a critical control element (boxed) within the transcribed region of the 5S gene. Photograph kindly provided by Professor D.D. Brown, from Sakonju *et al.*, *Cell* **19**, 13–25 (1980), by permission of Cell Press.

Transcription

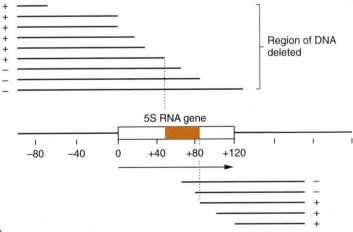

(b)

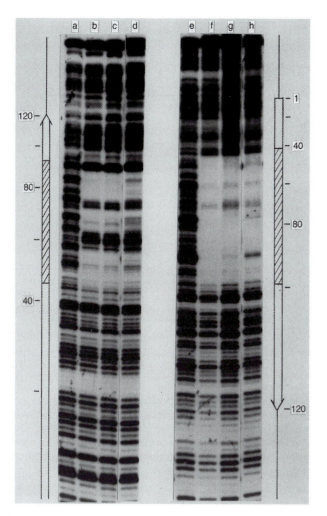

Figure 3.5

Binding of the TFIIIA transcription factor to the internal control region of 5S
DNA in a footprint assay in which binding of a protein protects the DNA from
digestion by DNase I and produces a clear region lacking the ladder of bands
produced by DNase I digestion of the other regions. Tracks a and e show the
two DNA strands of the 5S gene in the absence of added TFIIIA while tracks
b–d and f–h show the same DNA in the presence of TFIIIA. Photograph kindly
provided by Professor D.D. Brown, from Sakonju and Brown, *Cell* **31**, 395–405
(1982), by permission of Cell Press.

presence of the stable transcription complex and not on the precise
sequence of the DNA to which it binds since, as discussed above, the region
to which the polymerase normally binds can be deleted and replaced by
other sequences without drastically reducing transcription.

Although the assembly of transcription complexes on RNA polymerase
III genes was first defined on the 5S ribosomal RNA gene, other RNA polymerase III transcription units differ in the details of transcription complex
assembly. Thus, for example, the genes encoding the transfer RNAs which
play a key role in translation (see Section 3.3) also have an internal

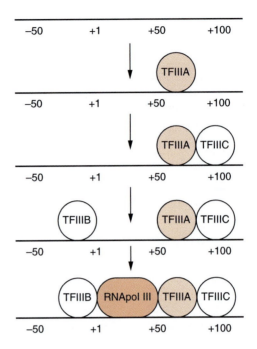

Figure 3.6

Transcription initiation by RNA polymerase III at the 5S rRNA gene promoter. Sequences downstream of the transcription initiation site (+1) are indicated by the + signs, sequences upstream are indicated by the – signs. Following binding of the transcription factor TFIIIA to the internal control sequence, TFIIIC and TFIIIB bind with TFIIIB then acting to recruit RNA polymerase III, allowing transcription to begin. Note that within the three-dimensional structure, TFIIIB interacts directly with TFIIIC, although this is not indicated in the figure.

promoter. However, in this case, due to differences in the sequence of the promoter TFIIIA is not required. Rather, TFIIIC binds directly to sequences within the promoter and subsequently recruits TFIIIB. As in the 5S RNA promoters, TFIIIB then recruits the RNA polymerase itself. Similarly, TFIIIB plays a critical role in transcription of the RNA polymerase III genes which have an upstream promoter such as that encoding the small nuclear RNA U6 which, unlike the other small nuclear RNAs of the splicesosome (see Section 3.3), is transcribed by RNA polymerase III rather than RNA polymerase II (Table 3.1). (For review of small nuclear RNA gene transcription see Hernandez, 2001.)

It is clear, therefore, that TFIIIB plays an essential role in the transcription of the three different types of RNA polymerase III gene promoters (for review see Hernandez, 1993) and is the functional equivalent of the RNA polymerase I SL1 factor which acts directly to recruit the RNA polymerase via protein–protein interaction (for a detailed comparison of transcriptional initiation by RNA polymerases I and III see Paule and White, 2000).

RNA polymerase II

The TATA box

Inspection of the region immediately upstream of the transcriptional start site reveals that a very wide variety of different genes transcribed by RNA polymerase II contain an AT-rich sequence which is found approximately 30 bases upstream of the transcription start site. This TATA box plays a critical role in promoting transcriptional initiation and in positioning the start site of transcription for RNA polymerase II. Its destruction by mutation or deletion effectively abolishes transcription of such genes (for review see Breathnach and Chambon, 1981).

Most importantly, this TATA box acts as the initial DNA target site for the progressive assembly of the basal transcription complex for RNA polymerase II which is considerably more complex than that for RNA polymerase I or III (for reviews see Nikolov and Burley, 1997; Woychick and Hampsey, 2002; Roeder, 2003). Thus, initially, the TATA box is bound by the transcription factor TFIID whose binding is facilitated by the presence of another transcription factor TFIIA (Fig. 3.7a). Interestingly, structural analysis of TFIID has revealed it to have a molecular clamp structure, which consists of four globular domains around an accessible groove, which can accommodate the DNA to which TFIID binds (Brand *et al.*, 1999) (color plate 3).

Subsequently, the TFIID/DNA complex is recognized by another transcription factor TFIIB (Fig. 3.7b). Structural analysis has shown that TFIIB binds on the opposite side of TFIID to that which is bound by TFIIA (Andel *et al.*, 1999) (color plates 4 and 5). This binding of TFIIB is an essential step in the formation of the initiation complex since, as well as binding to TFIID, TFIIB can also recruit RNA polymerase II itself. Hence the binding of TFIIB allows the subsequent recruitment of RNA polymerase II to the initiation complex in association with another factor TFIIF (Fig. 3.7c). Subsequently, two other factors TFIIE and TFIIH associate with the complex (Fig. 3.7d).

In particular, the recruitment of TFIIH plays a critical role in allowing the RNA polymerase to initiate transcription. Thus, TFIIH is in fact a multicomponent complex whose molecular structure has been determined (Chang and Kornberg, 2000; Schultz *et al.*, 2000) and which appears to play a key role in both transcription and the repair of damaged DNA (for reviews see Hoeijmakers *et al.*, 1996; Zurita and Merino, 2003).

Thus, TFIIH contains a kinase activity which is capable of phosphorylating the C-terminal domain of the largest subunit of RNA polymerase II. This C-terminal domain contains multiple copies of the sequence Tyr-Ser-Pro-Thr-Ser-Pro-Ser which is highly evolutionarily conserved (for a review see Stiller and Hall, 2002). The kinase activity of TFIIH phosphorylates serine 5 within this repeat and this allows transcription initiation to occur. However, transcription only proceeds for 20–30 bases and then the polymerase pauses and does not continue transcribing. Subsequent phosphorylation of serine 2 within the C-terminal domain of RNA polymerase II by the pTEF-b kinase is necessary to release this block and allow transcriptional elongation to occur (for review see Orphanides and Reinberg, 2002; Sims *et al.*, 2004) (Fig. 3.8).

Hence, phosphorylation of the C-terminal domain plays a critical role in allowing the polymerase to initiate and continue transcription. Thus,

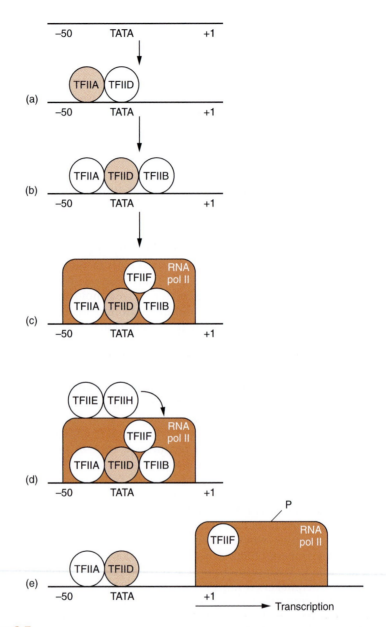

Figure 3.7

Transcription initiation at an RNA polymerase II promoter. Initially, the TFIID factor binds to the TATA box together with another factor TFIIA (panel a). Subsequently TFIIB is recruited by interaction with TFIID (panel b) and TFIIB then recruits RNA polymerase II and its associated factor TFIIF (panel c). TFIIE and TFIIH then bind and TFIIH phosphorylates the C-terminal domain of RNA polymerase II (panel d). This converts RNA polymerase II into a form which is capable of initiating transcription and RNA polymerase II and TFIIF then move off down the gene producing the RNA transcript, leaving TFIIA and TFIID bound at the promoter (panel e).

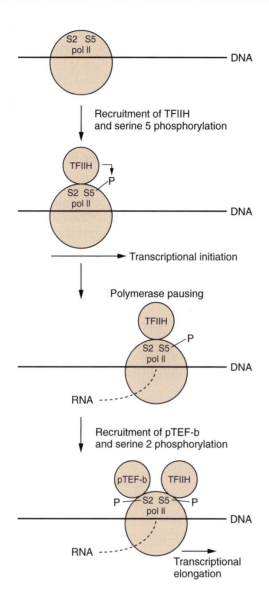

Figure 3.8

Recruitment of TFIIH results in the phosphorylation of RNA polymerase II on serine 5 of its C-terminal domain and allows transcriptional initiation to occur. However, the polymerase pauses and ceases transcribing after producing a short RNA transcript. Subsequently, recruitment of the pTEF-b kinase results in the phosphorylation of polymerase II on serine 2 of the C-terminal domain, allowing transcriptional elongation to produce the complete RNA product.

although the dephosphorylated form of RNA polymerase II is recruited to the DNA, its phosphorylation is necessary for transcription to produce the RNA product (Fig. 3.7e).

Both RNA polymerase II itself and an active transcription complex have been crystallized allowing structural analysis (for reviews see Klug, 2001; Landick, 2001; Asturias and Craighead, 2003) (color plate 6). This has revealed that, as the DNA is transcribed into RNA in the interior of the polymerase molecule, it encounters a wall of protein within the RNA polymerase (labeled A in Fig. 3.9). This forces it to make a right angled turn exposing the end of the nascent RNA and allowing ribonucleoside triphosphates to be added to it as transcription occurs. Subsequently, the newly-formed DNA–RNA hybrid produced as a consequence of transcription encounters another part of RNA polymerase, known as the rudder (labeled B in Fig. 3.9), which forces the separation of the RNA from the DNA. This allows the newly-formed part of the RNA molecule to exit and double-stranded DNA to reform (Fig. 3.9).

Interestingly, this complex interaction of the DNA and RNA polymerase is facilitated by TFIIB. Thus, recent structural studies of the RNA polymerase/TFIIB complex (Bushnell *et al.*, 2004; Chen and Hahn, 2004) have shown that TFIIB does more than simply recruit the polymerase. Thus, by interacting with both DNA-bound TFIID and the polymerase itself in a very precise structure, TFIIB ensures that the DNA and the polymerase molecule are correctly positioned and orientated relative to one another for the DNA to enter the interior of the polymerase, allowing transcription to occur.

As the polymerase moves off down the gene, TFIIF remains associated with it whilst TFIIA and TFIID remain bound at the promoter, allowing further cycles of recruitment of TFIIB, RNA polymerase II etc., leading to repeated rounds of transcription. Such a role of TFIIA and TFIID in allow-

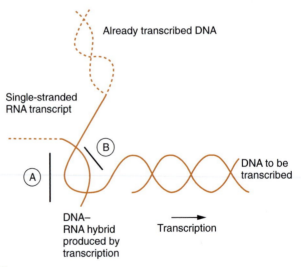

Figure 3.9

Movement of the DNA being transcribed through the RNA polymerase molecule. "A" indicates the wall within the polymerase protein which forces the DNA–RNA hybrid to make a right-angled turn, thereby allowing transcription to occur by addition of ribonucleoside triphosphates to the end of the RNA chain. "B" indicates the rudder region of the polymerase which forces the DNA–RNA hybrid to melt releasing the newly-formed RNA and allowing double-stranded DNA to reform. DNA which is about to be transcribed is shown by the solid lines and DNA which has already been transcribed by the dotted lines. The arrow indicates the direction of transcription.

ing repeated rounds of transcription is of particular interest in view of the finding that some RNA polymerase is found within the cell associated with a large number of proteins including TFIIB, TFIIF and TFIIH to form a so-called RNA polymerase holoenzyme (for reviews see Greenblatt, 1997; Myer and Young, 1998). Hence, in addition to the stepwise pathway involving progressive recruitment of TFIIB, RNA polymerase and TFIIH individually, it appears that an alternative pathway exists in which TFIIA and TFIID can recruit a complex containing TFIIB, TFIIE, TFIIF, TFIIH and RNA polymerase itself (Fig. 3.10). In addition, the holoenzyme also contains a number of other protein components which appear to be involved either in opening up the chromatin structure so as to allow transcription to occur (see Section 6.6) or in allowing the polymerase complex to be stimulated by transcriptional activators (see Section 8.3).

Common features of transcription by the three different polymerases

Although the three different polymerases each have a different function and different associated transcription factors, the three polymerases themselves are all multi-subunit proteins with several subunits being shared

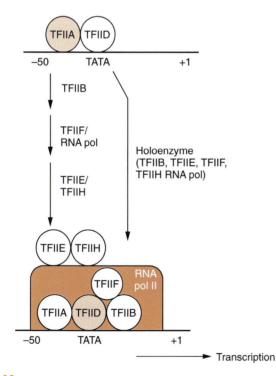

Figure 3.10

Following binding of TFIIA and TFIID to the promoter, the formation of the basal transcription complex for RNA polymerase II may take place via the sequential recruitment of TFIIB, TFIIF/RNA polymerase II and TFIIE/TFIIH as indicated in Fig. 3.7 or by the recruitment of a holoenzyme containing all of these factors.

between the different polymerases. Moreover, the largest subunits of each of the three enzymes show considerable sequence homology to one another.

In addition, each of the three polymerases shows a similar pattern of recruitment to the DNA with one specific transcription factor binding to a target sequence in the gene promoter followed by binding of one or more other proteins which then recruit the polymerase itself. As discussed above in the case of RNA polymerase II promoters which contain a TATA box, the original binding to DNA is achieved by the TFIID factor which binds to the TATA box. Interestingly, however, although TFIID was originally identified as a single factor it is now clear that it is composed of multiple protein components (for reviews see Burley and Roeder, 1996; Tansey and Herr, 1997). One of these proteins known as TBP (TATA-binding protein) is responsible for binding to the TATA box whilst the other components of the complex, known as TAFs (TBP-associated factors) do not bind directly to the TATA box but appear to allow TFIID to respond to stimulation by transcriptional activators (see Section 8.3).

Although most of the genes transcribed by RNA polymerase II contain a TATA box, a subset of RNA polymerase II genes do not contain the TATA box (for review see Weis and Reinberg, 1992; Smale, 2001; Bashirullah *et al.*, 2001). Paradoxically, however, TBP plays a key role also in the transcription of this class of RNA polymerase II genes. In this case, TBP does not bind to the DNA but is recruited by another DNA binding protein which binds to the initiator-element overlapping the transcriptional start site. TBP then binds to this initiator binding protein allowing the recruitment of TFIIB and the RNA polymerase itself, as occurs for promoters containing a TATA box. Hence TBP plays a central role in the assembly of the transcription complex for RNA polymerase II, joining the complex by binding to DNA in the case of TATA box-containing promoters and being recruited via protein–protein interactions in the case of promoters which lack a TATA box (Fig. 3.11).

These findings indicate therefore that TBP may be the basic transcription factor which is essential for transcription by RNA polymerase II, paralleling the role of SL1 for RNA polymerase I and of TFIIIB for RNA polymerase III. This idea is supported by the amazing finding that TBP is actually also a component of both SL1 and TFIIIB (for a review see White and Jackson, 1992). Thus SL1 is not a single protein but is actually a complex of four factors one of which is TBP (for a review see Grummt, 2003). Hence the recruitment of SL1 to an RNA polymerase I promoter by UBF (see above) actually results in the delivery of TBP to the DNA exactly as in the case of RNA polymerase II promoters which lack a TATA box. Similarly, in the 5S and tRNA genes, TBP is delivered to the promoter as part of the TFIIIB complex following prior binding of either TFIIIA or TFIIIC (see above) (for review see Rigby, 1993). Moreover, some RNA polymerase III genes such as that of the U6 RNA gene contain an upstream promoter with a TATA box and hence in this case TBP can bind directly to the promoter (Fig. 3.12). Remarkably, following its direct or indirect recruitment to the DNA, TBP can then make a variety of further contacts within the transcriptional complexes for RNA polymerases I, II and III in order to enhance assembly and/or activity of the complexes and these have now been defined by structural analysis (Bric *et al.*, 2004; Schröder *et al.*, 2003) (Fig. 3.13).

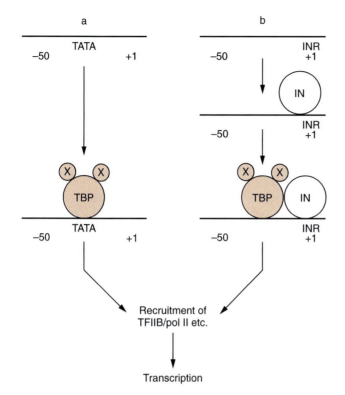

Figure 3.11

Transcription of promoters by RNA polymerase II involves the recruitment of TBP (and its associated factors (X) forming the TFIID complex) to the promoter. This may occur by direct binding of TBP to the TATA box where this is present (panel a) or by protein–protein interaction with a factor (IN) bound to the initiator element in promoters lacking a TATA box (panel b).

The multiple protein–protein and protein–DNA interactions which we have discussed in this section, therefore appear to serve merely to recruit TBP to the DNA which, in the case of all three polymerases, then leads, directly or indirectly, to the recruitment of the RNA polymerase itself. It has therefore been suggested that TBP represents an evolutionarily ancient transcription factor whose existence precedes the evolution of three independent eukaryotic RNA polymerases and which therefore plays a universal and essential role in eukaryotic transcription (for a review see Hernandez, 1993).

Although the three RNA polymerases have common features paralleling the fact that they all copy DNA into RNA, they each have their own specific role. As RNA polymerase II which transcribes protein-coding genes is the primary target of gene regulatory processes, post-transcriptional modifications affecting the RNA transcripts produced by this enzyme will be discussed in the remainder of this chapter and its regulation is the primary subject of the rest of this book, However, the regulation of genes transcribed by RNA polymerases I and III will be described where appropriate (see in particular Sections 7.5 and 9.5).

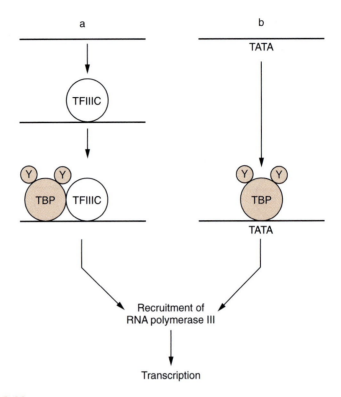

Figure 3.12

Transcription of promoters by RNA polymerase III involves the recruitment of TBP (and its associated factors (Y) forming the TFIIIB complex) to the promoter. This may be achieved by protein–protein interactions with either TFIIIA and TFIIIC or TFIIIC alone in the case of promoters lacking a TATA box (panel a) or by direct binding to the TATA box where this is present (panel b).

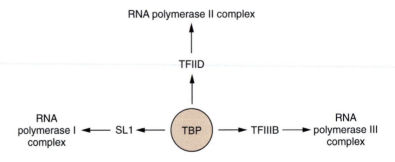

Figure 3.13

As a component of SL1, TFIID and TFIIIB, the TBP protein plays a key role in the transcription initiation complex for all three RNA polymerases.

3.3 Post-transcriptional events

Following transcription, a number of processes modify the primary transcript of genes which are transcribed by RNA polymerase II before it can be translated into protein (see Fig. 3.1). These modifications will now be discussed in turn.

Capping

Following initiation of the primary transcript, transcription proceeds by the progressive addition of ribonucleotides to the RNA chain so that an RNA molecule is produced with bases complementary to those present in the DNA being transcribed. Even before this process is complete, the 5′ end of the nascent mRNA is modified by the addition of a guanine residue which is not encoded by the DNA. The bond joining this guanine to the 5′ end of the RNA differs from the bonds linking the nucleotides in the RNA chain in two ways. Firstly, three phosphate molecules separate the G residue from the first base in the chain whereas only a single phosphate separates each residue in the rest of the chain. Second, the phosphate residues are joined via a 5′ to 5′ bond rather than the standard 3′ to 5′ bond which links the nucleotides in the chain (Fig. 3.14) (for a review see Banerjee, 1980).

Capping occurs only in eukaryotes and not in prokaryotes. It is believed to be necessary due to a difference in the process of translation between eukaryotic and prokaryotic mRNAs. Prokaryotic mRNAs are in general polycistronic, which means that several different proteins are translated from the same mRNA. Thus, the ribosome can initiate translation to produce protein internally within an mRNA molecule being guided by specific sequences (Shine–Dalgarno sequences) within the mRNA which are complementary to the 16S RNA of the small ribosomal subunit. In contrast, eukaryotic mRNAs are generally monocistronic with only one protein being derived from each mRNA. These mRNAs therefore lack the guide sequences present in prokaryotic mRNAs and translation is initiated by the ribosome binding the 5′ end of the mRNA by recognizing the methylated cap (Fig. 3.15).

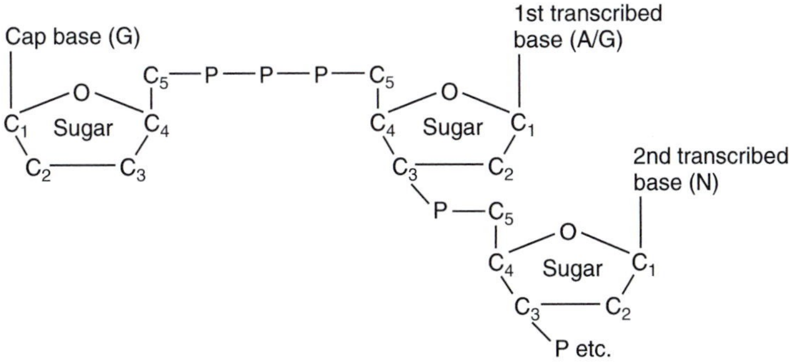

Figure 3.14

Structure of the cap which is present at the 5′ end of eukaryotic mRNAs. Note that the cap structure consists of a guanosine (G) residue which is linked via a 5′ to 5′ bond and three phosphate residues to the first transcribed base which is normally an A or a G residue. This linkage is in contrast to the normal 3′ to 5′ link with a single phosphate residue which joins the sugar residues in the nucleotide chain.

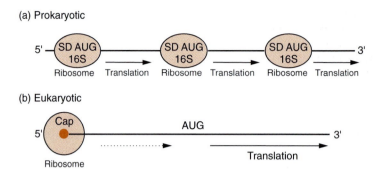

(a) Prokaryotic

(b) Eukaryotic

Figure 3.15

Translational initiation in prokaryotes (panel a) and eukaryotes (panel b). Note that prokaryotic mRNAs are polycistronic with several proteins being translated from a single mRNA. Initiation is achieved by the 16S RNA of the ribosome binding to a complementary sequence known as the Shine–Dalgarno (SD) sequence at various points within the mRNA molecule. Translation is then initiated at an AUG codon adjacent to the Shine–Dalgarno sequence. In contrast eukaryotic mRNAs are monocistronic with only a single protein being translated from the mRNA. Translation is initiated by the ribosome binding to the cap structure at the 5′ end of the mRNA and moving along the mRNA to initiate translation at an AUG codon which is located in an appropriate surrounding sequence.

Interestingly, in prokaryotes the so-called Shine–Dalgarno sequence which is recognized by the 16S ribosomal RNA is located within 10 bases upstream of the AUG sequence at which translation is initiated. In contrast, in eukaryotes the AUG initiation codon may be located several hundred bases away from the 5′ end of the RNA. Therefore, following binding at the cap, the ribosome migrates along the mRNA until it encounters the appropriate AUG initiation codon.

However, translation is not initiated at the first AUG which is encountered by the ribosome. Rather, the AUG triplet must be set within the appropriate consensus sequence which is related to GCCA/GCCAUGG. This so-called scanning hypothesis, in which the ribosome moves along the mRNA looking for this sequence was originally propounded by Kozak and the rules which determine whether a particular sequence surrounding an AUG codon will allow the ribosome to initiate transcription are known as Kozak's rules (Kozak, 1986).

As well as its function in allowing recognition by the ribosome, the cap structure also has an important role in protecting the 5′ end of the mRNA from attack by exonuclease enzymes which would otherwise recognize a free 5′ end and would digest the RNA in a 5′ to 3′ direction. Indeed, the removal of the cap structure is one of the steps in mRNA degradation by which the mRNA molecule is ultimately broken down (for review see Parker and Song, 2004).

More recently, a further role for capping has been described involving it acting as a checkpoint regulating elongation of the RNA transcript (for review see Orphanides and Reinberg, 2002). Thus, as noted in Section 3.2, following initiation and production of the first few nucleotides of the RNA, the RNA polymerase pauses and it is at this point that the RNA is capped. Once capping has occurred the pTEF-b kinase is recruited and it phosphorylates RNA polymerase II so that transcription continues and the full RNA transcript is produced. Hence, a checkpoint mechanism operates in which a full RNA transcript is not produced, unless the RNA is correctly capped (Fig. 3.16).

Polyadenylation

As discussed above, the 5′ end of the mRNA differs from that of the primary transcript only by the addition of the single modified G residue forming the cap. In contrast, however, at the 3′ end the mRNA molecule is much shorter than the original transcript and differs from it in terminating with a run of up to 200 adenosine residues which are added post-transcriptionally in a process known as polyadenylation (Fig. 3.17, for reviews see Barabino and Keller, 1999; Keller, 1995; Proudfoot, 1996).

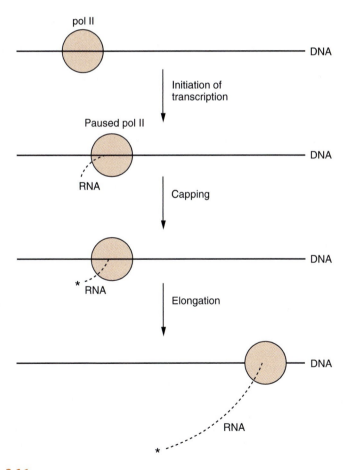

Figure 3.16

Following initiation of transcription and production of the first few bases of the RNA, the polymerase pauses and only continues transcription once the nascent RNA has been capped. Compare with Fig. 3.8 and note that capping is essential for recruitment of the pTEF-b kinase which is required for transcriptional elongation.

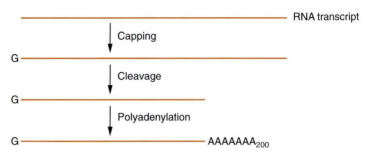

Figure 3.17

Modifications at the 5' and 3' end of RNA polymerase II transcripts. Note that the 5' end of the RNA is modified by the addition of a single G residue in the capping process whereas the 3' end is cleaved with removal of a large stretch of RNA and subsequent addition of up to 200 A residues to the free 3' end in the process of polyadenylation.

The process involves firstly the cleavage of the RNA transcript and then the subsequent addition of the A residues at the resulting free 3' end.

The site at which cleavage occurs (the polyadenylation site) is flanked by two conserved sequence elements (Fig. 3.18). Upstream of the poly(A) site, the RNA contains the essentially invariant sequence AAUAAA and downstream of the poly(A) site it contains a less well conserved region which is rich in G and U residues. These two sequence elements are recognized respectively by two protein complexes known as the cleavage and polyadenylation specificity factor (CPSF) and the cleavage stimulation factor CstF (see Fig. 3.18). Following the binding of these complexes they are thought to interact with one another and the cutting of the RNA occurs in the region between them. Note that this cutting is an endonucleolytic cleavage where cutting occurs within the RNA chain rather than the exonucleolytic cleavage which would occur following removal of the cap (see above) where degradation begins at a free end of the RNA molecule. Subsequent to cleavage, the enzyme poly (A) polymerase then adds the run of A residues to the free 3' end (Fig. 3.18).

Thus the RNA transcript is modified at both ends, firstly by capping at the 5' end and subsequently by cleavage and polyadenylation at the 3' end. It is believed that the primary role of the polyadenylation process is to protect the mRNA from degradation by exonucleases which would otherwise attack its free 3' end and rapidly degrade it. Thus, in many cases, an mRNA which is to be degraded is first deadenylated allowing its 3' end to be attacked (for a review see Parker and Song, 2004). In addition, it appears that the presence of the poly A tail can also regulate the efficiency by which an mRNA is translated (see Section 5.6). Hence, like the cap, the poly A tail appears to have a dual function in regulating both translation of the mRNA into protein and its degradation.

Interestingly, not all mRNAs encoding proteins are polyadenylated. In particular, the mRNAs encoding the histone molecules which play a critical role in chromatin structure (see Section 6.3) are not polyadenylated. In this case, however, the 3' end of the mRNA is protected from degradation since it can fold itself into a double-stranded stem loop structure which protects the mRNA from degradation. Interestingly, the folding and

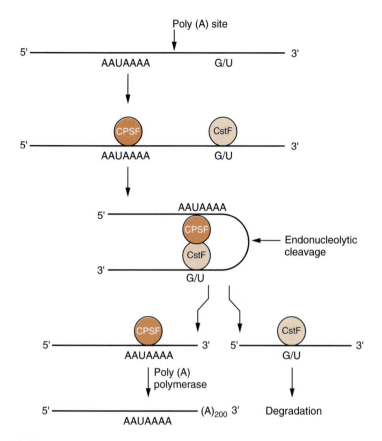

Figure 3.18

Polyadenylation. This process involves the binding of protein factors to the AAUAAA signal upstream of the poly(A) site (CPSF) and to the G/U rich sequence downstream of the poly(A) site (CstF). CPSF and CstF then interact with one another leading to cleavage in between them at the polyadenylation site. Subsequently the sequence downstream of the poly(A) site is degraded whilst the free 3′ end of the upstream region is polyadenylated by the enzyme poly(A) polymerase.

unfolding of this stem loop plays a critical role in regulating the stability of the histone mRNA resulting, for example, in its becoming highly stable during S phase of the cell cycle when DNA is being synthesized and more histones are required (see Section 5.5). Such an example is, however, the exception and the vast majority of RNA species transcribed by RNA polymerase II are polyadenylated at their 3′ ends following capping at the 5′ end.

RNA splicing

Probably the most unexpected finding to have been made during the modern era of gene cloning was the discovery that the protein coding regions of genes transcribed by RNA polymerase II are interrupted by DNA sequences which do not encode parts of the protein. These intervening sequences are known as introns because, as we shall see, although they are transcribed into RNA, their RNA product remains *in* the nucleus whereas the coding regions or exons, *exit* the nucleus.

Thus, when the DNA is transcribed both the exons and the introns which lie between them are transcribed into a single primary RNA transcript. As well as being modified at its 5′ and 3′ ends by capping and polyadenylation, this primary transcript must therefore also have its introns removed before it is transported to the cytoplasm. Otherwise, if it is translated, irrelevant amino acid sequences will appear in the middle of the protein or translation will be prematurely terminated due to the presence of a translation stop codon within the intervening sequence. The process whereby these introns are removed is known as RNA splicing (Fig. 3.19) (for reviews see Adams *et al.*,1996; Staley and Guthrie, 1998; Tollervey and Caceres, 2000).

Initial studies on RNA splicing concentrated on a DNA sequence analysis of the introns and the adjacent regions of the exons. Such studies revealed that virtually all eukaryotic introns for genes transcribed by RNA polymerase II begin with the bases GT and end with the bases AG (see Fig. 3.20). Although other features of the intron and the adjacent regions of the exon have been defined, such as a run of pyrimidine residues (the polypyrimidine tract) located adjacent to the AG sequence within the intron, these are much less well conserved than the GT/AG sequences.

Although this finding was of interest it did not of itself provide any functional insights into the mechanisms by which splicing occurred. These were initially provided by studies using nuclear extracts which could carry out the splicing reaction in the test tube. These experiments showed

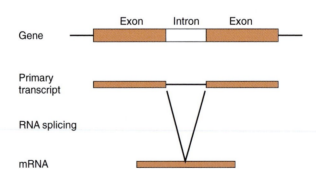

Figure 3.19

Removal of an intervening sequence (intron) from the primary RNA transcript by RNA splicing.

Figure 3.20

Exon–intron junctions in RNA polymerase II transcripts. Note that the first two bases of the intron are normally GU whilst the last two bases are normally AG.

that, when a simple RNA molecule containing two exons interrupted by an intron was added to the splicing reaction, it was possible to produce the spliced product with the two exons joined together and a free intron. Most interestingly, however, the intron emerged in a lariat structure in which its 5' end had folded back on itself and formed a 5' to 2' bond with an A residue located between 18 and 40 bases upstream of the AG sequence at the 3' end of the intron (Fig. 3.21).

This indicated that splicing proceeds in a series of steps (Fig. 3.21) in which the transcript is first cleaved at the junction between the upstream exon and the intron (the 5' splice site) resulting in a free upstream exon and the intron joined to exon II. In order to prevent a back reaction in which the intron rejoins with the upstream exon, the free 5' end of the exon rapidly folds back on itself and forms the 5' to 2' bond with the A residue within the intron. This A residue is known as the branch point. Following this process, the intron is then freed from the downstream exon by cleavage immediately after the next AG sequence following the branch

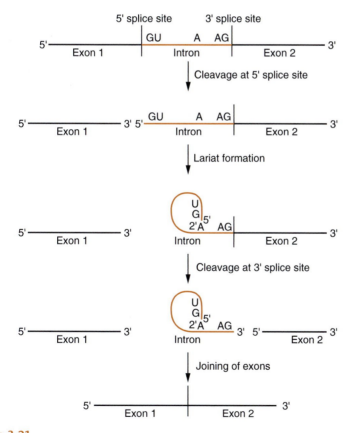

Figure 3.21

Mechanism of RNA splicing. Splicing is initiated by cleavage at the 5' splice site generating an intron–exon 2 structure with a free 5' end. This free 5' end then forms a 5' to 2' bond with an A residue within the intron (the branch point). This is followed by cleavage at the 3' splice site and subsequent joining of exon 1 to exon 2.

point. Subsequently the downstream exon then joins with the upstream exon to form the spliced product (Fig. 3.21). Hence the reaction has achieved its purpose by removing the intron from between the two exons and joining them together in a splicing reaction.

In considering this process, one might ask why the upstream exon does not simply diffuse away whilst the downstream exon is still linked to the intron and cannot be joined to it. The answer to this question is that this reaction does not take place with the RNA isolated within the nucleus. Rather, it takes place in a complex structure known as the spliceosome which consists of a number of different RNA and protein components which together catalyse this process and hold the various intermediates in the splicing process in the correct orientation relative to one another (for reviews see Staley and Guthrie, 1998; Jurica and Moore, 2003). The three-dimensional structure of the spliceosome has recently been defined. This shows that the spliceosome consists of large and small subunits, with a tunnel or channel in between that can accommodate the RNA which is about to undergo splicing (Azubel *et al.*, 2004) (color plate 7).

The RNA components of the spliceosome are small RNA molecules ranging from 56 to 217 bases in size and known as the U RNAs because they are rich in uridine residues (for review see Neilsen, 1994). Each of these U RNAs has a specific role in the splicing process and each of them is associated with specific proteins forming small nuclear ribonucleoprotein particles (snRNP) which consist of a U RNA with its associated proteins. Some of these proteins, such as the Sm proteins, are common to all the different U snRNPs whereas others, such as the U1snRNP A protein, are associated with only one specific RNP particle (for review see Lamond, 1999).

The splicing process is initiated by the binding of the U1 snRNP to the 5′ splice site (Fig. 3.22a). Subsequently the U2 accessory factor (U2AF) binds to the polypyrimidine tract located between the branch point A and the AG at the end of the intron. This then results in the U2 snRNP binding at the adjacent branch point (Fig. 3.22b). Next, the RNA is bound by the U5 snRNP which binds to the upstream exon and by the single snRNP which contains the U4 and U6 RNAs bound to one another. This binding results in the displacement of the U1 snRNP from the 5′ splice site where it is replaced by the U6 snRNP (Fig. 3.22c). This results in the release of the U1 snRNP and of the U4 RNA which is now no longer base paired to U6.

The U6 and the U2 snRNPs then interact with one another, bringing the 5′ splice site close to the branch point. This is followed by the cleavage of exon 1 from intron 1, lariat formation and the joining of exon 1 to exon 2 as outlined in Fig. 3.21. The U2 snRNP, the U6 snRNP and the U5 snRNP (which has moved from the upstream exon to the intron during this process) are released with the intron (Fig. 3.22d).

Hence, the entire process of splicing takes place in the tightly ordered structure of the spliceosome and this explains, for example, why exon 1 cannot simply diffuse away after its cleavage from the intron–exon 2. Indeed, it has been demonstrated that a specific component of the spliceosome hSLu7 binds tightly to exon 1 and holds it in close proximity to the AG of the correct 3′ splice site so allowing selection of the correct 3′ splice site to occur (Chua and Reed, 1999).

In addition to the U RNAs and the proteins associated with them in the UsnRNPs, the spliceosome also contains other proteins which are not directly associated with the U RNAs. The most intensively studied of these

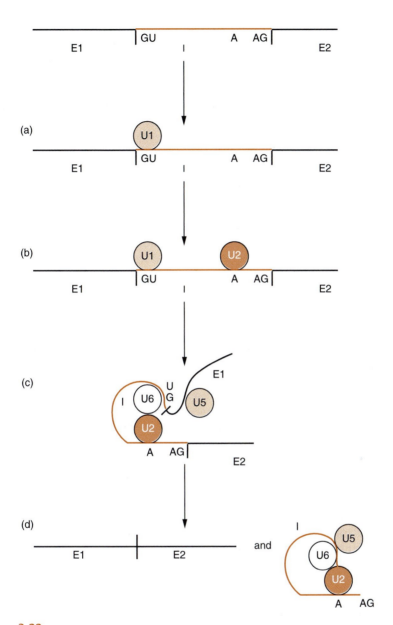

Figure 3.22

Involvement of snRNP particles bearing different U RNAs in RNA splicing.
Splicing is initiated by binding of the U1 particle to the 5′ splice site (panel a)
followed by binding of the U2 particle to the branch point (panel b). This is
followed by binding of the U5 particle to the upstream exon and of the U4/U6
particle to the 5′ splice site with release of the U1 particle and U4 RNA. U6 and
U2 particles then interact with one another (panel c), bringing the 5′ splice site
adjacent to the branch point. This is followed by cleavage at this site, lariat
formation and subsequent cleavage at the 3′ site with exon joining. The free
intron is released in association with the U2, U5 and U6 particles (panel d), with
the U5 particle having moved from the upstream exon to a position within the
intron.

are a series of proteins which are rich in serine and arginine residues and which are therefore known as SR proteins (S = serine, R = arginine in the one letter code for amino acids) (for reviews see Manley and Tacke, 1996; Tollervey and Caceres, 2000).

These proteins are believed to serve both in recruiting the snRNPs to the RNA to be spliced and in determining which splice sites are joined to one another. As the vast majority of RNAs contain multiple exons and introns (rather than the simple two exon, one intron model we have been considering) this is obviously highly critical in order to ensure that the correct exons are joined to one another so that exon I is joined to exon II which in turn is joined to exon III, rather than, for example, exon I being joined to exon III, resulting in exon II being missing from the mRNA and therefore producing a protein lacking a particular region. As will be described in Section 5.2 this process can be regulated so that different combinations of exons are joined to one another in different situations in a process known as alternative splicing and it appears that the SR proteins play a critical role in the regulation of this process.

Coupling of transcription and RNA processing within the nucleus

Although we have so far discussed transcription, capping, polyadenylation and splicing as if they were entirely separate processes, it is clear that they are in fact tightly coupled together within the nucleus (for reviews see Orphanides and Reinberg, 2002; Calvo and Manley, 2003). The first evidence that this was the case was the finding that deletion of the C-terminal domain (CTD) of RNA polymerase II (see Section 3.2) not only reduces transcription but also interferes with post-transcriptional processes such as capping, splicing and polyadenylation indicating that RNA polymerase II is also involved in these events.

As described in Section 3.2, phosphorylation of the CTD of RNA polymerase II leads to the polymerase together with TFIIF initiating transcription and moving off down the gene, leaving behind the other basal transcription factors such as TFIIB and TFIID. Interestingly, these basal factors are then progressively replaced by factors which are involved in capping, splicing and polyadenylation and which bind specifically to the phosphorylated form of the RNA polymerase II CTD.

Thus, as soon as transcription begins, capping enzymes bind to the CTD of RNA polymerase II allowing the 5′ end of the RNA to be capped as described above (Fig. 3.23b). Subsequently, the SR proteins and other components of the splicing complex (see above) bind to the CTD allowing splicing to occur (Fig. 3.23c). Finally, components of the polyadenylation complex (see above) interact with the transcribing complex and polyadenylate the RNA (Fig. 3.23d).

Interestingly, the recruitment of these various factors is closely coupled to the phosphorylation of the CTD of RNA polymerase II. Thus, as noted in Section 3.2, this region contains multiple copies of the amino acid sequence Tyr-Ser-Pro-Thr-Ser-Pro-Ser. It has been shown that phosphorylation of serine 5 in this sequence is essential for recruitment of capping factors (Mandal et al., 2004) whilst subsequent phosphorylation of serine 2 is necessary for recruitment of factors involved in 3′ end processing and polyadenylation (Ahn et al., 2004; Meinhard and Cramer, 2004) (Fig. 3.23).

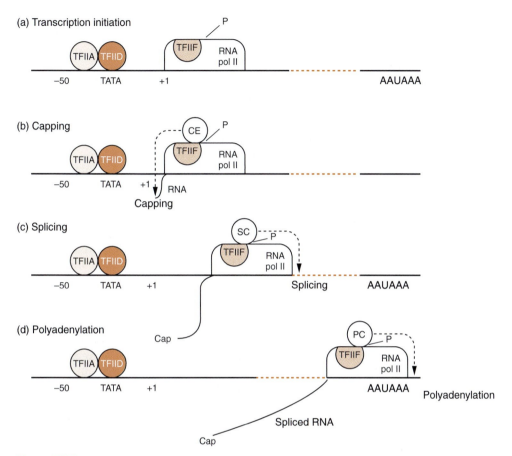

Figure 3.23

As the RNA polymerase II complex transcribes the gene (panel a), it becomes progressively associated with the proteins that mediate capping (panel b, CE = capping enzymes), splicing (panel c, SC = splicing complex) or polyadenylation (panel d, PC = polyadenylation complex). This allows the RNA to be capped, spliced and polyadenylated as it is being transcribed. Note the phosphorylation (P) of the RNA polymerase in the transcribing complex (see also Fig. 3.7) which is essential for the recruitment of the capping, splicing and polyadenylation factors.

Hence, the dual role of CTD phosphorylation in transcriptional elongation and processing of the RNA transcript couples the two events ensuring that the elongating transcript is correctly processed (for discussion see Ni *et al.*, 2004; Sims *et al.*, 2004). Indeed, this linkage provides the explanation for the relationship between capping and transcriptional elongation which was discussed above. Thus, following phosphorylation of the CTD on serine 5 by the kinase activity of TFIIH, capping of the RNA transcript occurs and this stimulates phosphorylation of the CTD on serine 2 by pTEF-b, allowing not only the recruitment of 3′ processing factors but also stimulating transcriptional elongation itself (for review see Orphanides and Reinberg, 2002) (Fig. 3.24).

Hence, the transcription complex, as it moves down the gene, attracts the factors which are needed to modify the RNA post-transcriptionally and this occurs whilst the RNA is being transcribed, paralleling the translation of the RNA whilst it is being transcribed which occurs in prokaryotes (see Section 3.1). Once the coupled processes of transcription, capping, splicing and polyadenylation have taken place, the RNA is fully mature and ready to be transported to the cytoplasm.

RNA transport

Although the processes of capping, polyadenylation and RNA splicing take place in the nucleus, the process of translation whereby the mature mRNA is converted into protein takes place in the cytoplasm. Thus the RNA must be transported through the nuclear membrane from the nucleus to the cytoplasm before it can be translated into protein (for reviews see Nakielny and Dreyfuss, 2000; Dimaano and Ullman, 2004).

A number of proteins which associate with the RNA in the nucleus and accompany it into the cytoplasm have been identified and are believed to play a critical role in mediating the transport of the RNA from the nucleus to the cytoplasm (for reviews see Daneholt, 2001; Dimaano and Ullman, 2004). One of these proteins, hnRNP A1, is a member of the heterogeneous nuclear RNP (hnRNP) proteins which associate with the primary transcripts of protein coding genes even before splicing complexes form on these RNAs (for review of hnRNP proteins see Dreyfuss *et al.*, 1993). Although some of these proteins such as hnRNP C disassociate from the mRNA before it is transported to the cytoplasm hnRNP A1 remains associated with the mRNA as it moves into the cytoplasm (Fig. 3.25). Indeed in the insect *Chironomus tentans* hnRNP A1 can be directly visualized in the electron microscope associated with mRNA which is passing through the nuclear pore into the cytoplasm (Daneholt, 2001) (Fig. 3.26).

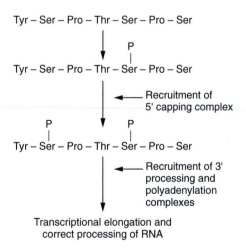

Tyr – Ser – Pro – Thr – Ser – Pro – Ser

P
|
Tyr – Ser – Pro – Thr – Ser – Pro – Ser

Recruitment of 5' capping complex

P   P
| |
Tyr – Ser – Pro – Thr – Ser – Pro – Ser

Recruitment of 3' processing and polyadenylation complexes

Transcriptional elongation and correct processing of RNA

Figure 3.24

The C-terminal domain of RNA polymerase II contains multiple copies of the sequence Tyr-Ser-Pro-Thr-Ser-Pro-Ser. Phosphorylation of this sequence on serine 5 is essential for recruitment of the 5′ capping complex whilst phosphorylation on serine 2 is essential for recruitment of the 3′ processing and polyadenylation complexes.

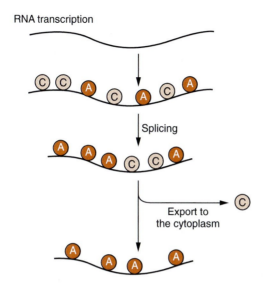

RNA transcription

Splicing

Export to
the cytoplasm

Figure 3.25

Prior to splicing the RNA transcript binds hnRNP proteins such as hnRNP A1 and
C. Although hnRNPC is released from the RNA prior to export to the cytoplasm
hnRNPA1 moves with the RNA into the cytoplasm.

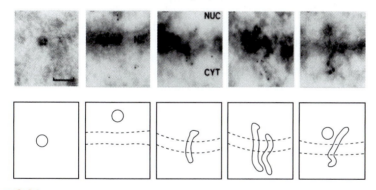

Figure 3.26

Electron micrographs (top row) and schematic diagrams (bottom row) showing
the passage of the mRNA/hnRNP particle from nucleus (NUC) to cytoplasm
(CYT) via the nuclear envelope (ne and dotted line) in the salivary glands of
Chironomus tentans. The mRNA/protein particle is shown within the nucleus (left
hand panel), approaching the nuclear envelope (second panel) and partially
unfolded particles passing through pores in the nuclear membrane are shown in
the central and right-hand panels. The scale bar indicates 100 nm. Photograph
kindly provided by Professor B. Daneholt from Visa *et al.*, *Cell* **84**, 253–254
(1996), by permission of Cell Press.

   Although hnRNP A1 is thus a candidate for a molecule mediating the
transport of the mRNA into the cytoplasm, no direct evidence exists that
it is actually essential for this process. However, such direct evidence does
exist in the case of another mRNA binding protein namely the Gle1
protein found in yeast. Thus this protein was originally identified as being
encoded by a gene whose mutation prevents the transport of polyadenylated mRNA from the nucleus to the cytoplasm in yeast, indicating that

it is essential for this purpose. Most interestingly, the Gle1 protein contains a short amino acid sequence known as a nuclear export signal (NES). If this sequence is transferred to another unrelated protein, it can direct the transfer of that protein from the nucleus to the cytoplasm. This indicates that, following binding of the Gle1 protein to an mRNA, the NES sequence directs the export of the protein and the associated mRNA from the nucleus to the cytoplasm (Fig. 3.27).

A similar NES sequence has also been identified in the Rev protein of the human immunodeficiency virus (HIV) which is involved in regulating the transport of different HIV mRNAs from the nucleus to the cytoplasm (see Section 5.4). Similarly, the TFIIIA transcription factor (see Section 3.2) also contains an NES sequence. As discussed in Section 3.2, the TFIIIA molecule binds to an internal region within the transcribed portion of the gene encoding the 5S RNA. This region is therefore also present in the resulting 5S RNA transcript and TFIIIA can bind to it at the RNA level also and promote transport of the 5S RNA from the nucleus to the cytoplasm.

Hence the internal promoter characteristic of the 5S RNA genes allows TFIIIA both to act as a transcription factor by binding to its recognition sequence at the DNA level and also to act as a transport protein by binding to the corresponding sequence at the RNA level. As discussed in Section 6.5 this binding of TFIIIA to both the 5S DNA and the 5S RNA is also used to regulate the level of transcription of the 5S RNA genes in oogenesis.

It is likely therefore that Gle1, Rev and TFIIIA act via similar mechanisms to promote the transport to the cytoplasm of different RNA molecules, with their action involving binding to the RNA by an RNA binding region of the protein followed by transport of the RNA/protein complex mediated by the NES signal sequence (Fig. 3.27).

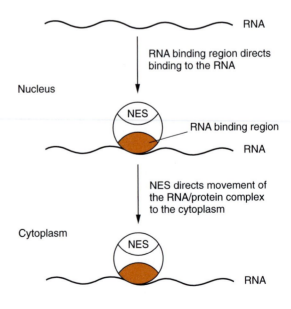

Figure 3.27

Proteins such as Gle1, Rev and TFIIIA bind to the RNA via an RNA binding region and then direct its movement to the cytoplasm via the Nuclear Export Signal (NES).

Although a variety of proteins are involved in the export of mRNAs from the nucleus to the cytoplasm, the process by which different proteins bind to different mRNAs is not a random one. Thus, studies in yeast showed 1000 distinct mRNAs associated with the Ycal export protein whilst 1150 associated with the Mex67 export protein, only 349 of which mRNAs overlap and can associate with both proteins. Moreover, the proteins encoded by these Mex67-associated mRNAs are distinct from those whose mRNAs associate specifically with Ycal encoding, for example, proteins involved in the cell cycle or carbohydrate metabolism, whilst mRNAs encoding translation factors or membrane proteins associate specifically with Mex67 (Fig. 3.28) (for a review see Keene, 2003). Hence, the mRNAs encoding functionally linked proteins can be exported together, facilitating their translation in parallel and the subsequent functional association of their corresponding proteins.

Evidently, only RNA that has been properly capped, spliced and polyadenylated should be transported to the cytoplasm (for a review see Maquat and Carmichael, 2001). This is of critical importance to prevent the transport of unspliced RNAs containing introns which could not be translated to produce functional protein. Just as proteins involved in capping, splicing and polyadenylation associate with the nascent RNA transcript and process it (see above), so proteins involved in RNA export associate with the RNA which is being processed and catalyse its export. For example, a protein known as Aly binds to the RNA during the splicing process and remains tightly associated with it (Zhou *et al.*, 2000). As Aly is essential for transport to the cytoplasm, this ensures that only properly spliced RNA leaves the nucleus (for reviews see Reed and Magni, 2001; Dimaano and Ullman, 2004). Interestingly, it has recently been shown that several SR proteins involved in splicing (see above) also promote the export of the mRNA, further strengthening the link between the processes of splicing and transport (Huang *et al.*, 2003).

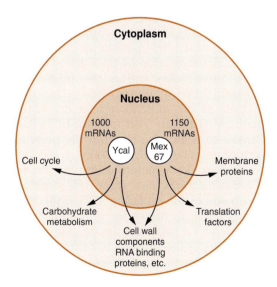

Figure 3.28

In yeast the export proteins Ycal and Mex67 are involved in the transport of distinct sets of mRNAs encoding proteins with distinct biological functions.

Translation

The process of translating the mRNA into protein takes place on defined cytoplasmic organelles known as ribosomes which, in eukaryotes, consist of a large subunit of 60S in size and a small subunit of 40S in size (for a review of translation see Pestova *et al.*, 2001). Each of these subunits contains both specific RNAs and proteins. Thus the large subunit consists of 28S, 5.8S and 5S ribosomal RNAs associated with 49 different proteins whilst the small subunit consists of an 18S ribosomal RNA and 33 associated proteins (for reviews see Lake, 1985; Moore, 1988).

Translation is initiated by the binding of the eukaryotic initiation factor (eIF), eIF-4E to the cap structure at the 5' end of the mRNA (see above) (Fig. 3.29a) (for reviews see Hentze, 1997; Dever, 1999). Subsequently other eIF factors such as eIF-4G and eIF-4A/B bind to the 5' end of the RNA (Fig. 3.29b). Together eIF-4A, eIF-4E, eIF-4G form a complex which is known as eIF-4F. The binding of the various components of eIF-4F is followed by the binding of a complex consisting of the small 40S ribosomal subunit, a transfer RNA molecule (tRNA), which carries the amino acid methionine, and the initiation factor eIF-2 (Fig. 3.29c). Subsequently, as described above, the 40S subunit migrates down the mRNA until it finds an AUG initiation sequence in the mRNA which is located in the appropriate context for the initiation of translation (Fig. 3.29d). At this point, eIF-2 is released from the complex and the 60S large ribosomal subunit binds to the small subunit to form the complete 80S ribosome which is able to initiate translation (Fig. 3.29e). Several other factors, including eIF-5, eIF-5b and eIF-6, play an essential role in the joining of the large and small ribosomal subunits (Pestova *et al.*, 2000; Ceci *et al.*, 2003).

A key role in the process of translation is played by the transfer RNA (tRNA) molecules. These RNAs are unique in that they can bind an amino acid at their 3' end (Fig. 3.30) and therefore function to bring the amino acids to the ribosome for incorporation into protein. The tRNAs are RNA molecules from 74–95 bases in length which can fold into a highly defined secondary structure (Fig. 3.30). In this structure one specific loop contains the so-called anticodon. Each tRNA molecule which binds a different amino acid has a distinct anticodon of three bases which can pair with the complementary three bases within the mRNA (for reviews see Normanly and Abelson, 1989; Schimmel, 1987). Thus the initiator tRNA described above contains the anticodon sequence CAU and therefore binds to the initiator AUG codon. This initiator tRNA carries a methionine amino acid linked to it and the chain of the protein molecule thus begins with a methionine residue (Fig. 3.31a).

Subsequently, a further tRNA molecule bearing the appropriate anticodon is recruited to bind to the next three base sequence in the mRNA immediately downstream of the initiating AUG sequence. As each particular tRNA containing a specific anticodon also binds to a specific amino acid, this tRNA will deliver a particular amino acid to the mRNA-bound ribosome, with the nature of the amino acid being specified by the nature of the three base codon immediately following the AUG (Fig. 3.31b).

This second tRNA is recruited to the ribosome by a specific eukaryotic elongation factor (eEF) eEF-1. The structure of the ribosome allows two specific sites within it to be occupied by tRNA molecules which are bound to adjacent three base sequences in the mRNA. Once this has occurred, an

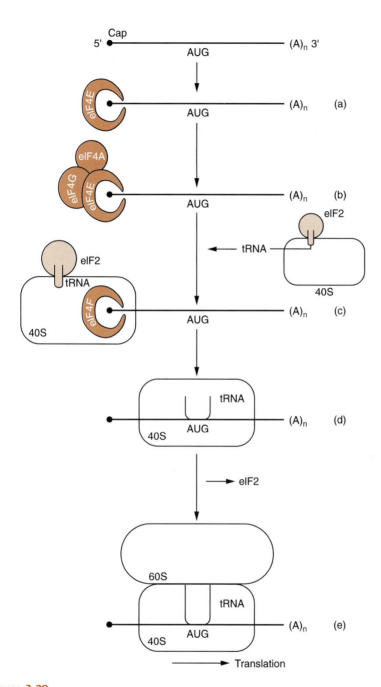

Figure 3.29

Mechanism of translational initiation. Initial binding of eIF4E to the cap structure (panel a) is followed by the binding of eIF4G and eIF4A (panel b). The complex of eIF4A E and G is known as eIF4F. The eIF4F complex is then recognized by a complex consisting of the 40S ribosomal subunit, the initiator tRNA and eIF2 (panel c). The 40S subunit then migrates along the RNA until it reaches the initiator AUG codon (panel d). This is followed by release of eIF2 and binding of the 60S large ribosomal subunit. Note that for clarity eIF4A, E and G are shown as a single eIF4F complex in panel c.

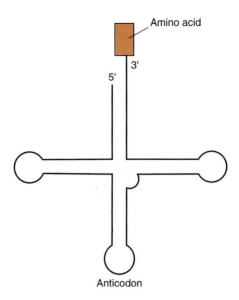

Figure 3.30

Cloverleaf structure of tRNA with stem structures containing paired bases and loops containing unpaired bases. Note the unpaired bases in the anticodon which will bind to the complementary bases in the codon of the mRNA, thereby delivering the appropriate amino acid bound to the 3′ end of the tRNA to the ribosome.

enzymatic activity associated with the large ribosomal subunit forms a peptide bond between the two amino acids which are bound to the two adjacent tRNA molecules. In this manner, the growth of the peptide chain of the protein molecule begins.

Subsequently, another elongation factor, eEF-2, catalyses the move of the ribosome three more bases down the mRNA molecule (Fig. 3.31c). The initiator tRNA is then released from the mRNA and a further tRNA molecule is recruited to the ribosome-bound mRNA by eEF-1 (Fig. 3.31d). This tRNA contains an anticodon which can bind to the next three bases in the mRNA sequence and has a specific amino acid bound to it. Once

Figure 3.31

Translation elongation. Following the arrival of the ribosome at the initiator AUG codon and binding of the initiator tRNA bearing a methionine amino acid to the AUG codon (panel a, see also Fig. 3.29), a second tRNA is recruited in association with the elongation factor eEF1 and binds to the next codon via its corresponding anticodon (panel b). Note that this codon may have any sequence depending on the next amino acid to be incorporated into the protein, a CUG codon which recruits tRNA with the anticodon sequence CAG and bearing a leucine amino acid is shown as an example. Subsequently a peptide bond is formed between the methionine and the next amino acid (leucine in this case) with translocation of the methionine to bond with the leucine residue. The elongation factor eEF1 then initiates translocation of the ribosome three bases along the mRNA and the first tRNA is released (panel c). The cycle then repeats itself with a new tRNA being recruited to the next three bases (panel d) with consequent peptide bond formation producing a three amino acid peptide.

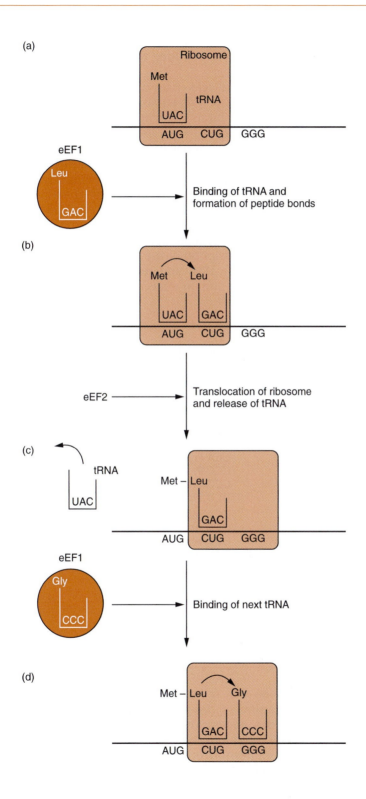

again, a peptide bond is formed between the two amino acid chain and this third amino acid, with the resulting three amino acid peptide being transferred to the new tRNA.

In this manner, the amino acid chain of the protein is gradually built up (Fig. 3.31) (for a review see Wilson and Noller, 1998). As each tRNA with a particular anticodon binds a specific amino acid, amino acids are added to the chain in accordance with the anticodon of the tRNA which is recruited. In turn this is determined by the codon present in the mRNA and so the information in the mRNA is gradually translated into protein in accordance with the genetic code (Table 3.2). Ultimately, the moving ribosome encounters one of the three stop codons which are not recognized by a specific tRNA and a single release factor catalyses the release of the completed polypeptide from the ribosome.

The production of the final protein molecule thus completes the process which began with the transcription of the gene and results in the genetic information encoded in the DNA and subsequently in the RNA being converted into a functional protein molecule.

Table 3.2   The genetic code

Base 1	Base 2			
	A	**C**	**G**	**U**
A	AAA Lys	ACA Thr	AGA Arg	AUA Ile
	AAC Asn	ACC Thr	AGC Ser	AUC Ile
	AAG Lys	ACG Thr	AGG Arg	AUG Met
	AAU Asn	ACU Thr	AGU Ser	AUU Ile
C	CAA Gln	CCA Pro	CGA Arg	CUA Leu
	CAC His	CCC Pro	CGC Arg	CUC Leu
	CAG Gln	CCG Pro	CGC Arg	CUG Leu
	CAU His	CCU Pro	CGU Arg	CUU Leu
G	GAA Glu	GCA Ala	GGA Gly	GUA Val
	GAC Asp	GCC Ala	GGC Gly	GUC Val
	GAG Glu	GCG Ala	GGG Gly	GUG Val
	GAU Asp	GCU Ala	GGU Gly	GUU Val
U	UAA Stop	UCA Ser	UGA Stop	UUA Leu
	UAC Tyr	UCC Ser	UGC Cys	UUC Phe
	UAG Stop	UCG Ser	UGG Trp	UUG Leu
	UAU Tyr	UCU Ser	UGU Cys	UUU Phe

3.4   Conclusions

The multi-step process which has been described in this chapter and results in the conversion of the genetic information in the DNA into protein, clearly offers numerous opportunities for the regulation of gene expression. The data leading to the conclusion that, in fact, the primary regulation of gene expression occurs at the level of transcription are described in the next chapter. The cases of regulation at other levels, such as RNA splicing, RNA transport, RNA stability and translation into protein, are described in Chapter 5.

References

Adams, M.D., Rudner, D.Z. and Rio, D.C. (1996). Biochemistry and regulation of pre-mRNA splicing. *Current Opinion in Cell Biology* **8**, 331–339.

Ahn, S. H., Kim, M. and Buratowski, S. (2004). Phosphorylation of serine 2 within the RNA polymerase II C-terminal domain couples transcription and 3-end processing. *Molecular Cell* **13**, 67–76.

Andel, F., Ladurner, A.G., Inouye, C., Tjian, R. and Nogales, E. (1999). Three-dimensional structure of the human TFIID-IIa-IIB complex. *Science* **286**, 2153–2156.

Asturias, F.J. and Craighead, J.L. (2003). RNA polymerase II at initiation. *Proceedings of the National Academy of Sciences USA* **100**, 6893–6895.

Azubel, M., Wolf, S.G., Sperling, J. and Sperling, R. (2004). Three-dimensional structure of the native spliceosome by Cryo-electron microscopy. *Molecular Cell* **15**, 833–839.

Banerjee, A.K. (1980). 5′ terminal cap structure in eukaryotic messenger ribonucleic acids. *Microbiological Reviews* **44**, 175–205.

Barabino, S.M.L. and Keller, W. (1999). Last but not least: regulated Poly(A) tail formation. *Cell* **99**, 9–11.

Bashirullah, A., Cooperstock, R.L. and Lipshitz, H.D. (2001). Spatial and temporal control of RNA stability. *Proceedings of the National Academy of Sciences USA* **98**, 7025–7028.

Brand, M., Leurent, C., Mallouh, V., Tora, L. and Schultz, P. (1999). Three-dimensional structures of the TAF$_{II}$-containing complexes TFIID and TFTC. *Science* **286**, 2151–2153.

Breathnach, R. and Chambon, P. (1981). Organisation and expression of eukaryotic split genes coding for proteins. *Annual Review of Biochemistry* **50**, 349–383.

Bric, A., Radebaugh, C.A. and Paule, M.R. (2004). Photocross-linking of the RNA Polymerase I preinitiation and immediate postinitiation complexes. *Journal of Biological Chemistry* **279**, 31259–31267.

Burley, S.K. and Roeder, R.G. (1996). Biochemistry and structural biology of transcription factor IID (TFIID). *Annual Review of Biochemistry* **65**, 769–799.

Bushnell, D.A., Westover, K.D., Davis, R.E. and Kornberg R.D. (2004). An RNA polymerase II-TFIIB cocrystal at 4.5 Angstroms. *Science* **303**, 983–988.

Calvo, O. and Manley, J.L. (2003). Strange bedfellows: polyadenylation factors at the promoter. *Genes and Development* **17**,1321–1327.

Ceci, M., Gaviraghi, C., Gorrini, C., Sala, L.A., Offenhauser, N., Marchisio, P.C. and Biffo, S. (2003). Release of eIF6 (p27BBP) from the 60S subunit allows 80S ribosome assembly. *Nature* **426**, 579–584.

Chang, W.-H. and Kornberg, R.D. (2000). Electron crystal structure of the transcription factor and DNA repair complex, core TFIIH. *Cell* **102**, 609–613.

Chen, H.T. and Hahn, S. (2004). Mapping the location of TFIIB within the RNA polymerase II transcription preinitiation complex; a model for the structure of the PIC. *Cell* **119**, 169–180.

Chua, K. and Reed, R. (1999). The RNA splicing factor hSlu7 is required for correct 3′ splice-site choice. *Nature* **402**, 207–210.

Daneholt, B. (2001). Assembly and transport of a premessenger RNP particle. *Proc. Natl. Acad. Sci. USA* **98**, 7012–7017.

Davidson, E.H. and Britten, R.J. (1979). Regulation of gene expression: possible role of repetitive sequences. *Science* **204**, 1052–1059.

Dever, T.E. (1999). Translation initiation: adept at adapting. *Trends in Biochemical Sciences* **24**, 398–403.

Dimaano, C. and Ullman, K.S. (2004). Nucleocytoplasmic transport: integrating mRNA production and turnover with export through the nuclear pore. *Molecular and Cellular Biology* **24**, 3069–3076.

Dreyfuss, G., Matunis, M.J., Pinol-Roma, S. and Burd, C.J. (1993). hnRNP proteins and the biogenesis of mRNA. *Annual Review of Biochemistry* **62**, 89–321.

Geiduschek, E.P. and Kassavetis, G.A. (2001). The RNA polymerase III transcription apparatus. *Journal of Molecular Biology* **310**, 1–26.

Greenblatt, J. (1997). RNA polymerase II holoenzyme and transcriptional regulation. *Current Opinion in Cell Biology* **9**, 310–319.

Grummt, I. (2003). Life on a planet of its own: regulation of RNA polymerase I transcription in the nucleolus. *Genes and Development* **17**, 1691–1702.

Hentze, M.W. (1997). eIF4G: a multipurpose ribosome adaptor. *Science* **275**, 500–501.

Hernandez, N. (1993). TBP: a universal transcription factor. *Genes and Development* **7**, 1291–1308.

Hernandez, N. (2001). Small nuclear RNA genes: a model system to study fundamental mechanisms of transcription. *Journal of Biological Chemistry* **276**, 26733–26736.

Hoeijmakers, J.H.J., Egly, J.-M. and Vermeulen, W. (1996). TFIIH: a key component in multiple DNA transactions. *Current Opinion in Genetics and Development* **6**, 26–33.

Huang, Y., Gattoni, R., Stévenin, J., and Steitz, J.A. (2003). SR splicing factors serve as adapter proteins for TAP-dependent mRNA export. *Molecular Cell* **11**, 837–843.

Jacob, F. and Monod, J. (1961). Genetic regulatory mechanism in the synthesis of proteins. *Journal of Molecular Biology* **3**, 318–356.

Jacob, S.T. (1995). Regulation of ribosomal gene transcription. *Biochemical Journal* **306**, 617–626.

Jeffreys, A.J. and Flavell, R.A. (1977). The rabbit beta globin gene contains a large insert in the coding sequence. *Cell* **12**, 1097–1108.

Jurica, M.S. and Moore, M.J. (2003). Pre-mRNA splicing: awash in a sea of proteins. *Molecular Cell* **12**, 5–14.

Keene, J.D. (2003). Organizing mRNA export. *Nature Genetics* **33**, 111–112.

Keller, W. (1995). No end yet to messenger RNA 3′ processing. *Cell* **81**, 829–832.

Klug, A. (2001). A marvellous machine for making messages. *Science* **292**, 1844–1846.

Kozak, M. (1986). Point mutations define a sequence flanking the AUG initiator codons that modulate translation by eukaryotic ribosomes. *Cell* **44**, 283–292.

Lake, J.A. (1985). Evolving ribosome structure: domains in Archaebacteria, Eubacteria, Eocytes and Eukaryotes. *Annual Review of Biochemistry* **54**, 507–530.

Lamond, A. I. (1999). Running rings around RNA. *Nature* **397**, 655–656.

Landick, R. (2001). RNA polymerase clamps down. *Cell* **105**, 567–570.

Mandal, S.S., Chu, C., Wada, T., Handa, H., Shatkin, A.J. and Reinberg, D. (2004). Functional interactions of RNA-capping enzyme with factors that positively and negatively regulate promoter escape by RNA polymerase II. *Proceedings of the National Academy of Sciences of the USA* **101**, 7572–7577.

Manley, J.L. and Tacke, R. (1996). SR proteins and splicing control. *Genes and Development* **10**, 1569–1579.

Maquat, L.E. and Carmichael, G.G. (2001). Quality control of mRNA function. *Cell* **104**, 173–176.

Meinhart, A. and Cramer, P. (2004). Recognition of RNA polymerase II carboxy-terminal domain by 3″-RNA-processing factors. *Nature* **430**, 223–226.

Moore, P.B. (1988). The ribosome returns. *Nature* **331**, 223–227.

Myer, V.E. and Young, R.A. (1998). RNA Polymerase II holoenzymes and subcomplexes. *Journal of Biological Chemistry* **273**, 27757–27760.

Neilsen, T.W. (1994). RNA–RNA interactions in the splicesosome: unravelling the ties that bind. *Cell* **78**, 1–4.

Ni, Z., Schwartz, B.E., Werner, J., Suarez, J.-R., and Lis, J.T. (2004). Coordination of transcription, RNA processing and surveillance by P-TEFb kinase on heat shock genes. *Molecular Cell* **13**, 55–65.

Nikolov, D.B. and Burley, S.K. (1997). RNA polymerase II transcription initiation: a structural view. *Proceedings of the National Academy of Sciences of the USA* **94**, 15–22.

Normanly, J. and Abelson, J. (1989). tRNA identity. *Annual Review of Biochemistry* **58**, 1029–1049.

Orphanides, G. and Reinberg, D. (2002). A unified theory of gene expression. *Cell* **108**, 439–451.

Parker, R. and Song, H. (2004). The enzymes and control of eukaryotic mRNA turnover. *Nature Structural and Molecular Biology* **11**, 121–127.

Paule, M.R. and White, R.J. (2000). Transcription by RNA polymerases I and III. *Nucleic Acids Research* **28**, 1283–1298.

Pestova, T.V., Lomakin, I.B., Lee, J.H., Choi, S.K., Dever, T.E. and Hellen, C.U.T. (2000). The joining of ribosomal subunits in eukaryotes requires eIF5B. *Nature* **403**, 332–335.

Pestova, T.V., Kolupaeva, V.G., Lomakin, I.B., Pilipenko, E.V., Shatsky, I.N., Agol, V.I. and Hellen, C.U.T. (2001). Molecular mechanisms of translation initiation in eukaryotes. *Proceedings of the National Academy of Sciences of the USA* **98**, 7029–7036.

Pieler, T. and Theunissen, O. (1993). TFIIIA: Nine fingers – three hands. *Trends in Biochemical Sciences* **18**, 226–230.

Proudfoot, N. (1996). Ending the message is not so simple. *Cell* **87**, 779–781.

Reed, R. and Magni, K. (2001). A new view of mRNA export: separating the wheat from the chaff. *Nature Cell Biology* **3**, E201–E204.

Rigby, P.W.J. (1993). Three in one and one in three: it all depends on TBP. *Cell* **72**, 7–10.

Roeder, R.G. (2003). The eukaryotic transcriptional machinery: complexities and mechanisms unforeseen. *Nature Medicine* **9**, 1239–1244.

Sakonju, S., Bogenhagen, D.F. and Brown, D.D. (1980). A control region in the centre of the 5S RNA gene directs specific initiation of transcription. *Cell* **19**, 13–25.

Schimmel, P. (1987). Aminoacyl tRNA synthetases: general scheme of structure function relationships in the polypeptides and recognition of transfer RNAs. *Annual Review of Biochemistry* **56**, 125–158.

Schramm, L. and Hernandez, N. (2002). Recruitment of RNA polymerase III to its target promoters. *Genes and Development* **16**, 2593–2620.

Schröder, O., Bryant, G.O., Geiduschek, E.P., Berk, A.J. and Kassavetis, G.A. (2003). A common site on TBP for transcription by RNA polymerases II and III. *EMBO Journal* **22**, 5115–5124.

Schultz, P., Fribourg, S., Poterszman, A., Mallouh, V., Moras, D. and Egly, J.M. (2000). Molecular structure of human TFIIH. *Cell* **102**, 599–607.

Sims, R.J., III, Belotserkovskaya, R. and Reinberg, D. (2004). Elongation by RNA polymerase II: the short and long of it. *Genes and Development* **18**, 2437–2468.

Smale, S.T. (2001). Core promoters: active contributors to combinatorial gene regulation. *Genes and Development* **15**, 2503–2508.

Staley, J.P. and Guthrie, C. (1998). Mechanical devices of the spliceosome: motors, clocks, springs and things. *Cell* **92**, 315–326.

Stiller, J.W. and Hall, B.D. (2002). Evolution of the RNA polymerase II C-terminal domain. *Proceedings of the National Academy of Sciences USA* **99**, 6091–6096.

Tansey, W.P. and Herr, W. (1997). TAFs: guilt by association? *Cell* **88**, 729–732.

Tilghman, S.M., Curtis, P.J., Tiemeier, D.C., Leder, P. and Weissman, C. (1978). The intervening sequence of a mouse beta globin gene is transcribed within the 15S beta globin mRNA precursor. *Proceedings of the National Academy of Sciences USA* **75**, 1309–1313.

Tollervey, D. and Caceres, J.F. (2000). RNA processing marches on. *Cell* **103**, 703–709.

Weis, L. and Reinberg, D. (1992). Transcription by RNA polymerase II initiator directed formation of transcription-competent complexes. *FASEB Journal* **6**, 3300–3309.

White, R.J. and Jackson, S.P. (1992). The TATA binding protein: a central role in transcription by RNA polymerases I and III. *Trends in Genetics* **8**, 284–288.

Wilson, K.S. and Noller, H.F. (1998). Molecular movement inside the translational engine. *Cell* **92**, 337–349.

Woychik, N. A. and Hampsey, M. (2002). The RNA polymerase II machinery: Structure illuminates function. *Cell* **108**, 454–463.

Zhou, Z., Luo, M.-J., Straesser, K., Katahira, J., Hurt, E. and Reed, R. (2000). The protein Aly links pre-messenger-RNA splicing to nuclear export in metazoans. *Nature* **407**, 401–405.

Zurita, M. and Merino, C. (2003). The transcriptional complexity of the TFIIH complex. *Trends in Genetics* , 578–584.

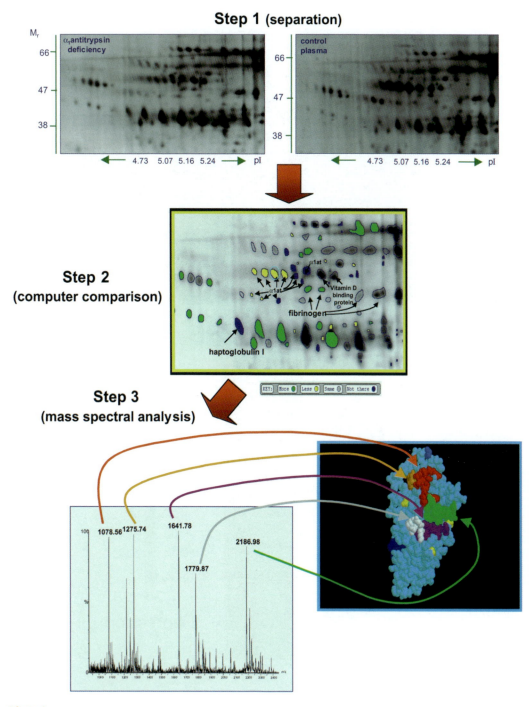

Step 1 (separation)

M$_r$

α$_1$antitrypsin deficiency

66 —
47 —
38 —

← 4.73 5.07 5.16 5.24 → pI

control plasma

66 —
47 —
38 —

← 4.73 5.07 5.16 5.24 → pI

Step 2 (computer comparison)

α1at

Vitamin D binding protein

α1at

fibrinogen

haptoglobulin I

KEY: More ● Less ○ Same ○ Not there ●

Step 3 (mass spectral analysis)

100 1078.56 1275.74 1641.78

2186.98

1779.87

%

Plate 1

Analysis of differences between two protein samples by two-dimensional gel electrophoresis to separate the proteins (step 1), computer comparison of the two gels to identify proteins which differ between the two samples (step 2) followed by excision of a protein spot which is altered and its identification by mass spectral analysis of the peptides produced by trypsin digestion (step 3). Plate kindly provided by Kevin Mills and Professor Brian Winchester.

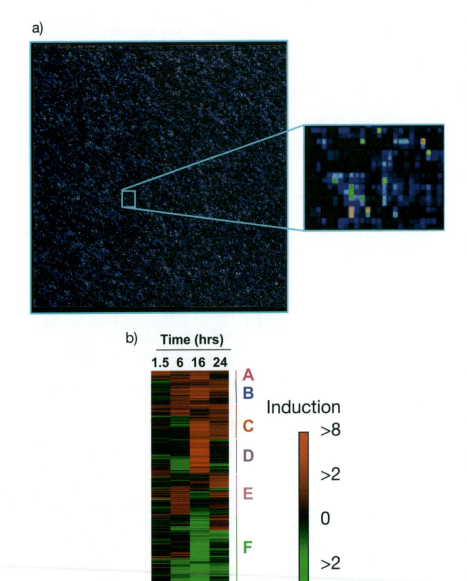

Plate 2

Gene chip analysis of gene expression patterns using the "Affymetrix gene chip system". Panel (a) shows a single gene chip hybridized to labeled RNA probe with a close-up of a specific region. Panel (b) shows the changes in expression of different genes during the activation of lung fibroblasts. Red and green show up-regulation and down-regulation respectively. Letters represent groups of genes showing similar patterns of expression over time. Plate kindly provided by Dr Rachel Chambers and Dr Mike Hubank.

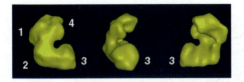

Plate 3

Three-dimensional structure of TFIID. Note the globular domains arranged around a groove into which the DNA fits. Plate kindly provided by Dr Patrick Schultz from Brand *et al.*, *Science* **286**, 2151–2153 (1999) by kind permission of the American Association for the Advancement of Science.

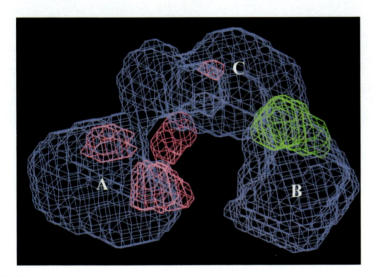

Plate 4

Position of TFIIB (green) and TFIIA (red) complexed with TFIID (blue). Plate kindly provided by Dr Eva Nogales from Andel *et al.*, *Science* **286**, 2153–2156 (1999) by kind permission of the American Association for the Advancement of Science.

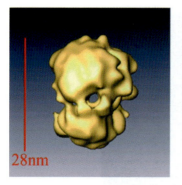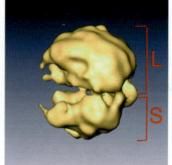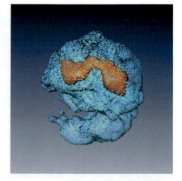

Plate 7

Left and middle panels: Two different views of the three-dimensional structure of the spliceosome at 20 Angstrom resolution. Note the large (L) and small (S) subunits and the cleft between them which accommodates the RNA to be spliced. Right-hand panel: Position of the RNAs involved in catalysing the splicing process (red) within the spliceosome (blue). Plate kindly provided by Professor Ruth Sperling.

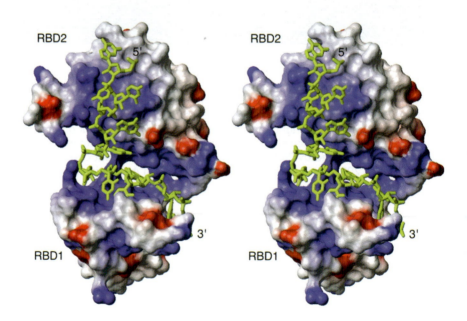

Plate 8

Two views of the structure of the Sxl splicing protein bound to an RNA molecule (green). Plate kindly provided by Dr Yutaka Muto and Professor Shigeyuki Yokoyama, from Handa *et al.*, *Nature* **398**, 579–585 (1999) by kind permission of Macmillan Magazines Ltd.

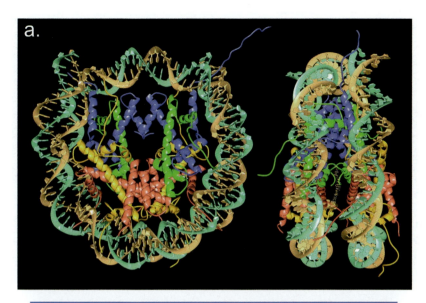

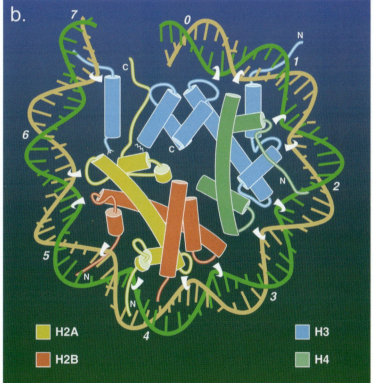

🟨 H2A	🟦 H3
🟧 H2B	🟩 H4

Plate 9

Two views of the structure of the nucleosome (panel a) and a schematic diagram indicating the positions of the different histones (panel b). Panel a kindly provided by Professor Tim Richmond, from Luger *et al.*, *Nature* **389**, 251–260 (1997), panel b kindly provided by Dr Daniela Rhodes, from Rhodes, *Nature* **389**, 251–260 (1997), both by kind permission of Macmillan Magazines Ltd.

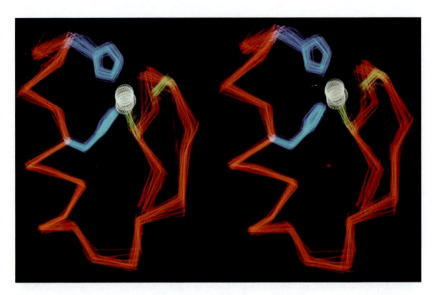

Plate 10

Structure of the Cys_2His_2 zinc finger of the Xfin factor. The cysteine residues are shown in yellow, the histidine side chains in blue and the zinc atom in white. Plate kindly provided by Dr Peter Wright, from Lee *et al.*, *Science* **245**, 635–637 (1989) by kind permission of the American Association for the Advancement of Science.

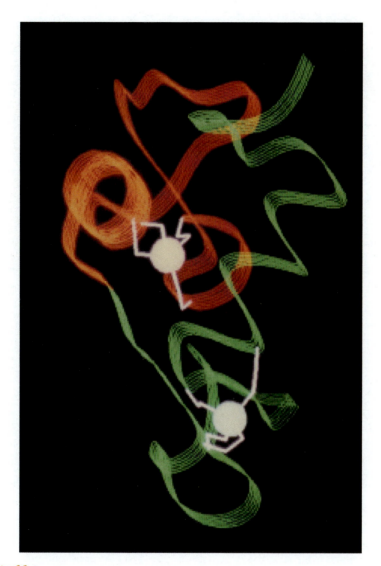

Plate 11

Model of the two Cys$_4$ zinc fingers of the glucocorticoid receptor. The two fingers are shown in red and green respectively with the two zinc atoms in white. Plate kindly provided by Professor Robert Kaptein from Hard *et al.*, *Science* **249**, 157–160 (1990) by kind permission of the American Association for the Advancement of Science.

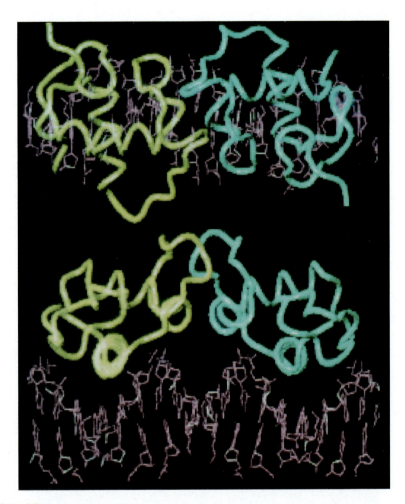

Plate 12

Two views of the estrogen receptor binding to DNA. The two zinc fingers are shown in green and blue with the DNA in purple. Plate kindly provided by Dr Daniela Rhodes from Schwabe *et al.*, *Cell* **75**, 567–578 (1993) by kind permission of Elsevier Science.

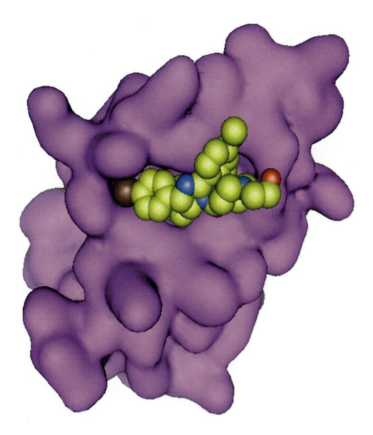

Plate 13

The binding of p53 to a pocket in the MDM2 molecule (pink) is mimicked by the synthetic small molecule nutlin-2. Plate kindly provided by Dr Bradford Groves and Dr Lyubomir Vassilev.

Regulation at transcription

4

SUMMARY

- Gene expression is primarily controlled at the level of transcription by regulating which genes should be copied into RNA in any situation.
- The major control point is at transcriptional initiation when the DNA begins to be copied into RNA.
- An increasing number of cases of regulation at transcriptional elongation also exist however, indicating that this is also an important control point.

4.1 Introduction

The existence of the many different potential regulatory stages discussed in the previous chapter has led to many studies to determine which of these are used in any particular situation. In general, such studies have shown that in higher eukaryotes, as in bacteria, the primary control of gene expression is at the level of transcription, and the evidence showing that this is the case is discussed in this chapter. A number of cases of post-transcriptional regulation do exist, however, and these are discussed in Chapter 5.

4.2 Evidence for transcriptional regulation

The evidence for the regulation of gene transcription comes from several types of study, which will be considered in turn.

Evidence from studies of nuclear RNA

If regulation of gene expression takes place at the level of transcription, the differences in the cytoplasmic levels of particular mRNAs which occur between different tissues should be paralleled by similar differences in the levels of these RNAs within the nuclei of different tissues. In contrast, regulatory processes in which a gene was transcribed in all tissues and the resulting transcript either spliced or transported to the cytoplasm in a minority of tissues would result in cases where differences in mRNA content occurred without any corresponding difference in the nuclear RNA (Fig. 4.1). Hence a study of the level of particular RNA species in the nuclear RNA of individual tissues or cell types serves as an initial test to distinguish transcriptional and post-transcriptional regulation.

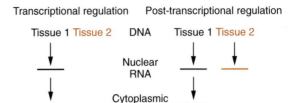

Figure 4.1

Consequences of transcriptional or post-transcriptional regulation on the level of a specific nuclear RNA in a tissue that expresses the corresponding cytoplasmic mRNA (tissue 1) and one that does not (tissue 2).

The earliest studies in this area focused on the highly abundant RNA species produced during terminal differentiation, which could be readily studied simply because of their abundance. Thus, even before the discovery of intervening sequences revealed the possibility of regulation at the level of RNA splicing, Gilmour *et al.* (1974) studied the processes regulating the accumulation of globin which occur when Friend erythroleukemia cells are treated with an inducer of globin production, dimethyl sulfoxide. In this system, the accumulation of cytoplasmic globin RNA, which produces the increase in synthesis of globin protein, was paralleled by increasing accumulation of globin RNA within the nucleus, suggesting that the primary effect of the inducer was at the level of globin gene transcription. Similarly, in the experiments of Groudine *et al.* (1974) globin-specific RNA was readily detectable in the nuclei of globin-synthesizing erythroblasts but not in the nuclear RNA of fibroblasts or muscle cells, hence indicating the involvement of transcriptional control in regulating the abundant production of globin RNA only in the red blood cell lineage and not in other cell types.

Subsequently, the use of Northern blotting (see Section 1.3) allowed the separation of different species within nuclear RNA by size, and their visualization by hybridization to an appropriate probe. In the case of genes with many intervening sequences, such as that encoding the egg protein ovalbumin, many different RNA species could be observed in the nucleus, including not only the primary transcript and the fully spliced RNA prior to transport to the cytoplasm but also a series of intermediate-sized RNAs from which some of the intervening sequences had still to be removed (Roop *et al.*, 1978; Fig. 4.2).

The identification of such potential precursors of the mature mRNA allowed a study of their expression in tissues either producing or not producing ovalbumin mRNA and protein. Thus cytoplasmic ovalbumin mRNA and protein are present only in the oviduct following stimulation with estrogen, and disappear when estrogen is withdrawn. Similarly, the mRNA and protein are absent in other tissues, such as the liver, and cannot be induced by treatment with the hormone in these tissues. Studies of the distribution of both the fully spliced nuclear RNA and the larger precursors (Roop *et al.*, 1978) showed that these species could only be detected in the nuclear RNA of the oviduct following estrogen stimulation and were absent in the liver nuclear RNA, or in oviduct nuclear RNA following estrogen withdrawal (Fig. 4.3). Hence the distribution of these precursors in the

Hybridization of oviduct RNA
to ovalbumin structural gene probe

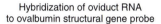

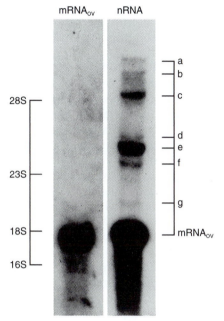

Figure 4.2

Northern blot showing that unspliced and partially spliced precursors (a–g) to the ovalbumin mRNA (mRNA$_{ov}$) are detectable in the nuclear RNA (nRNA) of estrogen-stimulated oviduct tissue. Photograph kindly provided by Professor B.W. O'Malley, from Roop *et al.*, *Cell* **15**, 671–685 (1978), by permission of Cell Press.

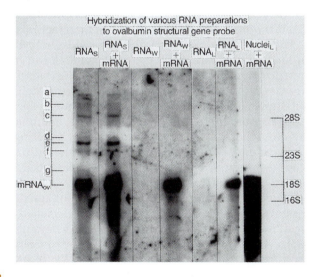

Figure 4.3

Northern blot showing that the nuclear precursors to ovalbumin mRNA (seen in Fig. 4.2) are detectable in the nuclear RNA of estrogen-stimulated oviduct (RNA$_s$) but are absent in the nuclear RNA of estrogen-withdrawn oviduct (RNA$_w$) and of liver (RNA$_L$). Moreover, ovalbumin mRNA mixed with withdrawn oviduct (RNA$_w$ + mRNA) or liver (RNA$_L$ + mRNA) nuclear RNA is not degraded, showing that the absence of RNA for ovalbumin in these nuclear RNAs is not due to a nuclease specifically degrading the ovalbumin RNA. Photographs kindly provided by Professor B.W. O'Malley, from Roop *et al.*, *Cell* **15**, 671–685 (1978), by permission of Cell Press.

nucleus exactly parallels that of the cytoplasmic mRNA, a finding entirely consistent with the transcriptional induction of the ovalbumin gene in the oviduct in response to estrogen. These observations are difficult to reconcile, however, with a model in which the hormone acts to relieve a block in RNA splicing or transport, which exists in untreated oviduct and other tissues, since such models would predict an accumulation of unspliced or untransported RNA for ovalbumin within the nucleus.

These early studies have now been abundantly supplemented by many others measuring the nuclear RNA levels of other specific genes whose expression changes in particular situations, such as those encoding a number of highly abundant mRNAs present only in the soya bean embryo (Goldberg *et al.*, 1981) or that encoding the developmentally regulated mammalian liver protein, α-fetoprotein (Latchman *et al.*, 1984). In general, such studies have led to the conclusion that in most cases alterations in specific mRNA levels in the cytoplasm are accompanied by parallel changes in the levels of the corresponding nuclear RNA species.

Interestingly, such studies carried out with cloned DNA probes for individual RNA species of relatively high abundance are in contrast to reports using R_0t curve analysis (see Section 1.3) to study variations in the total nuclear RNA population in different tissues. Such experiments, which examine mainly low-abundance RNAs, have suggested that in some organisms, such as sea urchins (Wold *et al.*, 1978) and tobacco plants (Kamaly and Goldberg, 1980), the nuclear RNA is a highly complex mixture of different RNAs, some of which can be retained in the nucleus in one tissue and transported to the cytoplasm in other tissues. These studies, which are indicative of post-transcriptional control in these organisms, are discussed in Section 5.1. It is noteworthy, however, that in mammals a different situation exists. Thus, in these organisms the increased number of different sequences in nuclear, compared with cytoplasmic, RNA can be accounted for by the presence of intervening sequences, which are transcribed but not transported to the cytoplasm, without the need to postulate the existence of whole transcripts which are confined to the nucleus in specific tissues.

Hence it appears from these studies that, at least in mammals and other higher eukaryotes, the regulation of transcription, resulting in parallel changes in nuclear and cytoplasmic RNA levels, is the primary means of regulating gene expression. However, the studies described so far suffer from the defect that they measure only steady-state levels of specific nuclear RNAs. It could be argued that the gene is being transcribed in the non-expressing tissue but that the transcript is degraded within the nucleus at such a rate that it cannot be detected in assays of steady-state RNA levels (Fig. 4.4). This degradation in the non-expressing tissue but not in the expressing tissue could be produced directly by specific regulation of the rate of turnover of a particular RNA in the different tissues, producing rapid degradation only in the non-expressing tissue. Alternatively, it might result from a block to splicing or transport of the RNA in the non-expressing tissue, followed by the rapid degradation of this unspliced or untransported RNA, which in the expressing tissue would be processed or transported before it could be degraded. These considerations necessitate the direct measurement of gene transcription itself in the different tissues in order to establish unequivocally the existence of transcriptional regulation. Methods to do this have been devised and these will now be discussed (Methods Box 4.1).

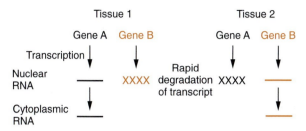

Figure 4.4

Model in which lack of expression of genes in particular tissues is caused by rapid degradation of their RNA transcripts.

METHODS BOX 4.1

Measurement of transcription rates

(a) Pulse labeling to measure transcription (Fig. 4.5)
  - Add radioactively labeled nucleotides to cells.
  - After 5–10 min harvest cells and isolate RNA (including radioactively labeled newly synthesized RNA).
  - Hybridize labeled RNA to dot blot containing DNA from genes whose transcription is being measured.

(b) Nuclear run on to measure transcription (Fig. 4.7)
  - Isolate nuclei from cells.
  - Add radioactively labeled nucleotide to the isolated nuclei.
  - After 1–2 h isolate RNA from nuclei (including labeled RNA produced by transcribing polymerase "running on" to the end of the gene).
  - Hybridize labeled RNA to dot blot containing DNA from genes whose transcription is being measured.

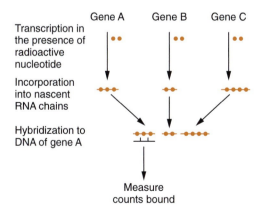

Figure 4.5

Pulse-labeling assay to assess the transcription rate of a specific gene (gene A) by measuring the amount of radioactivity (dots) incorporated into nascent transcripts.

Evidence from pulse-labeling studies

The synthesis of RNA from DNA by the enzyme RNA polymerase involves the incorporation of ribonucleotides into an RNA chain. Therefore the synthesis of any particular RNA can be measured by adding a radioactive ribonucleotide (usually uridine labeled with tritium) to the cells and measuring how much radioactivity is incorporated into RNA specific for the gene of interest. Clearly, if the degradation mechanisms discussed in the last section do exist, they will, given time, degrade the radioactive RNA molecule produced in this way and drastically reduce the amount of labeled RNA detected. In order to prevent this, the rate of transcription is measured by exposing cells briefly to the labeled uridine in a process referred to as pulse labeling. The labeled uridine is incorporated into nascent RNA chains that are being made at this time and, even before a complete transcript has had time to form, the cells are lysed and total RNA is isolated from them. This RNA, which contains labeled partial transcripts from genes active in the tissue used, is then hybridized to a piece of DNA derived from the gene of interest, the number of radioactive counts that bind providing a measure of the incorporation of labeled precursor into its corresponding RNA (Fig. 4.5) (see Methods Box 4.1, Section a).

This method provides the most direct means of measuring transcription and has been used, for example, to show that the induction of globin production which occurs in Friend erythroleukemia cells in response to treatment with dimethyl sulphoxide is mediated by increased transcription of the globin gene (Lowenhaupt et al., 1978). This increased transcription produces the increased levels of globin-specific RNA in the nucleus and cytoplasm of the treated cells, which was discussed earlier in this chapter.

Although pulse labeling provides a very direct measure of transcription rates, the requirement to use very short labeling times to minimize any effects of RNA degradation limits its applicability. Thus in the experiments of Lowenhaupt et al. (1978) it was possible to measure the amount of radioactivity incorporated into globin RNA in the very brief labeling times used (5 or 10 min) only because of the enormous abundance of globin RNA and the very high rate of transcription of the globin gene. With other RNA species, the rates of transcription are insufficient to provide measurable incorporation of label in the short pulse time. More label will, of course, be incorporated if longer pulse times are used, but such pulse times allow the possibility of RNA turnover and are therefore subject to the same objections as the measurement of stable RNA levels.

Hence, although pulse labeling can be used to establish unequivocally that transcriptional control is responsible for the massive synthesis of the highly abundant RNA species present in terminally differentiated cells, it cannot be used to demonstrate the generality of transcriptional control processes and, in particular, their applicability to RNAs which, although regulated in different tissues, never become highly abundant.

This limitation of the pulse-labeling method is especially relevant in view of the existence of the Davidson and Britten model for gene regulation (Davidson and Britten, 1979) which specifically postulates that highly abundant RNA species will be regulated in a different manner to the bulk of RNA species, which are of moderate or low abundance. Thus, in this model (Fig. 4.6) all genes are postulated to be transcribed in all tissues at

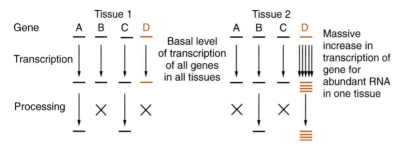

Figure 4.6

Davidson and Britten model of post-transcriptional regulation, in which all genes are transcribed at a low basal rate and regulation is achieved by controling whether the resulting transcript is processed to mature mRNA. Transcriptional control is confined to genes (such as gene D) where the level of RNA required cannot be achieved by processing all the primary transcript produced by the low basal rate of transcription.

a low basal rate and regulation takes place at a post-transcriptional level by deciding which transcripts are processed to functional mRNA and which are degraded. In the case of most genes the level of RNA and protein required in any particular tissue would be met by processing correctly all of the primary transcript produced by this low basal rate of transcription. The level of the abundant RNA species would be such, however, that they could not be produced with this low rate of gene transcription, even by correctly processing all of the primary transcript. Hence, for the genes encoding these RNA species, a special mechanism would operate and their transcription would be dramatically increased in some tissues, as observed by pulse labeling.

This theory postulates that the regulation of gene expression by changes in transcription is confined to the few genes encoding highly abundant RNA species, and that post-transcriptional control processes will regulate the expression of less abundant RNAs whose genes will be transcribed even in tissues where no mRNA is synthesized. In order to test this theory it is necessary to use a method of measuring transcription which, although less direct than pulse labeling, is more sensitive and hence can be applied to a wider variety of cases, including non-highly abundant mRNAs. This method is discussed in the next section.

Evidence from nuclear run-on assays

The primary limitation on the sensitivity of pulse labeling is the existence within the cell of a large pool of non-radioactive ribonucleotides, which are normally used by the cell to synthesize RNA. When labeled ribonucleotide is added to the cell, it is considerably diluted in this pool of unlabeled precursor. The amount of label incorporated into RNA in the labeling period is therefore very small, since the majority of ribonucleotides incorporated are unlabeled. The sensitivity of this method is thus severely reduced, resulting in its observed applicability only to genes with very high rates of transcription. Interestingly, however, although transcription takes place in the nucleus, most of the pool of precursor ribonucleotides is present in the cytoplasm. Hence it is possible, by removing the cytoplasm and isolating nuclei, to remove much of the pool of unlabeled ribonucleotide. When this occurs the RNA polymerase ceases transcribing due to lack of ribonucleotide but remains associated with the DNA (Fig. 4.7). Labeled ribonucleotides can then be added directly to the isolated nuclei in a test-tube and the RNA polymerase resumes transcribing and runs on to the end of the gene incorporating labeled ribonucleotide into the RNA transcript. Because the labeled ribonucleotides are not diluted in the unlabeled cytoplasmic pool, considerably more label is incorporated into any particular RNA transcript than is observed in pulse-labeling experiments. The label incorporated into any particular transcript is detected by hybridization to its corresponding DNA exactly as in pulse-labeling experiments (see Methods Box 4.1, Section b).

This method, which is known as a nuclear run-on assay, is therefore much more widely applicable than pulse labeling, and can be used to quantify the transcription of genes that are never transcribed at levels detectable by pulse labeling. Moreover, very many studies have now established that the RNA synthesized by isolated nuclei in the test-tube is similar to that made by intact whole cells, and that the method is therefore not only sensitive but also provides an accurate measure of transcription, free from artefact.

In initial studies, nuclear run-on assays were used to measure the transcription of highly abundant RNA species. Thus, for example, nuclei isolated from the erythrocytes of adult chickens (which, unlike the equivalent cells in mammals, do not lose their nuclei) were shown to transcribe the gene encoding the adult β-globin protein, whereas nuclei prepared from similar cells isolated from embryonic chickens failed to transcribe this gene and instead transcribed the gene encoding the form of β-globin made in the embryo (Groudine et al., 1981). Such a finding parallels the observation discussed earlier that the RNA specific for the adult form of β-globin is present only in adult and not in embryronic erythroid cell nuclei (see above) and indicates that the developmentally regulated production of different forms of globin protein is under transcriptional control.

A similar parallelism between the results of nuclear RNA studies and nuclear run-on assays of transcription is also found in the case of the ovalbumin gene. Thus the presence of ovalbumin-specific RNA only in the nuclear RNA of hormonally stimulated oviduct tissue is paralleled by the ability of nuclei prepared from stimulated oviduct cell nuclei to transcribe the ovalbumin gene at high levels in run-on assays. In contrast, nuclei from other tissues or unstimulated oviduct cells failed to transcribe this gene, paralleling the observed absence of ovalbumin RNA in the nuclei and cytoplasm of these tissues (Swaneck et al., 1979). Hence the observed increase in nuclear and cytoplasmic RNA for ovalbumin in response to estrogen is indeed caused by increased transcription of the ovalbumin gene. Interestingly, in this study the incorporation of label into ovalbumin RNA in the run-on assay was observed to peak after 15 min and did not decrease at longer labeling times of up to 1 h. This suggested that, unlike intact cells, isolated nuclei do not degrade or process the RNA that they synthesize, and hence allowed the use of longer labeling times, further increasing the sensitivity of this technique (see Methods Box 4.1 and compare Section a with Section b).

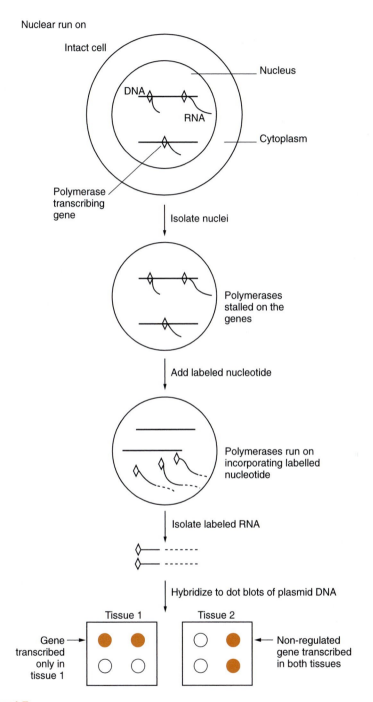

Figure 4.7

Nuclear run-on assay. Following isolation of nuclei, RNA polymerase ceases transcribing due to lack of ribonucleotides but remains associated with the DNA. When radioactively labeled ribonucleotide is added, the polymerase resumes transcription and runs on to the end of the gene incorporating the labeled ribonucleotide into the RNA transcript. The labeled RNA made in this way in different tissues can then be used to probe filters containing DNA from genes whose transcription is to be measured in the different tissues.

This increased sensitivity has allowed the use of nuclear run-on assays to demonstrate that transcriptional control of many genes encoding specific proteins is responsible for the previously observed differences in the levels of these proteins in different situations. Cases where such transcriptional control has been demonstrated in this manner are far too numerous to mention individually but involve a range of different tissues and organisms, such as the synthesis of α-fetoprotein in fetal, but not adult, mammalian liver, production of insulin in the mammalian pancreas, the expression of the *Drosophila melanogaster* yolk protein genes only in ovarian follicle cells, the expression of aggregation stage-specific genes in the slime mould *Dictyostelium discoideum*, and the expression of soya bean seed proteins such as glycinin in embryonic, but not adult, tissues.

These studies on individual genes for particular proteins have been supplemented by the more general studies of Darnell and colleagues (Derman *et al.*, 1981; Powell *et al.*, 1984). In these experiments the authors studied 12 different genes whose corresponding mRNAs were present in mouse liver cytoplasm but were absent in brain cytoplasm. These included both genes encoding previously isolated liver-specific proteins, such as albumin or transferrin, and those which had been isolated simply on the basis of the presence of their corresponding cytoplasmic RNA in the liver and not in other tissues and for which the protein product had not yet been identified. Measurement of the transcription rate of these genes in nuclei isolated from brain or liver showed that such transcription was detectable only in the liver nuclei (Fig. 4.8), indicating that the difference in cytoplasmic mRNA level was produced by a corresponding difference in gene transcription. In these experiments the rate of transcription of the 12 genes was also measured in nuclei prepared from kidney tissue. In this tissue mRNAs corresponding to two of the genes were present at a considerably lower level than that observed in the liver, while RNA corresponding to the other 10 genes was undetectable. As with the brain nuclei, the level of transcription detectable in the kidney nuclei exactly paralleled the level of RNA present. Thus, only the two genes producing cytoplasmic mRNA in the kidney were detectably transcribed in this tissue and the level of transcription of these genes was much lower than that seen in the liver nuclei.

These studies on a relatively large number of different liver RNAs of different abundances, when taken together with the studies of many genes encoding specific proteins, indicate that for genes expressed in one or a few cell types, increased levels of RNA and protein in a particular cell type are brought about primarily by increases in gene transcription and that, at least in mammals, post-transcriptional control mechanisms, such as that postulated by Davidson and Britten (1979), are not the primary means of regulating gene expression, although they may be more important in other animals, such as the sea urchin (see Section 5.1).

Evidence from polytene chromosomes

Although pulse-labeling and nuclear run-on studies of many genes have conclusively established the existence of transcriptional regulation, it is necessary to discuss another means of demonstrating such regulation, in which increased transcription can be directly visualized. As described in

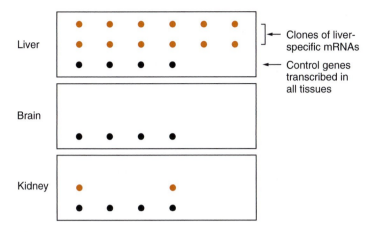

Figure 4.8

Nuclear run-on assay to measure the transcription rates of genes encoding liver-specific mRNAs and of control genes expressed in all tissues. Note that the liver-specific genes are not transcribed at all in brain, while in the kidney the only two of these genes to be transcribed are those that are known to produce a low level of mRNA in the kidney.

Section 2.3, the chromosomal DNA in the salivary glands of *Drosophila* is amplified many times, resulting in a giant polytene chromosome. Such chromosomes exhibit along their length areas known as puffs in which the DNA has decondensed into a more open state, resulting in the expansion of the chromosome (Fig. 4.9). If cells are allowed to incorporate labeled ribonucleotides into RNA and the resulting RNA is then hybridized back to the polytene chromosomes, it localizes primarily to the positions of the puffs. Hence these puffs represent sites of intense transcriptional activity which, because of the large size of the polytene chromosomes, can be directly visualized.

Most interestingly, many procedures which, in *Drosophila*, result in the production of new proteins, such as exposure to elevated temperature (heat shock) or treatment with the steroid hormone ecdysone, also result in the production of new puffs at specific sites on the polytene chromosomes, each treatment producing a different specific pattern of puffs. This suggests that these sites contain the genes encoding the proteins whose synthesis is increased by the treatment, and that this increased synthesis is mediated via increased transcription of these genes, which can be visualized in the puffs. In the case of ecdysone treatment this has been confirmed directly by showing that the radioactive RNA synthesized immediately after ecdysone treatment hybridizes strongly to the ecdysone-induced puffs but not to a puff which regresses upon hormone treatment. In contrast, RNA prepared from cells prior to ecdysone treatment hybridizes only to the hormone-repressed puff and not to the hormone-induced puffs (Fig. 4.10; Bonner and Pardue, 1977). Similarly, RNA labeled after heat shock hybridizes intensely to puff 87C, which appears following exposure to elevated temperature and is now known to contain the gene encoding the 70 kDa heat-shock protein (hsp70) which is the major protein made in *Drosophila* following heat shock (Spradling *et al.*, 1975).

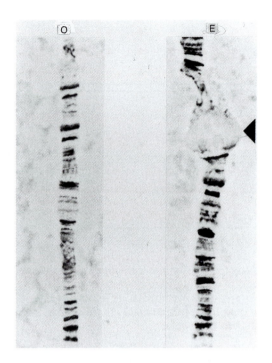

Figure 4.9

Transcriptionally active puff (arrowed) in a polytene chromosome of *Drosophila melanogaster*. The puff appears in response to treatment with the steroid hormone ecdysone (E) and is not present prior to hormone treatment (O). Photograph kindly provided by Dr M. Ashburner.

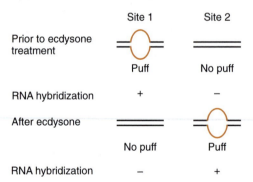

Figure 4.10

The newly synthesized RNA made following ecdysone stimulation can be labeled with ^{3}H uridine and shown to hybridize to the puffs that form following ecdysone treatment. Conversely, a puff that regresses after hormone treatment binds only the labeled RNA synthesized before addition of the hormone.

Thus the large size of polytene chromosomes allows a direct visualization of the transcriptional process and indicates that, as in other situations, gene activity in the salivary gland is regulated at the level of transcription.

4.3 Regulation at transcriptional elongation

Initiation of transcription

In the majority of cases where increased transcription of a particular gene has been demonstrated, it is likely that such increased transcription is mediated by an increased rate of initiation of transcription by RNA polymerase which occurs at the region of the gene known as the promoter (see Section 3.2). Hence in a tissue in which a gene is being transcribed actively, a large number of polymerase molecules will be moving along the gene at any particular time, resulting in the production of a large number of transcripts. Such a series of nascent transcripts being produced from a single transcription unit can be visualized in the lampbrush chromosomes of amphibian oocytes, the nascent transcripts associated with each RNA polymerase molecule increasing in length the further the polymerase has proceeded along the gene, resulting in the characteristic nested appearance (Fig. 4.11).

By contrast, in tissues where a gene is transcribed at very low levels, initiation of transcription will be a rare event and only one or a very few

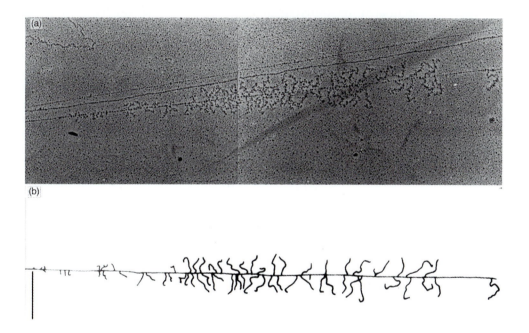

Figure 4.11

Electron micrograph (A) and summary diagram (B) of a lampbrush chromosome in amphibian oocytes, showing the characteristic nested appearance produced by the nascent mRNA chains attached to transcribing RNA polymerase molecules. The bar indicates 1 µm. Photograph kindly provided by Dr R.S. Hill, from Hill and Macgregor, *J. Cell Sci.* **44**, 87–101 (1980), by permission of the Company of Biologists Ltd.

polymerase molecules will be transcribing a gene at any particular time. Similarly, the absence of transcription of a particular gene in some tissues will result from a failure of RNA polymerase to initiate transcription in that tissue (Fig. 4.12). A large number of sequences upstream of the point at which initiation occurs and which are involved in its regulation have now been described, and these will be discussed in Chapter 7.

Transcriptional elongation

Although the majority of cases of transcriptional regulation are likely to occur at the level of initiation, some cases have been described where regulation occurs following the initiation of transcription by RNA polymerase and the production of a truncated RNA of less than full length. In such cases, transcriptional control may operate by releasing a block to elongation of the nascent transcript (for reviews see Conaway *et al.*, 2000; Arndt and Kane, 2003; Sims *et al.*, 2004).

This form of regulation is responsible, for example, for the tenfold decline in mRNA levels encoding the cellular oncogene c-*myc* (for discussion of cellular oncogenes see Chapter 9) which occurs when the human pro-myeloid cell line HL-60 is induced to differentiate into a granulocyte type cell. If nuclear run-on assays are carried out using nuclei from undifferentiated or differentiated HL-60 cells, the results obtained vary depending on the region of the c-*myc* gene whose transcription is being measured. Thus the c-*myc* gene consists of three exons, which appear in the messenger RNA and which are separated by intervening sequences that are removed from the primary transcript by RNA splicing. If the labeled products of the nuclear run-on procedure are hybridized to the DNA of the second exon, the levels of transcription observed are approximately tenfold higher in nuclei derived from undifferentiated cells than in nuclei from differentiated cells. Hence the observed differences in c-*myc* RNA levels in these cell types are indeed produced by differences in transcription rates. However, if the same labeled products are hybridized to DNA from the first exon of the c-*myc* gene, virtually no difference in the level

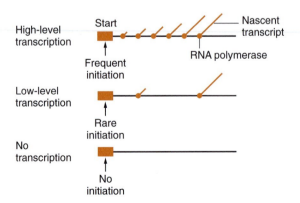

Figure 4.12

Regulation of transcriptional initiation results in differences in the number of RNA polymerase molecules transcribing a gene and therefore in the number of transcripts produced.

of transcription of this region in the differentiated compared with the undifferentiated cells is observed.

Comparison of the rates of transcription of the first and second exons in undifferentiated and differentiated cells indicates that regulation takes place at the level of transcriptional elongation, rather than initiation. Thus, although similar numbers of polymerase molecules initiate transcription of the c-*myc* gene in both cell types, the majority terminate in differentiated cells near the end of exon 1, do not transcribe the remainder of the gene, and hence do not produce a functional RNA. In contrast, in undifferentiated cells most polymerase molecules that initiate transcription, transcribe the whole gene and produce a functional RNA. Hence the fall in c-*myc* RNA in differentiated cells is regulated by means of a block to elongation of nascent transcripts (Fig. 4.13). A 180 bp sequence from the 3′ end of the first exon of c-*myc* has been shown to mediate this effect and to block transcriptional elongation if placed within the transcribed region of another gene (Wright and Bishop, 1989). Interestingly, the rapid inhibition of transcriptional elongation following differentiation is supplemented, several days after differentiation, by an inhibition of c-*myc* transcription at the level of initiation.

Similar effects on transcriptional elongation have been seen in several other cellular oncogenes such as c-*myb*, c-*fos* and c-*mos*, indicating that this mechanism is not confined to a single oncogene and may be quite widespread.

Interestingly, a protein affecting the rate of transcriptional elongation has been identified in the case of the human immunodeficiency virus (HIV-1). Thus early in infection with this virus, a low level of transcription occurs from the HIV-1 promoter and many of the transcripts terminate very close to the promoter producing very short RNAs. Subsequently the viral Tat protein binds to these RNAs at a region known as Tar, which is located at +19 to +42 relative to the start site of transcription at nucleotide +1. This binding produces two effects. Firstly, there is a large increase in transcriptional initiation leading to many more RNA

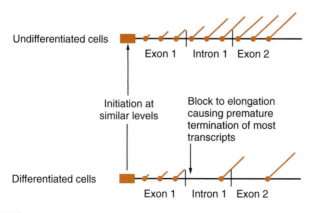

Figure 4.13

Regulation of transcriptional elongation in the c-*myc* gene. A block to elongation at the end of exon 1 results in most transcripts terminating at this point in differentiated HL-60 cells.

transcripts being initiated. In addition, however, Tat also overcomes the block to elongation leading to the production of a greater proportion of long transcripts which can encode the viral proteins.

Hence Tat produces a large increase in the production of HIV transcripts able to encode viral proteins both by stimulating transcriptional initiation and by overcoming a block to transcriptional elongation close to the promoter (Fig. 4.14). As well as the viral Tat protein, cellular proteins which can stimulate transcriptional elongation have also been identified. Indeed such an effect is involved in the increased expression of the gene encoding the heat inducible hsp70 protein that occurs following exposure of cells to elevated temperature and which results from the binding of the heat-shock transcription factor (HSF) to the promoter of this gene (see Sections 4.2 and 7.2) (for a review see Lis and Wu, 1993) and other transcriptional activators have also been shown to act by stimulating both transcriptional initiation and elongation (Yankulov et al., 1994).

A key target for factors which regulate transcriptional elongation is the phosphorylation of the C-terminal domain (CTD) of RNA polymerase II. Thus, as described in Section 3.2, such phosphorylation of the CTD is a key event required for the RNA polymerase II transcription complex to move off down the gene allowing transcriptional elongation to occur. Interestingly, the HIV Tat protein has been shown to be able to interact with a kinase protein that can phosphorylate the CTD. Hence, Tat attracts this kinase to the HIV promoters ensuring phosphorylation of RNA polymerase II and producing transcriptional elongation (for a review see Conaway et al., 2000).

As well as regulating transcriptional elongation, phosphorylation of the CTD of RNA polymerase II is also linked to the recruitment of factors involved in capping, splicing and polyadenylation of the transcript,

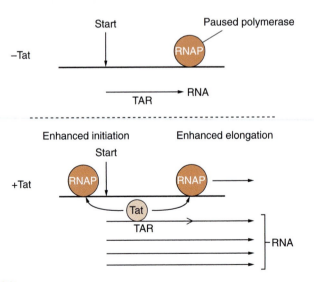

Figure 4.14

The HIV Tat protein acts both by enhancing the rate of transcriptional initiation by RNA polymerase (RNAP) and by enhancing the rate of elongation by overcoming polymerase pausing close to the promoter. Both these effects are achieved by the Tat protein binding to a specific region (TAR) of the nascent RNA.

thereby ensuring that the elongating transcript is correctly processed (see Section 3.3). This linkage is seen also in the HIV/Tat case. Thus, recent studies have shown that phosphorylation of the CTD of RNA polymerase II on serine 5 by the Tat-regulated cellular kinase also stimulates the capping of the HIV RNA (Zhou *et al.*, 2003). Similarly, the P-TEFb cellular kinase has been shown to stimulate 3′ end processing as well as transcriptional elongation in the case of the heat-shock genes by phosphorylating serine 2 of the CTD (Ni *et al.*, 2004).

The involvement of CTD phosphorylation in regulating transcriptional elongation is also seen in the case of the zebrafish protein, Foggy (Guo *et al.*, 2000). Thus, the Foggy protein interacts with the non-phosphorylated form of RNA polymerase II to block transcriptional elongation. However, when RNA polymerase II is phosphorylated it is no longer inhibited by Foggy allowing transcriptional elongation to proceed (Fig. 4.15). When this activity of Foggy is inactivated by mutation, the resulting zebrafish fail to produce the correct number of dopamine synthesizing neurones during development. This indicates that the regulation of transcriptional elongation by Foggy is likely to be critical for the correct expression of genes involved in the production of this cell type and more generally demonstrates that regulation at the level of transcriptional elongation can play a key role in controling gene expression during development.

The viral Tat protein and the cellular Foggy protein illustrate that regulators of transcriptional elongation can act either to stimulate or inhibit this process. Indeed, in yeast a pair of antagonistic regulators of transcriptional elongation have been identified. Thus, the Fkh2p factor stimulates transcriptional elongation and processing of the RNA, by enhancing phosphorylation of the CTD of RNA polymerase II on serine 2 and serine 5, whilst Fkhlp inhibits such phosphorylation and thereby inhibits transcriptional elongation and RNA processing (Morillon *et al.*, 2003) (Fig. 4.16).

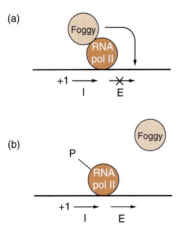

Figure 4.15

(a) The Foggy protein acts by interacting with the RNA polymerase II transcription complex to prevent transcriptional elongation (E) following transcriptional initiation (I). (b) Phosphorylation of RNA polymerase II prevents this inhibitory effect of Foggy, allowing transcriptional elongation to proceed.

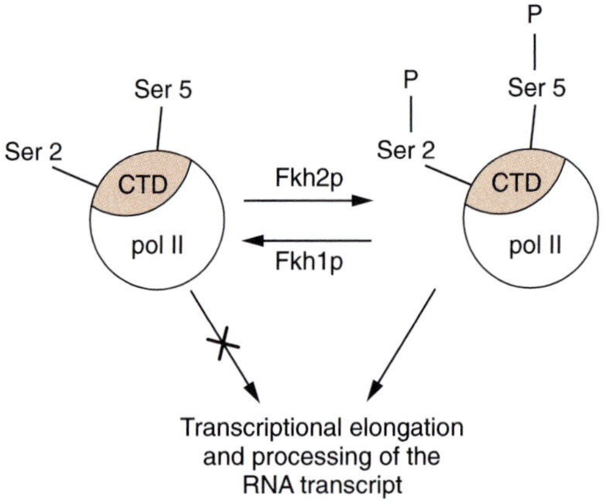

Figure 4.16

The yeast transcription factors Fkhlp and Fkh2p have opposite effects on the phosphorylation of the C-terminal domain (CTD) of RNA polymerase II and thereby produce opposite effects on transcriptional elongation and processing of the resulting RNA.

Hence, transcriptional elongation appears to be regulated by the balance between factors which stimulate this process and those which inhibit it. As many of these factors regulate transcriptional elongation by controlling the phosphorylation of the CTD of RNA polymerase II, they also regulate post-transcriptional processes which are regulated by such phosphorylation, thereby linking RNA transcript production with its proper processing.

Such regulation of transcriptional elongation is therefore likely to be of importance in a number of different situations. It is clear, however, that in most cases transcription is regulated at the level of initiation with specific regulatory processes acting to enhance the rate at which RNA polymerases initiate transcription of the DNA into RNA.

4.4 Conclusions

The experiments described in this chapter provide conclusive evidence that control of transcriptional initiation is the major means used to regulate gene expression in eukaryotic organisms. The mechanism by which such transcriptional control is achieved will be discussed in Chapters 6, 7 and 8. Some cases of post-transcriptional control have been described, however, and these will be discussed in the next chapter.

References

Arndt, K.M. and Kane, C.M. (2003). Running with RNA polymerase: eukaryotic transcript elongation. *Trends in Genetics* **19**, 543–550.

Bonner, J.J. and Pardue, M.L. (1977). Ecdysone-stimulated RNA synthesis in salivary glands of *Drosophila melanogaster* assay by *in situ* hybridisation. *Cell* **12**, 219–225.

Conaway, J.W., Shilatifard, A., Dvir, A. and Conaway, R.C. (2000). Control of elongation by RNA polymerase II. *Trends in Biochemical Sciences* **25**, 375–380.

Davidson, E.H. and Britten, R.J. (1979). Regulation of gene expression: possible role of repetitive sequences. *Science* **204**, 1052–1059.

Derman, E., Krauter, K., Walling, L., Weinberger, C., Ray, M. and Darnell, J.E. (1981). Transcriptional control in the production of liver specific mRNAs. *Cell* **23**, 731–739.

Gilmour, R.S., Harrison, P.R., Windass, J.D., Affara, N.A. and Paul, J. (1974). Globin messenger RNA synthesis and processing during haemoglobin induction in Friend cells. 1. Evidence for transcriptional control in clone M2. *Cell Differentiation* **3**, 9–22.

Goldberg, R.B., Hosheck, G., Ditta, G.S. and Breidenbach, R.W. (1981). Developmental regulation of cloned superabundant mRNAs in soybean. *Developmental Biology* **83**, 218–231.

Groudine, M., Hoitzer, H., Scherner, K. and Therwath, A. (1974). Lineage dependent transcription of globin genes. *Cell* **3**, 243–247.

Groudine, M., Peretz, M. and Weintraub, H. (1981). Transcriptional regulation of haemoglobin switching in chicken embryos. *Molecular and Cellular Biology* **1**, 281–288.

Guo, S., Yamaguchi, Y., Schillbach, S., Wada, T., Lee, J., Goddard, A., French, D., Handa, H. and Rosenthal, A. (2000). A regulator of transcriptional elongation controls vertebrate neuronal development. *Nature* **408**, 366–369.

Kamaly, J.C. and Goldberg, R.B. (1980). Regulation of structural gene expression in tobacco. *Cell* **19**, 935–946.

Latchman, D.S., Brzeski, H., Lovell-Badge, R.H. and Evans, M.J. (1984). Expression of the alpha-foetoprotein gene in pluripotent and committed cells. *Biochimica et Biophysica Acta* **783**, 130–136.

Lis, J. and Wu, C. (1993). Protein traffic on the heat shock promoter: parking stalling and trucking along. *Cell* **74**, 1–4.

Lowenhaupt, K., Trent, C. and Lingrel, J.B. (1978). Mechanisms for accumulation of globin mRNA during dimethyl sulfoxide induction of mouse erythroleukaemia cells: synthesis of precursors and mature mRNA. *Developmental Biology* **63**, 441–454.

Morillon, A., O'Sullivan, J., Azad, A., Proudfoot, N. and Mellor, J. (2003). Regulation of elongating RNA polymerase II by forkhead transcription factors in yeast. *Science* **300**, 492–495.

Ni, Z., Schwartz, B.E., Werner, J., Suarez, J.-R., and Lis, J.T. (2004). Co-ordination of transcription, RNA processing and surveillance by P-TEFb kinase on heat shock genes. *Molecular Cell* **13**, 55–65.

Powell, D.J., Freidman, J.M., Oulethe, A.J., Krauter, K.S. and Darnell, J.E. (1984). Transcriptional and post-transcriptional control of specific messenger RNAs in adult and embryonic liver. *Journal of Molecular Biology* **179**, 21–35.

Roop, D.R., Nordstrom, J.L., Tsai, S.-Y., Tsai, M.-J. and O'Malley, B.W. (1978). Transcription of structural and intervening sequences in the ovalbumin gene and identification of potential ovalbumin mRNA precursors. *Cell* **15**, 671–685.

Sims, R.J., III, Belotserkovskaya, R. and Reinberg, D. (2004). Elongation by RNA polymerase II: the short and long of it. *Genes and Development* **18**, 2437–2468.

Spradling, A., Penman, S. and Pardue, M.L. (1975). Analysis of *Drosophila* mRNA by *in situ* hybridization: sequences transcribed in normal and heat shocked cultured cells. *Cell* **4**, 395–404.

Swaneck, G. E., Nordstrom, J.L., Kreuzaler, F., Tsai, M.-J. and O'Malley, B.W. (1979). Effect of estrogen on gene expression in chicken oviduct, evidence for transcriptional control of the ovalbumin gene. *Proceedings of the National Academy of Sciences of the USA* **76**, 1049–1053.

Wold, B.J., Klein, W.H., Hough-Evans, B.R, Britten, R.J. and Davidson, E.H. (1978). Sea urchin embryo mRNA sequences expressed in the nuclear RNA of adult tissues. *Cell* **14**, 941–950.

Wright, S. and Bishop, J.M. (1989). DNA sequences that mediate attenuation of transcription from the mouse proto-oncogene c-*myc*. *Proceedings of the National Academy of Sciences of the USA* **86**, 505–509.

Yankulov, K., Blau, J., Purton, T., Roberts, S. and Bentley, D.L. (1994). Transcriptional elongation by RNA polymerase II is stimulated by transactivators. *Cell* **77**, 749–759.

Zhou, M., Deng, L., Kashanchi, F., Brady, J.N., Shatkin, A.J. and Kumar, A. (2003). The Tat/TAR-dependent phosphorylation of RNA polymerase II C-terminal domain stimulates cotranscriptional capping of HIV-1 mRNA. *Proceedings of the National Academy of Sciences of the USA* **100**, 12666–12671.

Post-transcriptional regulation

5

SUMMARY

- Although transcriptional control is the predominant means of regulating gene expression, post-transcriptional control processes also occur, often as a supplement to transcriptional control.
- Such post-transcriptional control can occur at a number of different points between initial transcription and protein synthesis.
- Alternative splicing and RNA editing are post-transcriptional processes which are often used as a supplement to transcriptional control to produce distinct but related proteins from a single gene.
- Regulation of RNA stability or of translation are often used as supplements to transcriptional control in situations where a rapid change in protein levels is required.
- Post-transcriptional regulation frequently occurs for the genes encoding transcription factors ensuring that the proteins which regulate transcription are not themselves regulated at this level.

5.1 Regulation after transcription?

Although the evidence discussed in the preceding chapter indicates that, in mammals at least, the primary control of gene expression lies at the level of transcription, a number of cases exist where changes in the rate of synthesis of a particular protein occur without a change in the transcription rate of the corresponding gene. Indeed, in some lower organisms post-transcriptional regulation may constitute the predominant form of regulation of gene expression.

In the sea urchin, for example, the nuclear RNA contains many more different RNA species than are found in the cytoplasmic messenger RNA. Hence a large proportion of the genes transcribed give rise to RNA products that are not transported to the cytoplasm and do not function as a messenger RNA. Interestingly, however, this process is regulated differently in different tissues; an RNA species which is confined to the nucleus in one tissue being transported to the cytoplasm and functioning as a messenger RNA in another tissue. Thus, up to 80% of the cytoplasmic mRNAs found in the embryonic blastula are absent from the cytoplasmic RNA of adult tissues, such as the intestine, but are found in the nuclear RNA of such tissues (Wold *et al.*, 1978).

Although post-transcriptional regulation in mammals does not appear to be as generalized as in the sea urchin, some cases exist where changes in cytoplasmic mRNA levels occur without alterations in the rate of gene transcription. Such post-transcriptional regulation may be more important in controlling variations in the level of mRNA species expressed in all tissues than in the regulation of mRNA species that are expressed in only one or a few tissues. For example, in the experiments of Darnell and colleagues (Powell *et al.*, 1984), which demonstrated the importance of transcriptional control in the regulation of liver-specific mRNAs (see Section 4.2), tissue-specific differences in the levels of the mRNAs encoding actin and tubulin (which are expressed in all cell types) were observed in the absence of differences in transcription rates. Clearly, these and other cases where mRNA levels alter in the absence of changes in transcription rates indicate the existence of post-transcriptional control processes and require an understanding of their mechanisms.

In principle, such post-transcriptional regulation could operate at any of the many stages between gene transcription and the translation of the corresponding mRNA in the cytoplasm. Indeed, the available evidence indicates that in different cases regulation can occur at any one of these levels. Each of these will now be discussed in turn.

5.2 Regulation of RNA splicing

RNA splicing

The finding that the protein-coding regions of eukaryotic genes are split by intervening sequences (introns) which must be removed from the initial transcript by RNA splicing of the protein-coding exons (see Section 3.3) led to much speculation that this process might provide a major site of gene regulation. Thus, in theory, an RNA species transcribed in several tissues might be correctly spliced to yield functional RNA in one tissue and remain unspliced in another tissue. An unspliced RNA would either be degraded within the nucleus or, if transported to the cytoplasm, would be unable to produce a functional protein due to the interruption of the protein-coding regions (Fig. 5.1). Several cases of such processing versus discard decisions have now been described in *Drosophila* and a similar pathway in which unspliced RNAs are specifically degraded within the nucleus has been characterized in yeast (Bousquet-Antonelli *et al.*, 2000). Hence, in lower organisms, there appear to be regulatory pathways in which a particular transcript is spliced to produce a functional mRNA in one situation whilst remaining unspliced and then being degraded within the nucleus in another situation.

Alternative RNA splicing

Although such processing versus discard decisions are not widespread in mammals, numerous cases of alternative RNA splicing have been described, both in mammals and other organisms. In this process (for reviews see Black, 2000; Graveley, 2001) a single gene is transcribed in several different tissues, the transcripts from this gene being processed differentially to yield different functional messenger RNAs in the different tissues (Fig. 5.2). In many cases these RNAs are translated to yield different protein products. It is noteworthy that this mechanism of gene regulation involves not only regulation of processing but also regulation

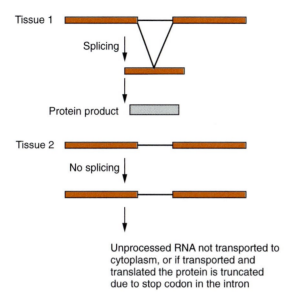

Figure 5.1

The absence of RNA splicing of a transcript in a particular tissue results in a lack of production of the corresponding protein.

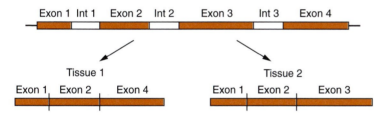

Figure 5.2

Alternative splicing of the same primary transcript in two different ways results in two different mRNA molecules.

of transcription, in that the alternatively processed RNAs are transcribed in only a restricted range of cell types and not in many other cells.

Cases of alternative RNA processing occur in the genes involved in a wide variety of different cellular processes, ranging from genes which regulate embryonic development or sex determination in *Drosophila* to those involved in muscular contraction or neuronal function in mammals. Indeed, a recent survey of the whole human genome concluded that at least 74% of human genes with multiple exons are alternatively spliced (Johnson *et al.*, 2003). A representative selection of such cases is given in Table 5.1.

For convenience, cases of alternative RNA processing can be divided into three groups (Leff *et al.*, 1986): (a) situations where the 5′ end of the differentially processed transcripts is different, (b) situations where the 3′ end of the differentially processed transcripts is different, and (c) situations where both the 5′ and 3′ ends of the differentially processed transcripts are identical.

Table 5.1 Cases of alternative splicing which are regulated developmentally or tissue specifically

Protein	Species	Nature of transcripts which undergo alternative splicing	Cell types carrying out alternative splicing
(a) Immune system			
Immunoglobulin heavy-chain IgD. IgE, IgG, IgM	Mouse	3′ end differs	B cells
Lyt-2	Mouse	Same transcript	T cells
(b) Enzymes			
Alcohol dehydrogenase	*Drosophila*	5′ end differs	Larva and adult
Aldolase A	Rat	5′ end differs	Muscle and liver
α-Amylase	Mouse	5′ end differs	Liver and salivary gland
(2′5′) Oligo A synthetase	Human	3′ end differs	B cells and monocytes
(c) Muscle			
Myosin light chain	Rat/mouse/human/ chicken	5′ end differs	Cardiac and smooth muscle
Myosin heavy chain	*Drosophila*	3′ end differs	Larval and adult muscle
Tropomyosin	Mouse/rat/human/ *Drosophila*	Same transcript	Different muscle cell types
Troponin T	Rat/quail/chicken	Same transcript	Different muscle cell types
(d) Nerve cells			
Calcitonin/CGRP	Rat/human	3′ end differs	Thyroid C cells or neural tissue
Myelin basic protein	Mouse	Same transcript	Different glial cells
Neural cell adhesion molecule	Chicken	Same transcript	Neural development
Preprotachykinin	Bovine	Same transcript	Different neurones
(e) Others			
Fibronectin	Rat/human	Same transcript	Fibroblasts and hepatocytes
Early retinoic acid-induced gene 1	Mouse	Same transcript	Stages of embryonic cell differentiation
Thyroid hormone receptor	Rat	Same transcript	Different tissues

Situations where the 5′ end of the transcripts is different

In these cases, two alternative primary transcripts are produced by transcription from different promoter elements and these are then processed differentially (for review of the use of different promoters in the same gene see Landry *et al.*, 2003). In several situations, such as the mouse α-amylase gene (Fig. 5.3), differential splicing is controlled simply by the presence or absence of a particular exon in the primary transcript. Thus in the salivary gland, where transcription takes place from an upstream promoter, the exon adjacent to this promoter is included in the processed RNA and a downstream exon is omitted. In the liver, where the transcripts are initiated 2.8 kb downstream and do not contain the upstream exon, the processed RNA includes the downstream exon. However, other cases of this type, for example that of the myosin light chain gene (Fig. 5.4), are

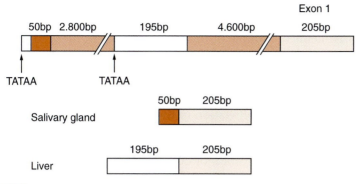

Figure 5.3

Alternative splicing at the 5′ end of α-amylase transcripts in the liver and salivary gland. The two alternative start sites for transcription are indicated (TATAA) together with the 5′ region of the mRNAs produced in each tissue.

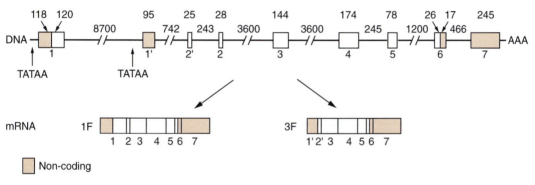

Non-coding

Figure 5.4

Alternative splicing of the myosin light chain transcripts in different muscle cell types produces two mRNAs (1F and 3F) differing at their 5′ ends. The two alternative start sites of transcription used to produce each of the RNAs are indicated (TATAA), together with the intron–exon structure of the gene.

more complex with each of the alternative primary transcripts containing both the alternatively spliced exons. In such cases it is assumed that the different primary transcripts fold into different secondary structures which favour the different splicing events. Whether this is the case or not, it is clear that cases of alternative splicing arising from differences in the site of transcriptional initiation represent further examples of transcriptional regulation in which the variation in RNA splicing is secondary to the selection of the different promoters in the different tissues. This is not so for the remaining two categories of alternative processing event.

Situations where the 3′ end of the transcripts is different

After the primary transcript has been produced, it is rapidly cleaved and a poly(A) tail is added (see Section 3.3). In many genes the process of cleavage and polyadenylation occurs at a different position within the primary transcript in different tissues, and the different transcripts are then differentially spliced.

The best-defined example of this process occurs in the genes encoding the immunoglobulin heavy chain of the antibody molecule, and plays an important role in the regulation of the antibody response to infection. Thus, early in the immune response, the antibody producing B cell synthesizes membrane-bound immunoglobulin molecules whose interaction with antigen triggers proliferation of the B cell and results in the production of more antibody-synthesizing cells. The immunoglobulin produced by these cells is secreted, however, and can interact with antigen in tissue fluids, triggering the activation of other cells in the immune system. The production of membrane-bound and secreted immunoglobin molecules is controlled by the alternative splicing of different RNA molecules differing in their 3′ ends (Fig. 5.5). The longer of these two molecules contains two exons encoding the portion of the protein that anchors it in the membrane. When this molecule is spliced, both these two exons are included, but a region encoding the last 20 amino acids of the secreted form is omitted. In the shorter RNA, the two transmembrane domain-encoding exons are absent and the region specific to the secreted form is included in the final messenger RNA.

If the polyadenylation site used in the production of the shorter immunoglobulin RNA is artificially removed, preventing its use (Danner and Leder, 1985), the expected decrease in the production of secreted immunoglobulin is paralleled by a corresponding increase in the synthe-

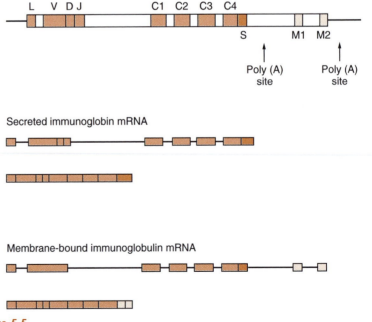

Figure 5.5

Alternative splicing of the immunoglobulin heavy chain transcript at different stages of B-cell development. The two unspliced RNAs produced by use of the two alternative polyadenylation sites in the gene are shown, together with the spliced mRNAs produced from them.

sis of the membrane-bound form of the protein (Fig. 5.6a). This indicates that the choice of splicing pattern is controlled by which polyadenylation site is used; removal of the upstream site resulting in increased use of the downstream site and increased production of the messenger RNA encoding the membrane-bound form.

This switch in the polyadenylation site is dependent on an increase in concentration of one of the subunits of the polyadenylation factor CstF (see Section 3.3) which occurs during B-cell development (Takagaki *et al.*, 1996). Thus CstF binds preferentially to the poly(A) site of the mRNA encoding the membrane form compared with that of the secreted form. Hence, early in B-cell development when CstF levels are low, it binds to the poly(A) site for the membrane form and so this mRNA is produced. As the levels of CstF rise binding and cleavage occurs at the poly(A) site for the secreted mRNA (Fig. 5.7).

This finding indicates that, in at least some cases of this type, the primary regulatory event is that determining the site of cleavage and polyadenylation, and that, as with cases of differential promoter usage,

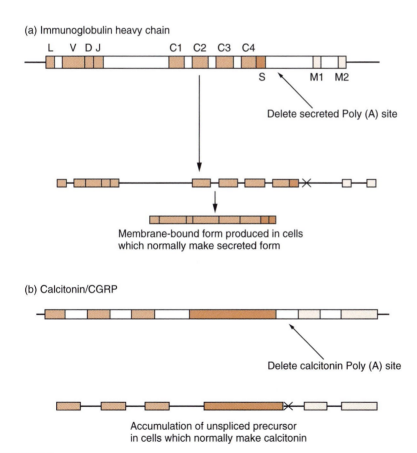

(a) Immunoglobulin heavy chain

Delete secreted Poly (A) site

Membrane-bound form produced in cells
which normally make secreted form

(b) Calcitonin/CGRP

Delete calcitonin Poly (A) site

Accumulation of unspliced precursor
in cells which normally make calcitonin

Figure 5.6

Effect of deleting the more upstream of the two polyadenylation sites in the immunoglobulin heavy chain (a) and calcitonin/CGRP genes (b) on the production of the alternatively spliced RNAs derived from each of these genes.

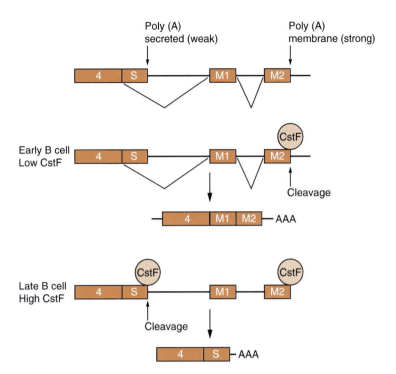

Figure 5.7

Role of the polyadenylation factor CstF in the resulted polyadenylation of the immunoglobulin transcript. At low levels of CstF it binds preferentially to the membrane poly (A) site. Following cleavage at this site, splicing joins exons 4, M1 and M2. At higher levels of CstF it also binds at the weaker secreted Poly(A) site leading to loss of exons M1 and M2 by cleavage of the mRNA.

alternative RNA splicing is regulated by differences in the structure of the transcript produced in different tissues.

Not all cases of alternative RNA splicing, where the 3′ end of the alternatively spliced RNAs varies, are of this type, however. This conclusion has emerged from the intensive study of the gene encoding the calcium regulatory protein, calcitonin, carried out by Rosenfeld and colleagues. When the gene encoding the calcitonin protein (which is a small peptide of 32 amino acids) was isolated it was found that it had the potential to produce an RNA encoding an entirely different peptide of 36 amino acids, which was named calcitonin-gene-related peptide (CGRP). Unlike calcitonin, which is produced in the thyroid gland, CGRP is produced in specific neurons within the brain and peripheral nervous system. These two peptides are produced (Fig. 5.8) by alternative splicing of two distinct transcripts differing in their 3′ ends.

This case differs from that of the immunoglobulin heavy chain, however, in that deletion of the polyadenylation site used in the shorter, calcitonin-encoding RNA does not result in an increase in CGRP expression in cells normally expressing calcitonin. Instead, large unspliced transcripts utilizing the downstream (CGRP) polyadenylation site accumulate in these cells (Fig. 5.6b). Although in CGRP-producing cells such

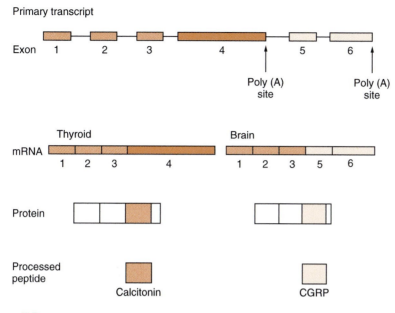

Primary transcript

Figure 5.8

Alternative splicing of the calcitonin/CGRP gene in brain and thyroid cells. Alternative splicing followed by proteolytic cleavage of the protein produced in each tissue yields calcitonin in the thyroid and CGRP in the brain.

transcripts would normally be spliced to yield CGRP messenger RNA, this does not occur in these experiments in cells normally producing calcitonin, and hence these unspliced precursors accumulate. This suggests that in the calcitonin/CGRP gene the use of different polyadenylation sites is secondary to the difference in RNA splicing, suggesting the existence of tissue-specific splicing factors, whose presence or absence in a specific tissue determines the pattern of calcitonin/CGRP RNA splicing.

Situations where both the 5′ and 3′ ends of the differently processed transcripts are identical

The existence of tissue-specific splicing factors which regulate alternative splicing is also indicated by the existence of cases where a transcript with identical 5′ and 3′ ends is spliced differently in different tissues, and which therefore cannot be explained by differential usage of promoters or polyadenylation sites.

Although the initial reports of such cases were confined to the eukaryotic DNA viruses, such as SV40 and adenovirus, many cases involving the cellular genes of higher eukaryotes were described subsequently. In one such case (Fig. 5.9) involving the skeletal muscle troponin T gene (Breitbart *et al.*, 1987) the same RNA can be spliced in up to 64 different ways in different muscle cell types. The existence of tissue-specific splicing factors acting on this gene is indicated by the finding that the artificial introduction and expression of this gene in non-muscle cells or myoblasts results in the complete removal of exons 4–8, whereas in muscle cells

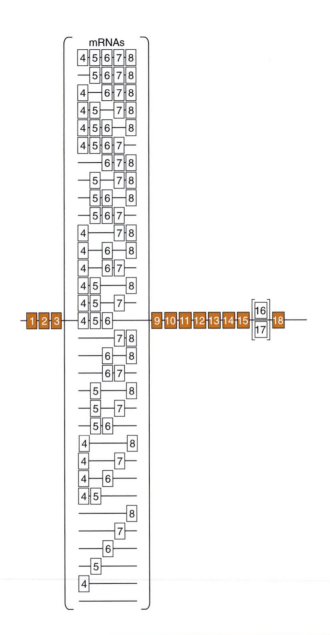

Figure 5.9

Alternative splicing of the four combinatorial exons (4–8) and the two mutually exclusive exons (16 and 17) can result in up to 64 distinct mRNAs from the rat troponin T gene. Redrawn from R.E. Breitbart and B. Nadal-Ginard, *Cell* **49**, 793–803 (1987), by permission of Professor B. Nadal-Ginard and Cell Press.

(myotubes) the correct pattern of alternative splicing seen with the endogenous gene is reproduced faithfully.

It should be noted, however, that the 64 possible mRNAs produced in this case, is certainly not the most dramatic example of multiple mRNAs being produced by alternative splicing. Thus, the vast number of different alternative exons found in the Dscam gene of *Drosophila* could result in over 38 000 different mRNAs being produced, which is greater than the number of different genes in this organism! (for review see Black, 2000; Graveley, 2001). This illustrates the extraordinary power of alternative splicing to produce multiple mRNAs encoding related but distinct proteins from a single gene.

Mechanism of alternative RNA splicing

RNA sequences involved in alternative splicing

The idea that certain factors are necessary for each particular pattern of splicing in genes such as calcitonin/CGRP or troponin, clearly begs the question of how such factors act. It is likely that these factors recognize *cis*-acting sequences within the RNA transcript itself. Clearly, the interaction of these factors with such sequences could produce alternative splicing either by promoting splicing at the site of the *cis*-acting sequence at the expense of the alternative splice site (Fig. 5.10a) or by inhibiting splicing at the site of binding and thereby promoting the use of the alternative splice site (Fig. 5.10b).

Both of these types of mechanism appear to be used in different cases. Indeed, examples of each of these mechanisms can be seen in the hierarchy of alternatively spliced genes which regulates sex determination in *Drosophila* (for review see Baker, 1989). In this hierarchy each gene product controls the alternative splicing of the next gene in the pathway resulting in different protein products in males and females. In turn these products differentially regulate the splicing of the next gene in the hierarchy leading ultimately to the production of a male or female fly (Fig. 5.11).

Thus the *Sxl* gene is differentially spliced in males and females, with the product of the female-specific mRNA not only controlling the splicing of the next gene in the hierarchy, *tra*, but also promoting its own female-specific splicing. Mutations which affect this autocatalytic function of Sxl on its own RNA map in the intron between exon 2 and the male-specific exon 3, indicating that the Sxl protein acts by inhibiting the male-specific

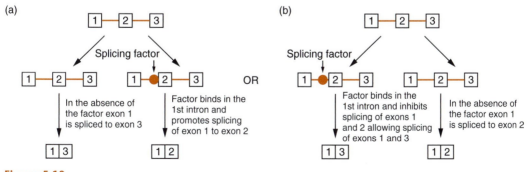

Figure 5.10

Possible models by which an alternative splicing factor can affect splicing by binding to a *cis*-acting sequence. In (a), the factor acts by promoting the use of the weaker of the two potential splicing sites while in (b) it acts by inhibiting use of the stronger of the two sites so that the other, weaker, site is used.

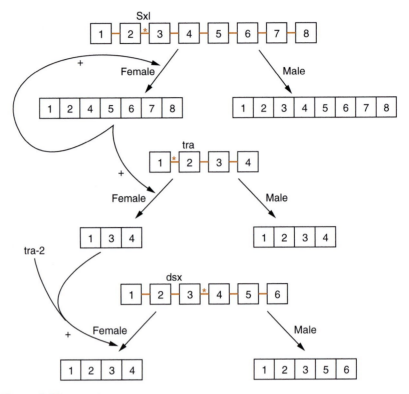

Figure 5.11

Schematic diagram (not to scale) of the hierarchy of alternatively spliced genes which controls sex determination in *Drosophila*. Female-specific splicing of the Sxl transcript produces a protein which promotes the female-specific splicing of its own RNA and that of the *tra* gene. In turn, the protein produced by the *tra* female-specific transcript in conjunction with the *tra-2* gene product promotes the female-specific splicing of the *dsx* transcript. The sites of mutations in the *Sxl* and *dsx* genes which affect their sex-specific splicing are indicated. Modified from Baker (1989).

splicing of exons 2 and 3. Similarly Sxl prevents the use of the male-specific exon 2 in the *tra* gene by binding within intron 1 and preventing the binding of the constitutive splicing factor U2AF which in turn recruits the U2snRNP particle which is essential for removal of all introns (see Section 3.3) (Valcarcel *et al.*, 1993).

In contrast, the action of the *tra* and *tra-2* gene products on the splicing of the *dsx* transcript appears to be mediated by promoting the use of the female-specific splice rather than the constitutive male-specific splicing event. In this case, mutations in *dsx* which affect its splicing map in the intron between exon 3 and the female-specific exon 4, indicating that the alternative splicing factor promotes the splicing of exons 3 and 4 (see below).

As well as responding to the presence or absence of a specific factor, specific RNA sequences can also regulate the pattern of alternative splicing in response to cellular signaling pathways. Thus, when the calcium/calmodulin-dependent protein kinase type IV (CaMKIV signaling) is activated, inclusion of a specific exon (STREX) in transcripts from the BK potassium channel gene is repressed and the exon is spliced out whereas in the absence of active CaMKIV, this exon is included in the RNA (Fig. 5.12a). This effect is mediated by a 54 bp sequence from the 3' splice site upstream of the regulated STREX exon (Xie and Black, 2001). Thus, mutation of this sequence abolishes regulation of splicing by CaMKIV whilst transferring it to a constitutionally spliced exon renders it sensitive to CaMKIV regulation (Fig. 5.12b). This sequence thus acts as a CaMKIV responsive RNA element (CaRRE) regulating splicing of the STREX exon in response to CaMKIV activation (for review see O'Donovan and Darnell, 2001).

Nature of alternative splicing factors

The evidence for the existence of alternative splicing factors and the identification of cis-acting sequences with which they interact have led to many attempts to identify these factors, (for reviews see Wang and Manley, 1997; Smith and Valcárcel, 2000).

In *Drosophila*, which is genetically very well characterized, the main approach to this problem has been a genetic one. Thus as discussed above,

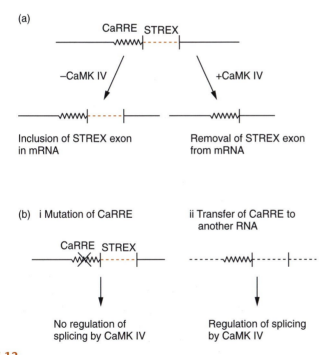

Figure 5.12

Regulation of splicing in BK potassium channel gene transcripts by the phosphorylation enzyme CaMK IV. (a) In the presence of active CaMK IV, the STREX exon is spliced out of the BK gene RNA whereas in the absence of active CaMK IV this splicing event is inhibited. (b) The response to CaMK IV is mediated by a 54 bp sequence from the 3' splice site upstream of STREX which is known as CaMK IV response RNA element (CaRRE) since inactivation of this sequence prevents regulation of splicing by CaMK IV (i) whereas its inclusion in a completely different RNA (dotted line) causes splicing of that RNA to be regulated by CaMK IV.

genes such as *Sxl* and *tra*, originally identified by the fact that mutations within them disrupted the process of sex determination, are now known to act by controlling alternative splicing. The products of these genes are alternative splicing factors and their study allows a unique insight into the nature of such factors.

The sequencing of the *Sxl* and *tra-2* genes has revealed that the proteins they encode contain one or more copies of a ribonucleoprotein (RNP) consensus sequence that is found in a wide variety of RNA binding proteins, such as those of the mammalian spliceosome, and constitutes an RNA binding domain (for reviews see Bandziulis *et al.*, 1989; Burd and Dreyfuss, 1994). Indeed, structural analysis of *Sxl* bound to the *tra* mRNA has shown that the RNA binds to a V-shaped cleft within the *Sxl* protein (Handa *et al.*, 1999) (color plate 8). Hence Sxl and Tra-2 influence alternative splicing by binding directly to *cis*-acting sequences in the spliced RNAs discussed above.

Hence specific factors have been identified in *Drosophila* which are expressed only in a particular situation and which regulate specific alternative splicing events. Interestingly, however, in vertebrates it appears that many tissue-specific splicing events depend on quantitative variations in the levels of splicing factors which are present in all tissues. Thus the SF2 factor is a member of the SR proteins family discussed in Section 3.3 which is present in all cells and is essential for the basic process of splicing itself (for review of SR proteins see Manley and Tacke, 1996). However, its concentration has been shown to influence which of two competing upstream splice sites is joined to a downstream site. Thus high concentrations of SF2 favor the more proximal of the two sites whilst low concentrations favor the more distal site (Fig. 5.13; for review see Smith and Valcárcel, 2000). Interestingly the constitutively expressed factor hnRNPA1 which binds to RNA before it is spliced (see Section 3.3) has the opposite effect, favoring the use of the more distal site. Hence, in this situation, the outcome of a specific alternative splicing event could be different in two different tissues, depending on the relative concentration of SF2 and hnRNPA1 in each tissue, even though both factors were present in both tissues (Fig. 5.13).

Such a system, in which the different outcome of alternative splicing events in each tissue is controlled by quantitative differences in the relative levels of constitutively expressed factors, can readily be fitted into the mechanistic models illustrated in Fig. 5.10 simply by suggesting that the two factors have opposite effects on the relative strengths of the two competing splice sites. Indeed it is likely that SR proteins such as SF2, can influence splice site selection in two ways, both involving the recruitment of the UsnRNP particles which are essential for splicing to occur (see Section 3.3). Thus SR proteins can bind to the 5′ splice site and promote the binding of the U1snRNP to this site (Fig. 5.14a). In addition, they can also bind to sequences known as splicing enhancers within the exon downstream of the regulated splice site and stimulate the binding of the U2AF protein to the branch point, thereby recruiting the U2 snRNP particle (see Section 3.3) (for reviews see Wang and Manley, 1997; Blencowe, 2000) (Fig. 5.14b).

Thus both tissue-specific factors and variations in constitutively expressed factors can regulate alternative splicing. Indeed, as described above, both Sxl and SR proteins, such as SF2, can act by regulating the

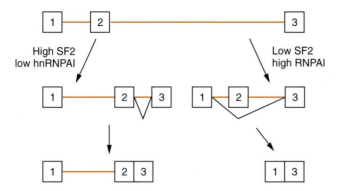

Figure 5.13

The pattern of splicing can be regulated by the balance in levels between the constitutively expressed SF2 and hnRNPA1 proteins. A high ratio of SF2 to hnRNPA1 favors use of the proximal exon (2) whilst a low ratio favors the distal exon (3).

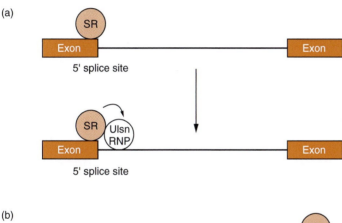

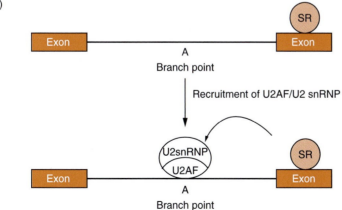

Figure 5.14

SR proteins, such as SF2, can stimulate the use of a particular splice site either by binding to the 5′ splice site and promoting binding of the U1snRNP to this site (panel a) or by binding to the downstream exon sequence and promoting recruitment of U2AF and hence of the U2 snRNP to the branch point (panel b).

recruitment of U2AF. It is not surprising therefore that some cases of alternative splicing are regulated by the interaction of SR proteins with specifically expressed factors. Thus in the case of the *dsx* gene, the splicing of exon 3 to the female-specific exon 4 does not occur in the absence of the female-specific proteins Tra and Tra-2 (Fig. 5.12). This is because U2AF binds only weakly to the branch point of this intron and the binding site of SR proteins in exon 4 is weak and too far away for SR to enhance U2AF binding (Fig. 5.15). In the presence of Tra and Tra-2 however, the interaction of SR with the exon 4 sequence is stabilized, allowing SR in turn to promote binding of U2AF and splicing of exon 3 to exon 4 (Lynch and Maniatis, 1995, Fig. 5.15). The SR proteins and tra/tra-2 contain both an RNA binding domain found in many RNA binding proteins with different functions (for a review see Burd and Dreyfuss, 1994) and a domain rich in serine (S) and arginine (R) residues which gives the SR proteins their name (for a review see Manley and Tacke, 1996).

Interestingly, as well as controlling specific splicing decisions, SR proteins can also determine the overall rate of splicing. Thus, during mitosis or when cells are exposed to elevated temperature or other stressful conditions, splicing is generally inhibited. This is dependent upon an SR protein known as SRp38. Thus, in mitotic cells or following exposure to elevated temperature, SRp38 is phosphorylated to produce its active form which interacts with the UlsnRNP (see Section 3.3) and inhibits 5' splice

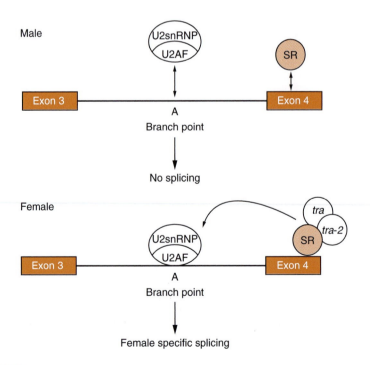

Figure 5.15

In male *Drosophila* SR protein binds weakly to the female-specific exon 4 sequence and hence cannot promote binding of the U2snRNP to the branch point. In females, however, *tra* and *tra-2* stabilize the binding of SR and hence allow it to stimulate U2snRNP binding with consequent splicing of exon 3 and exon 4.

site selection (Fig. 5.16) (Shin *et al.*, 2004). In the absence of SRp38, cells show reduced survival when exposed to elevated temperature and have a defect in mitosis indicating the functional importance of this effect.

In summary, alternative splicing is controlled by a series of proteins with similar functional domains, some of which are expressed in a highly specific manner whilst others are expressed at different levels in all cell types.

Generality of alternative RNA splicing

The cases discussed above indicate the use of alternative splicing in a wide variety of biological processes. In mammals such splicing has been shown to regulate the immune system's production of antibodies, the production of neuropeptides such as CGRP and the tachykinins, substance P and substance K, as well as the synthesis of the different forms of several of the major sarcomere muscle proteins. As noted above, around three-quarters of genes with multiple exons appear to be alternatively spliced in humans (Johnson *et al.*, 2003). Similarly, in *Drosophila* much of the posterior body plan is determined by developmentally regulated differential splicing of the ultrabithorax gene, while sex determination is also controlled by differential splicing of a hierarchy of genes in males and females (see above; for a review see Baker, 1989).

The widespread use of alternative splicing in mammals does not refute the initial conclusion that gene regulation occurs primarily at the level of transcription, however. Thus, alternative splicing represents a response to a requirement for the production of related but different forms of a gene product in different tissues. It therefore supplements the regulation of transcription of the gene responsible for producing the different forms.

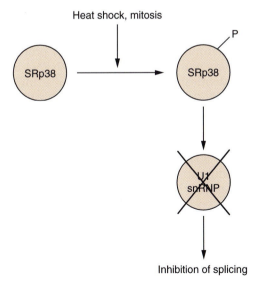

Figure 5.16

The SRp38 protein is converted to its active, phosphorylated form during mitosis or following exposure of cells to elevated temperature. It then interacts with the UIsnRNP and inhibits RNA splicing.

Thus the immunoglobulin heavy chain gene, which produces both membrane-bound and secreted forms of the protein at different stages of B-cell development, is transcribed only in B cells and not in other cell types, while the transcription of the troponin T gene, which produces multiple different isoforms in different muscle cell types, is confined to differentiated muscle cells.

Indeed, a genome wide analysis of the proteins produced by alternative splicing in species as diverse as humans and *Drosophila* indicated that such alternate splicing inserts or deletes whole functional domains of proteins more often than would be expected by chance (Kriventseva *et al.*, 2003). This general analysis reinforces the specific examples such as troponin and the immunoglobulins and indicates the key role of alternative splicing in producing distinct but related proteins.

Interestingly, alternative splicing is particularly common in non-dividing cells such as muscle cells and nerve cells (for a review see Grabowski, 1998). As discussed in Chapter 6, the reprogramming of cellular commitment and gene transcription often requires a cell division event allowing alterations in the chromatin structure of the DNA and its associated proteins to occur. Hence alternative splicing may represent a means of conveniently changing the pattern of protein production in cells where such reprogramming cannot be achieved by cell division.

5.3 RNA editing

The finding that two different protein products can be produced from the same RNA by alternative splicing has been supplemented by the observation that a similar result can be achieved by a post-transcriptional sequence change in the messenger RNA (for reviews see Reenan, 2001; Samuel, 2003). Thus apolipoprotein B, which plays an important role in lipid transport, is known to exist in two closely related forms. A large protein of 512 kDa known as apo-B100 is synthesized by the liver, while a smaller protein apo-B48 is made by the intestine. The smaller protein is identical to the amino-terminal portion of the larger protein. Analysis of the mRNA encoding these proteins revealed a 14.5 kb RNA in both tissues. These two RNAs were identical with the exception of a single base at position 6666, which is a cytosine in the liver transcript and a uracil in the intestinal transcript (Fig. 5.17). This change has the effect of replacing a CAA codon, which directs the insertion of a glutamine residue, with a UAA stop codon, which causes termination of translation of the intestine RNA and hence results in the smaller protein being made.

Only one gene encoding these proteins is present in the genome, and it is not alternatively spliced. In both intestinal and liver DNA this gene has a cytosine residue at position 6666. Hence the uracil in the intestinal transcript must be introduced by a post-transcriptional RNA editing mechanism.

Indeed a cytidine deaminase enzyme has been identified and shown to bind to the apoB mRNA at sequences adjacent to the edited site. This enzyme removes an amine (NH_2) group from the cytosine residue generating a uracil. This enzyme is expressed in the intestine where editing of the apoB mRNA occurs, but not in the liver, where editing does not occur. Interestingly, however, it is also present in other tissues such as the testis, ovary and spleen, where apoB mRNA is not expressed indicating that this

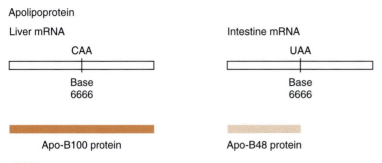

Apolipoprotein

Liver mRNA

CAA

Base
6666

Apo-B100 protein

Intestine mRNA

UAA

Base
6666

Apo-B48 protein

Figure 5.17

RNA editing of the apolipoprotein B transcript in the intestine produces an
mRNA encoding the truncated protein apo-B48.

enzyme is likely to edit other transcripts expressed in these tissues. Indeed,
several other transcripts which undergo C to U editing have now been
identified (for a review see Blanc and Davidson, 2003) and a number of
different cytidine deaminase enzymes capable of carrying out this form of
editing have been characterized (Wedekind *et al.*, 2003).

As well as C to U editing by cytidine deaminase, an adenosine deaminase
enzyme also exists in mammalian cells which removes an amino residue
from adenine to produce an inosine base which is read as a guanine residue
by the translation apparatus (for reviews see Reenan, 2001; Maas *et al.*, 2003).
This form of editing was first identified in the gene encoding a receptor for
the excitatory amino acid glutamate which is expressed in neuronal cells.
The editing of an A residue to an I (G) residue in the transcript of this gene
results in it encoding an arginine rather than the glutamine found in other
related receptors and alters its properties so that it is permeable to calcium.
Interestingly, this editing of the glutamate receptor mRNA is essential for
the survival of the animal. Thus, mice lacking the adenosine deaminase
enzyme (ADAR2) which carries out this editing are prone to seizures and
die young. However, their survival can be restored to normal by inserting
a glutamate receptor gene in which the alteration produced by editing has
been carried out at the DNA level, removing the need for editing at the RNA
level (Higuchi *et al.*, 2000).

Although the glutamate receptor is thus a critical target for RNA editing, editing by adenosine deamination is prevalent, especially in the
nervous system, and also occurs in other neuronally expressed receptors
such as the kainate receptors and the serotonin receptor (for reviews see
Reenan, 2001; Maas *et al.*, 2003). Indeed, a recent comprehensive search
for edited transcripts in *Drosophila* identified 16 previously unidentified
targets for A to I editing (Hoopengardner *et al.*, 2003). All of these were
involved in electrical or chemical neurotransmission and the editing event
targeted a functionally important residue. Hence, A to I editing plays a
critical role in the nervous system in producing different functional variants of molecules involved in neuronal signaling.

As well as affecting the nature of the protein produced from a particular mRNA, A to I editing by adenosine deaminase enzymes can also affect
the splicing of the edited RNA. This is seen in the case of the gene encoding the ADAR2 adenosine deaminase enzyme itself. Thus, the ADAR2
transcript contains an AA sequence 47 bases upstream of the AG sequence

which is normally used as a 3′ splice site. Editing of the AA sequence to AI (which mimics AG) creates a new 3′ splice site which is used for splicing, resulting in the 47 base sequence being retained in the mRNA (Fig. 5.18) (Rueter *et al.*, 1999).

This case is of particular interest in that the editing event is regulated so that it occurs frequently in the brain and lung where ADAR2 transcripts predominantly contain the 47 base extra exon and much less frequently in the heart where this exon is therefore absent from the mRNA. Moreover, it demonstrates that editing by adenosine deaminase can affect RNA splicing as well as altering the encoded protein and that it can affect the RNA encoding the adenosine deaminase itself as well as RNAs encoding other proteins such as the glutamate receptor.

The existence of two distinct types of editing enzymes (adenosine deaminase and cytidine deaminase) in mammalian cells, each with multiple substrates, indicates that this represents a widely used mechanism which, like alternative splicing, can produce related but distinct proteins from the same transcript. As with alternative splicing, however, this is likely to act as a supplement to transcriptional control, the apoB gene, for example, being transcribed in the liver and the intestine but not in other tissues, whilst many of the targets for A to I editing are transcribed only in the nervous system. Interestingly, as well as these similarities in the role/outcome of RNA editing and alternative splicing, there is evidence that the two processes may occur in parallel. Thus, it has recently been shown that the adenosine deaminase enzyme is found in large ribonucleoprotein particles, which also contain Sm and SR splicing factors (Raitskin *et al.*, 2001).

5.4   Regulation of RNA transport

Transport from nucleus to cytoplasm

The process of RNA splicing takes place within the nucleus, whereas the machinery for translating the spliced RNA is found in the cytoplasm. Hence the spliced mRNA must be transported to the cytoplasm if it is to direct protein synthesis (see Section 3.3). The first example of regulation at this level was described in the human immunodeficiency virus (HIV-1) (for a review see Sandri-Goldin, 2004).

Thus early in infection of cells with HIV, the virus produces a high level of very short non-functional transcripts and a small amount of full length functional RNA (see Section 4.3). The full length RNA is spliced with the removal of two introns so that the predominant transcript which appears in the cytoplasm is a fully spliced mRNA (Fig. 5.18; for a review see Cullen. 1991). This transcript encodes the regulatory proteins Tat and Rev. As discussed in Section 4.3, the Tat protein acts greatly to increase the rate of transcription of the viral genome, as well as promoting transcriptional elongation so that predominantly full length RNAs are produced in the nucleus. However, at this second stage of infection most of the mRNAs that appear in the cytoplasm are either unspliced or have had only the first intron removed (Fig. 5.19). As these transcripts encode the viral structural proteins, this allows the high level production of viral particles which is necessary late in infection.

This change in the nature of the HIV RNA in the cytoplasm is dependent on the action of the Rev protein which is made early in infection (for

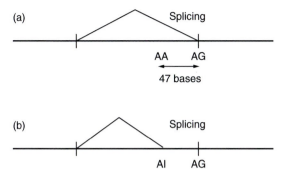

Figure 5.18

In the ADAR2 RNA, splicing normally occurs using an AG sequence as the 3′ splice site (a). RNA editing alters an AA sequence 47 bases upstream to AI. As AI resembles AG, this sequence is used as the 3′ splice site resulting in an extra 47 bases being retained in the final mRNA (b).

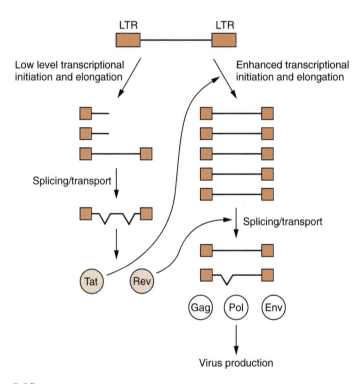

Figure 5.19

Regulation of HIV gene expression. Early in infection a low level of transcriptional initiation and elongation produces a small amount of the fully spliced viral RNA which encodes the viral Tat and Rev proteins. When these proteins are produced, the Tat protein enhances both the initiation and elongation of transcription, resulting in an increase in the production of the viral RNA. The Rev protein then acts post-transcriptionally to promote splicing/transport so the unspliced and singly spliced RNAs encoding the viral structural proteins (Gag, env) and the reverse transcriptase (pol) accumulate.

review see Cullen and Malim, 1991). Most interestingly, Rev does not affect the amounts of the different RNAs in the nucleus. Rather it is an RNA binding protein and binds to a specific site (the Rev response element) in the second intron of the HIV transcripts. This binding promotes the transport of these RNAs to the cytoplasm (Fig. 5.20). As the fully spliced RNA lacks the Rev binding site, its transport is not accelerated and so the proportion of unspliced or singly spliced RNA in the cytoplasm increases. Hence Rev acts at the level of RNA transport, the first regulatory protein to do so.

As noted in Section 3.3 the Rev protein resembles several cellular RNA transport proteins in containing a nuclear export signal (NES) which promotes transport of the RNA to which it has bound from the nucleus to the cytoplasm (for a review see Gerace, 1995). Moreover, Rev has been shown to bind to a cellular protein Rip 1 which is localized in the pores of the nuclear membrane and this interaction is likely to play a key role in its ability to mediate export of the HIV RNA (Fig. 5.21).These findings indicate that Rev mediates RNA export via a cellular pathway, involving the NES signal which it shares with cellular RNA transport proteins. It is likely therefore that, as well as the examples of constitutive RNA transport mediated by this system (see Section 3.3), there also exist situations in which this system is used to regulate the transport of specific cellular mRNAs in different tissues or in response to specific stimuli. This would, for example, provide a convenient explanation of the existence of RNA species that are confined to the nucleus in specific tissues of the sea urchin

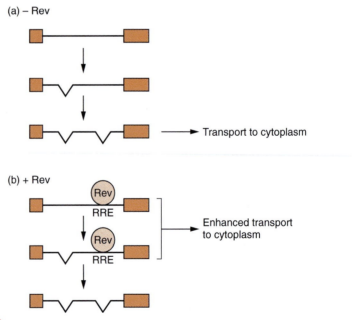

Figure 5.20

Binding of Rev to its response element (RRE) in the HIV-1 RNA promotes the transcript of unspliced or singly spliced RNAs which contain the binding site at the expense of doubly spliced RNA which lacks the binding site and therefore cannot bind Rev.

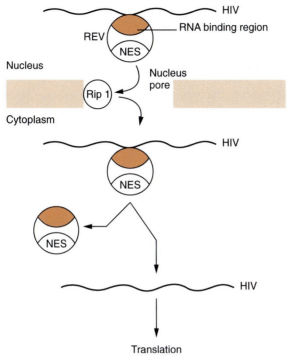

Figure 5.21

The Rev protein binds to the HIV RNA by means of its RNA binding domain and its nuclear export signal (NES) then binds to cellular proteins such as Rip 1 in the nuclear pore and promotes the export of the RNA/protein complex to the cytoplasm.

(Section 5.1). Similarly, a mechanism which prevented the transport of unspliced RNA in genes regulated by processing versus discard decisions (see Section 5.2) would represent a means of preventing wasteful translation of RNA species containing interruptions in the protein-coding sequence.

Transport within the cytoplasm

As well as controlling the transport of mRNA from the nucleus to the cytoplasm (Fig. 5.22A) it is also possible for regulatory processes to control the location of an mRNA species within the cytoplasm of a polarized cell, resulting in the corresponding protein being made only in certain parts of the cell (Fig. 5.22B) (for reviews see Hazelrigg, 1998; Mohr and Richter, 2001). This process is observed, for example, within the oocyte (egg) resulting in different proteins being localized in different regions of the egg. Following fertilization, this differential distribution of proteins in turn controls embryonic development so that different regions of the embryo develop from different regions of the fertilized egg. Thus, for example, in *Drosophila* the bicoid mRNA is located at the anterior end of the egg whilst the nanos mRNA is localized to the posterior end. The resulting opposite gradients of the Bicoid and Nanos proteins in turn control the polarity of the head, thorax and abdomen of the embryo so that different regions of

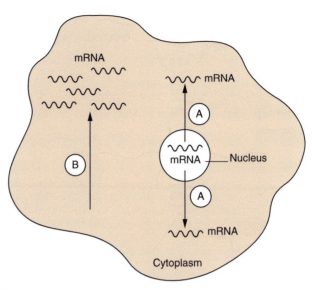

Figure 5.22

Transport of mRNA within the cell can involve both its export from the nucleus to the cytoplasm (A) as well as in some cases its preferential localization within a particular region of the cell cytoplasm (B).

the body are produced depending on the relative concentrations of the two proteins. (Fig. 5.23).

In some cases, such as the bicoid mRNA, specific sequences have been localized in the 3' untranslated region of the mRNA which are essential for correct localization and which can confer the bicoid pattern of localization on a non-localized mRNA. These sequences in the 3' untranslated region of the bicoid mRNA are essential for it to bind the staufen protein, to form a ribonucleoprotein complex. In turn, recruitment of the staufen protein is essential for movement of the bicoid mRNA to the anterior end of the egg (Fig. 5.24a). Interestingly, the staufen protein is also expressed in mammalian neurones and appears to play a critical role in directing specific mRNAs to the multiple dendritic processes of the neurone and excluding them from the single axon (Fig. 5.24b) (for a review see Roegiers and Jan, 2000). Hence staufen plays a key role in directing specific mRNAs to particular parts of the cell in very different cell types in widely different organisms.

As with the bicoid mRNA, initial studies also identified the 3' untranslated region of the oskar mRNA as being important for its localization in the egg, although the oskar mRNA localizes to the posterior rather than the anterior region. However, a more recent study has shown that correct localization of the oskar mRNA in the cytoplasm is also dependent on the splicing of the first two exons of the oskar transcript in the nucleus (Hachet and Ephrussi, 2004). Thus an artificial RNA with the first two exons already joined together (so that splicing is not required) is not correctly localized even though the mRNA is indistinguishable from one produced by splicing and contains the 3' untranslated region (Fig. 5.25).

This is likely to be explained by the findings described in Section 3.3, that RNA transport factors become associated with the RNA during the

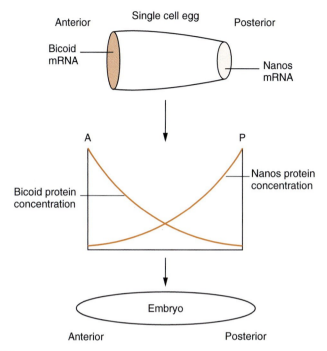

Figure 5.23

Localization of the bicoid and nanos mRNAs at opposite ends of the *Drosophila* egg sets up opposite gradients in the concentrations of their corresponding proteins which in turn specify the anterior/posterior polarity of the resulting embryo.

splicing process and one of these is presumably necessary for correct localization of the oskar mRNA. In agreement with this idea, the translation initiation factor eIF-4AIII has recently been shown to associate with the oskar mRNA during splicing and to be essential for its correct localization in the cytoplasm (Palacios *et al.*, 2004). Hence, as well as its role in translation of mRNA into protein, this factor is also involved in the nuclear splicing process and in correct cytoplasmic localization of the spliced mRNA.

5.5 Regulation of RNA stability

Cases of regulation by alterations in RNA stability

Once the mRNA has entered the cytoplasm, the number of times that it is translated, and hence the amount of protein it produces, will be determined by its stability. The more rapidly degraded an RNA is, the less protein it will produce. Indeed, it is now clear that the rate of turnover of an mRNA plays an important role in determining its level in the cell (for reviews see Khodursky and Bernstein, 2003; Wickens and Goldstrohm, 2003; Wilusz and Wilusz, 2004).

Hence an effective means of gene regulation could be achieved by changing the stability of an RNA species in response to some regulating

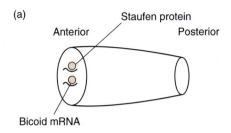

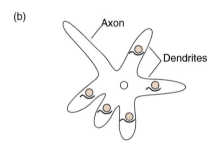

Figure 5.24

The Staufen protein plays a key role in localizing the bicoid mRNA to the anterior end of the *Drosophila* embryo (a) and in directing specific mRNAs to the dendrites and not the axon of mammalian neurones (b).

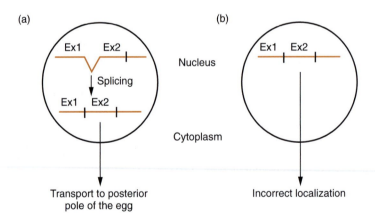

Figure 5.25

Correct localization of the oskar mRNA in the cytoplasm is dependent on the splicing of exons 1 and 2 in the nucleus (panel a). If exons 1 and 2 are joined together in an artificial RNA transcript so that splicing is not required, the mRNA does not correctly localize in the cytoplasm (panel b).

signal. A number of situations where the stability of a specific RNA species is changed in this way have been described (for reviews of this topic see Sachs, 1993; Ross, 1996; Bashirullah *et al.*, 2001). Thus, for example, the mRNA for the milk protein, casein, turns over with a half-life of around 1 h in untreated mammary gland cells. Following stimulation with the hormone prolactin, the half-life increases to over 40 h (Guyette *et al.*, 1979), resulting in increased accumulation of casein mRNA and protein production in response to the hormone (Fig. 5.26). Similarly, the increased production of the DNA-associated histone proteins in the S (DNA synthesis) phase of the cell cycle is regulated in part by a fivefold increase in histone mRNA stability that occurs in this phase of the cell cycle (for reviews of the regulation of histone gene expression see Marzluff, 1992; Stein *et al.*, 1992).

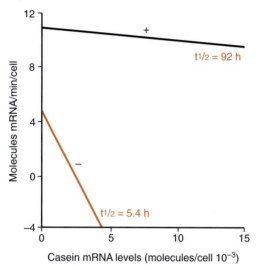

Figure 5.26

Difference in stability of the casein mRNA in the presence (+) or absence (–) of prolactin. Redrawn from W.A. Guyette *et al.*, *Cell* **17**, 1013–1023 (1979), by permission of Professor J. Rosen and Cell Press.

Interestingly, a recent genome-wide survey of all cellular mRNAs in yeast demonstrated that regulation at the level of mRNA stability was frequently observed for mRNAs whose corresponding proteins are involved in rRNA synthesis and ribosome production (Grigull *et al.*, 2004). Hence, this form of regulation may be particularly frequent for genes-encoding proteins involved in the process of protein synthesis itself. A representative selection of cases where mRNA stability is altered in a particular situation is given in Table 5.2.

Mechanisms of stability regulation

The first stage in defining the mechanism of changes in RNA stability is to identify the sequences within the RNA that are involved in mediating the observed alterations. This can be achieved by transferring parts of the gene encoding the RNA under study to another gene and observing the effect on the stability of the RNA expressed from the resulting hybrid gene. In a number of cases, short regions have been identified which can confer the pattern of stability regulation of the donor gene upon a recipient gene that is not normally regulated in this manner.

In many cases such regions are located in the 3′ untranslated region of the mRNA, downstream of the stop codon that terminates production of

Table 5.2 Regulation of RNA stability

mRNA	Cell type	Regulatory event	Increase or decrease in half-life
Cellular oncogene c-*myc*	Friend erythroleukemia cells	Differentiation in response to dimethyl sulphoxide	Decrease from 35 to less than 10 min
c-*myc*	B cells	Interferon treatment	Decreased
c-*myc*	Chinese hamster lung fibroblasts	Growth stimulation	Increased
Epidermal growth factor receptor	Epidermal carcinoma cells	Epidermal growth factor	Increased
Casein	Mammary gland	Prolactin	Increased from 1 to 40 h
Vitellogenin	Liver	Estrogen	Increased 30-fold
Type I pro-collagen	Skin fibroblasts	Cortisol	Decreased
Type I pro-collagen	Skin fibroblasts	Transforming growth factor β	Increased
Histone	HeLa	Cessation of DNA synthesis	Decreased from 40 to 80 min
Tubulin	CHO	Accumulation of free tubulin subunits	Decreased 10-fold

the protein (for review see Wickens *et al.*, 1997). Thus the cell-cycle-dependent regulation of histone H3 mRNA stability is controlled by a 30 nucleotide sequence at the extreme 3′ end of the molecule. Similarly, the destabilization of the mRNA encoding the transferrin receptor in response to the presence of iron can be abolished by deletion of a 60 nucleotide sequence within the 3′ untranslated region (for reviews see Hentze and Kuhn, 1996; Aisen *et al.*, 2001). Both of these sequences have the potential to form stem-loop structures by intra-molecular base pairing (Fig. 5.27), suggesting that changes in stability might be brought about by alterations in the folding of this region of the RNA in response to a specific signal (see Section 3.3).

The localization of sequences involved in the regulated degradation of specific mRNA species to the 3′ untranslated region is in agreement with the important role of this region in determining the differences in stability observed between different RNA species. Such sequences may act either by promoting endonucleolytic cleavage within the RNA transcript or by promoting loss of the poly (A) tail, opening up the RNA to exonucleolytic attack via its free 3′ end (see Section 3.3) (for a review see Parker and Song, 2004). This suggests that differences in RNA stability, whether between different RNA species or in a single RNA in different situations, may be controlled primarily by this region.

Despite this, cases where other regions of the RNA mediate the observed alterations in stability have also been described. The most extensively studied of such cases concerns the auto-regulation of the mRNA encoding the microtubule protein, β-tubulin, in response to free tubulin monomers (Pachter *et al.*, 1987; Yen *et al.*, 1988). This auto-regulation prevents the wasteful synthesis of tubulin when excess free tubulin, not polymerized into microtubules, is present, and is caused by a destabilization of the tubulin mRNA. A short sequence, only 13 bases in length, from the 5′ end

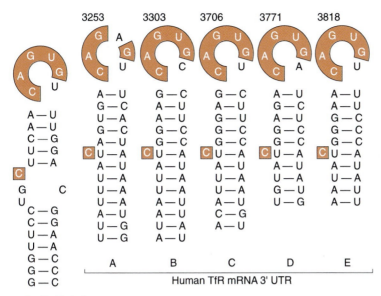

Figure 5.27

Similar stem-loop structures in the human ferritin and transferrin receptor mRNAs. Note the boxed conserved sequences in the unpaired loops and the absolute conservation of the boxed C residue, found within the stem, five bases 5' of the loop. Redrawn from J.L. Casey *et al.*, *Science* **240**, 924–928 (1988), by permission of Dr J.B. Harford and AAAS.

of the β-tubulin mRNA is responsible for this destabilization, and can confer the response on an unrelated mRNA. Most interestingly, these bases actually encode the first four amino acids of the tubulin protein, raising the possibility that the trigger for degradation of the tubulin mRNA might be the recognition of these amino acids in the tubulin protein, rather than the corresponding nucleotides in the tubulin RNA. In an elegant series of studies, Cleveland and colleagues (Yen *et al.*, 1988) showed that this was indeed the case. Thus changing the translational reading frame of this region, such that the identical nucleotide sequence encoded a different amino acid sequence, abolished the auto-regulatory response (Fig. 5.28a), whereas changing the nucleotide sequence in a manner which did not alter the encoded amino acids (due to the degeneracy of the genetic code) left the response intact (Fig. 5.28b).

Hence, stability of the RNA transcript can be regulated by sequences in different parts of the RNA, which can be recognized at the level of the RNA itself or the protein which it encodes.

Role of stability changes in regulation of gene expression

A consideration of the situations where changes in the stability of a particular RNA occur (see Table 5.2) suggests that the majority have two features in common. Firstly, changes in stability of a particular mRNA are very often accompanied by parallel alterations in the transcription rate of the corre-

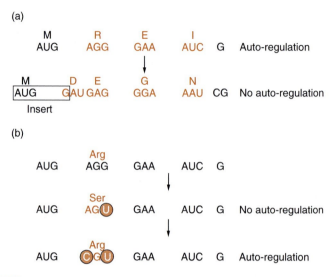

Figure 5.28

Effect of changes in the β-tubulin sequence on auto-regulation of tubulin mRNA stability.

sponding gene. Thus prolactin treatment of mammary gland cells results in a two- to fourfold increase in casein gene transcription, and the increased stability of histone mRNA in the S phase of the cell cycle is accompanied by a three- to fivefold increase in transcription of the histone genes.

Second, cases where RNA stability is regulated are very often those where a rapid and transient change in the synthesis of a particular protein is required. Thus synthesis of the histone proteins is necessary only at one particular phase of the cell cycle, when DNA is being synthesized. Following cessation of DNA synthesis a rapid shut-off in the synthesis of unnecessary histone proteins is required. Similarly, following the cessation of hormonal stimulation it would be highly wasteful to continue the synthesis of hormonally dependent proteins such as casein or vitellogenin. In the case of the cellular oncogene c-*myc*, whose RNA stability is transiently increased when cells are stimulated to grow, such continued synthesis would not only be highly wasteful but is also potentially dangerous to the cell. Thus this growth-regulatory protein is only required for a short period when cells are entering the growth phase, its continued inappropriate synthesis at other times carrying the risk of disrupting cellular growth regulatory mechanisms, possibly resulting in transition to a cancerous state (see Chapter 9).

These considerations suggest that alterations in RNA stability are used as a significant supplement to transcriptional control in cases where rapid changes in the synthesis of a particular protein are required. Thus, if transcription is shut off in response to withdrawal of a particular signal, inappropriate and metabolically expensive protein synthesis will continue for some time from pre-existing mRNA, unless that RNA is degraded rapidly. Similarly, rapid onset of the expression of a particular gene can be achieved by having a relatively high basal level of transcription with high RNA turnover in the absence of stimulation, allowing rapid onset of translation from pre-existing RNA following stimulation. Hence, as with alternative RNA processing, cases where RNA stability is regulated represent an adaptation to the requirements of a particular situation and do not affect the conclusion that regulation of gene expression occurs primarily at the level of transcription.

5.6 Regulation of translation

Cases of translational control

The final stage in the expression of a gene is the translation of its messenger RNA into protein (see Section 3.3). In theory, therefore, the regulation of gene expression could be achieved by producing all possible mRNA species in every cell and selecting which were translated into protein in each individual cell type. The evidence that different cell types have very different cytoplasmic RNA populations (see Chapter 1) indicates, however, that this extreme model is incorrect. Nonetheless, the regulation of translation, such that a particular mRNA is translated into protein in one situation and not another, does occur in some special cases (for reviews see Curtis *et al.*, 1995; Hentze, 1995; Pain, 1996).

The most prominent of such cases is that of fertilization. In the unfertilized egg protein synthesis is slow, but upon fertilization of the egg by a sperm a tremendous increase in the rate of protein synthesis occurs. This increase does not require the production of new mRNAs after fertilization. Rather, it is mediated by pre-existing maternal RNAs which are present in the unfertilized egg but are only translated after fertilization. Although in many species such translational control produces only quantitative changes in protein synthesis, in others it can affect the nature as well as the quantity of the proteins being made before and after fertilization. Thus in the clam *Spisula solidissima* some new proteins appear after fertilization, while others which are synthesized in large amounts before fertilization are repressed thereafter. However, the RNA populations present before and after fertilization are identical (Fig. 5.29), indicating that translational control processes are operating. Translational control also operates in the egg to ensure that mRNAs which are localized to particular regions of the egg cytoplasm (see Section 5.4) are not translated prior to their arrival at the correct region of the cell.

As well as producing parallel changes in the translation of many RNA species, translational control processes may also operate on individual RNAs in a particular cell type. Thus, for example, the rate of translation of the globin RNA in reticulocytes is regulated in response to the availability of the heme co-factor which is required for the production of hemoglobin. Similarly, the translation of the RNA encoding the iron-binding protein, ferritin, is regulated in response to the availability of iron.

Mechanism of translational control

In principle, translational regulation could operate via modifications in the cellular translational apparatus affecting the efficiency of translation of particular RNAs, or by modifications in the RNA itself which affect the way in which it is translated by the ribosome. Evidence is available indicating that both these types of mechanism are used in different cases.

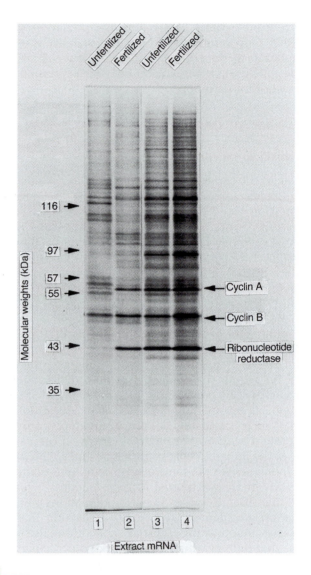

Figure 5.29

Translational control in the clam *Spisula solidissima*. Different proteins are synthesized *in vivo* before fertilization (track 1) and after fertilization (track 2). If,
however, RNA is isolated either before (track 3) or after fertilization (track 4) and
translated *in vitro* in a cell-free system, identical patterns of proteins are
produced. Hence the difference in the proteins produced *in vivo* from identical
RNA populations must be due to translational control. Photograph kindly
provided by Dr N. Standart and Dr T. Hunt, from Standart, *Semin. Dev. Biol.*
3, 367–380 (1992).

Thus, in the absence of heme a cellular protein kinase in the reticulocyte becomes active and phosphorylates the protein initiation factor eIF-2
(see Section 3.3) resulting in its inactivation. Since this factor is required
for the initiation of protein synthesis, translation of the globin RNA ceases
until heme is available. However, the use of such a mechanism, in which
total inactivation of the cellular translational apparatus is used to regulate

the translation of a single RNA, is possible only in the reticulocyte, where the globin protein constitutes virtually the only translation product.

In other cell types, where a large number of different RNA species are expressed, such a mechanism is normally used only where large-scale repression of many different RNA species occurs. Thus phosphorylation of eIF-2 (produced by a different kinase from that activated by the absence of heme) also occurs following exposure of cells to stress, when the translation of most cellular mRNAs is repressed (for a review see Silverman and Williams, 1999).

Interestingly, following exposure to specific stresses, some mRNAs continue to be translated since their protein products are required to mediate response to the stress. Thus, for example, the mRNAs encoding the heat-shock proteins are translated even following exposure of cells to elevated temperature and these proteins then produce a protective effect against the damaging effects of stress. This evidently raises the question as to how such translation can occur when general translation initiation factors have been inactivated.

A mechanism by which this is achieved has been elucidated in the case of the cationic amino acid transporter gene (Cat-1) whose mRNA continues to be translated following amino acid starvation in yeast. The Cat-1 mRNA contains an internal ribosome entry site (IRES) which allows the ribosome to bind to the mRNA internally and translate it rather than binding to the 5′ cap as normally occurs (see Section 3.3). Such IRES-mediated translation is activated after amino acid starvation, allowing the Cat-1 protein to carry out its function of mediating enhanced amino acid uptake (Fig. 5.30).

Recent studies have determined the mechanism by which translation via the IRES is induced after amino acid starvation (Yaman *et al.*, 2003). Thus, prior to amino acid starvation, the structure of the RNA does not allow the ribosome access to the IRES. However, following amino acid starvation, a short upstream open reading frame is translated into a small peptide of 48 amino acids and the ribosome reading through this region has the effect of changing the structure of the RNA, thereby exposing the IRES (Fig. 5.31). Moreover, unlike Cap-dependent translation, the IRES in the Cat-1 mRNA actually requires phosphorylated eIF-2 for its activity and is therefore stimulated by the phosphorylation of this factor during stresses such as amino acid starvation (Fig. 5.31).

IRES sequences are not confined to yeast, and have been found in a variety of viral and cellular mRNAs, including a number of mammalian mRNAs. It appears that in many of these mRNAs, as in the Cat-1 case, the IRES may have a role in allowing the RNA to continue to be translated when Cap-dependent translation is inhibited, for example during mitosis or following stress (for reviews see Hellen and Sarnow, 2001; Stoneley and Willis, 2004).

As well as widespread repression of translation, regulation of translation via initiation factors can also produce translational control of specific mRNAs by acting in conjunction with sequences within the mRNA itself which are often located within the 5′ untranslated region upstream of the translational start site (for a review see Jackson and Wickens, 1997). This is seen in the case of specific mRNAs whose translation is activated by treatment of cells with insulin or several growth factors (for a review see Gingras *et al.*, 2001). In this case, in the absence of insulin or growth factor

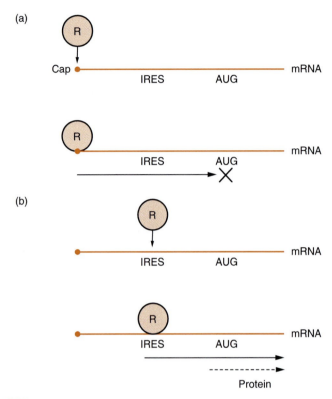

Figure 5.30

Following amino acid starvation, protein synthesis by ribosomes (R) binding to the 5' cap is decreased due to inactivation of specific initiation factors (a). However, translation initiated by ribosomes binding to an internal ribosome entry site (IRES) continues so allowing specific proteins to be made even though most protein synthesis is repressed (b).

the cap binding translational initiation factor eIF-4E (see Section 3.3) is associated with another protein eIF-4E binding protein (eIF-4Ebp). This association prevents the other initiation factors eIF-4G and eIF-4A from binding to the RNA. Such binding is particularly necessary for the translation of these insulin-regulated mRNAs since they have a high degree of secondary structure in their 5' region which must be unwound by the helicase activity of eIF-4A before translation can begin. Following insulin or growth factor treatment, eIF-4Ebp is phosphorylated, releasing it from eIF-4E and allowing the other factors to bind and unwind the mRNA so that translation can occur (Fig. 5.32).

Interestingly, the mRNA encoding insulin-like growth factor II itself exists in two different forms which are 4.8 kb and 6.0 kb long. The longer mRNA contains sequences upstream of the AUG initiator codon which render it dependent on the helicase activity of eIF-4A and its translation is therefore regulated by this pathway. In contrast, the 4.8 kb mRNA lacks these 5' sequences and is hence translated in an insulin/growth factor independent manner (Nielsen et al., 1995). Hence in this case, alternative

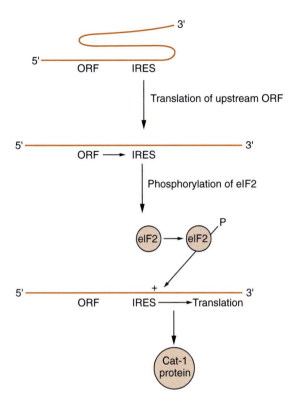

Figure 5.31

Prior to amino acid starvation, the IRES in the cat-1 mRNA is inaccessible. Following amino acid starvation, however, a small upstream open reading frame (ORF) is translated into protein and this changes the structure of the mRNA so exposing the IRES. Moreover, the activity of the IRES is stimulated by the phosphorylation of eIF-2 which also occurs after stresses such as amino acid starvation.

splicing generates two different mRNAs which encode the same protein but are subject to distinct translational regulation.

Several other cases where translational regulation of particular mRNA species is mediated by sequences in their 5′ untranslated region have been defined. Thus, the enhanced translation of the ferritin mRNA in response to iron is mediated by a sequence in this region which can fold into a stem-loop structure (for reviews see Hentze and Kuhn, 1996; Aisen *et al.*, 2001). The structure of this stem-loop is very similar to that found in the 3′ untranslated region of the transferrin receptor mRNA, whose stability is negatively regulated by the presence of iron (see Fig. 5.27 and Table 5.3).

Table 5.3   Regulation of the transferrin receptor and ferritin genes

	Effect of iron on protein production	Mechanism	Position of stem-loop structure
Ferritin	Increased	Increased mRNA translation	5′ untranslated region
Transferrin receptor	Decreased	Decreased mRNA stability	3′ untranslated region

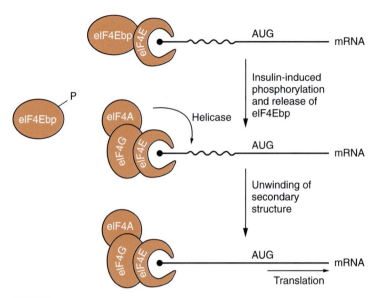

Figure 5.32

In the absence of insulin the translation initiation factor eIF-4E is bound by eIF-4Ebp and translation does not occur due to the presence of secondary structure in the mRNA. Following exposure to insulin, eIF-4Ebp is phosphorylated leading to its release from eIF-4E, eIF-4A and G then bind to eIF-4E and the helicase activity of eIF-4A then unwinds the RNA allowing translation to occur.

This has led to the suggestion that such loops may represent functionally equivalent iron response elements whose opposite effects on gene expression are dependent upon their position (5′ or 3′) within the RNA molecule. This idea was confirmed by transferring the transferrin receptor stem-loop to the 5′ end of an unrelated RNA, resulting in the iron-dependent enhancement of its translation. The identical structure is therefore capable of mediating opposite effects on RNA stability and translation, depending on its position within the RNA molecule.

Such an apparent paradox can be explained if it is assumed that the stem-loop structure is an iron response element (IRE) which unfolds in the presence of iron (Fig. 5.33). In the ferritin mRNA, where the element is in the 5′ end, this will allow the binding and movement of the 40S ribosomal subunit along the mRNA until it reaches the initiation codon and translation begins. In contrast, in the transferrin receptor mRNA, where this element is in the 3′ untranslated region, such unfolding renders the RNA susceptible to nuclease degradation at an increased rate. In agreement with this model, an IRE binding protein (IRE-BP) has been identified which binds to the stem-loop element in both the ferritin and transferrin receptor mRNAs. The RNA binding activity of IRE-BP increases dramatically in cells that have been deprived of iron, suggesting that its binding normally stabilizes the stem-loop structure.

Not all cases of translational regulation mediated by sequences in the 5′ untranslated region operate via such stem-loop structures, however. Increased expression of the yeast transcriptional regulatory protein GCN4 in response to amino acid starvation is caused by increased translation of

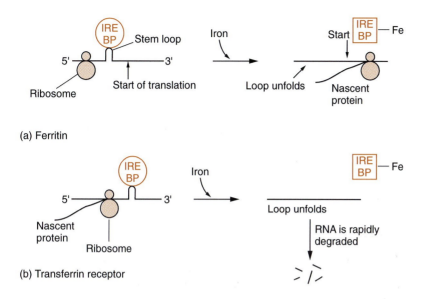

(a) Ferritin

(b) Transferrin receptor

Figure 5.33

Role of iron-induced unfolding of the stem-loop structure in producing increased translation of the ferritin mRNA (panel a) and increased degradation of the transferrin receptor mRNA (panel b). In each case the unfolding of the stem loop is dependent on the dissociation of a binding protein (IRE-BP) whose ability to bind to the stem loop decreases dramatically in the presence of iron.

its RNA (for a review see Morris and Geballe, 2000). The translational regulation of this molecule is mediated by short sequences within the 5′ untranslated region of the RNA, upstream of the start point for translation of the GCN4 protein. As in the case of the upstream ORF in the Cat-1 mRNA (see above), such sequences are capable of being translated to produce peptides, although in this case only of two or three amino acids (Fig. 5.34). Unlike the Cat-1 case however, translational initiation at the second, third or fourth of these upstream sequences in GCN4 to produce these small peptides, means that the ribosome fails to reinitiate at the translational start point for GCN4 production and hence this protein is not synthesized.

Following amino acid starvation, the production of these small peptides is suppressed and production of GCN4 correspondingly increased. Once again, this switch in initiation also involves the phosphorylation of the eIF-2 factor which is involved in translational regulation in response to heme or stress (see above). Thus following amino acid starvation, transfer RNA molecules lacking a bound amino acid accumulate and this activates an enzyme which phosphorylates eIF-2. This decreased activity of eIF-2 caused by its phosphorylation then results in the ribosome not initiating at three of the upstream sites in the 5′ untranslated region, which allows increased initiation at the downstream site leading to enhanced GCN4 production.

Interestingly, alternative initiation codons are also found in the genes encoding two mammalian liver transcriptional regulatory proteins C/EBPα and C/EBPβ. As discussed in Section 8.4, however, in this case the two

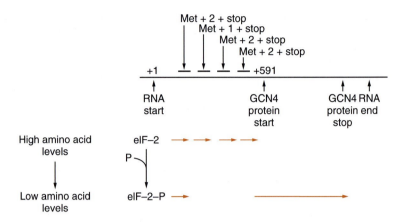

Figure 5.34

Presence of short open reading frames capable of producing small peptides in the 5′ untranslated region of the yeast GCN4 RNA. Translation of the RNA to produce these small proteins suppresses translation of the GCN4 protein. The position of the methionine residue beginning each of the small peptides is indicated together with the number of additional amino acids incorporated before a stop codon is reached. When high amino acid levels are present, the small proteins are made and the production of GCN4 is therefore suppressed. When amino acid levels fall, the eIF-2 translational initiation factor is phosphorylated. This prevents the translation of three of the four short peptides and thereby promotes GCN4 production.

initiation codons are used to produce two different forms of the proteins. One of these contains an N-terminal sequence which allows it to activate transcription, whereas the other form does not and so lacks this ability and can therefore interfere with the stimulatory activity of the activating form. Such alternative translation of the same mRNA thus parallels the use of alternative splicing to produce different protein products from the same gene.

Study of the cases of translational control where the mechanisms have been defined thus makes it clear that, although sequences in the 5′ untranslated region of particular mRNAs are frequently involved in the translational regulation of their expression, the mechanism by which they do so may differ dramatically in different cases.

Although the 5′ untranslated region is an obvious location for sequences involved in mediating translational control, cases where sequences in the 3′ untranslated region play a role in the regulation of translation have been reported. Thus, sequences in this region are involved in modulating the efficiency of translation of specific mRNAs which occurs upon fertilization of the egg (see above) as well as in mediating the increased translation of the mRNA encoding lipoxygenase which occurs during erythroid differentiation (Ostareck *et al.*, 2001). In this case, a sequence in the 3′ untranslated region of the lipoxygenase mRNA directly controls the ability of the 60S ribosomal subunit to bind to the mRNA. Thus, early in erythroid differentiation, the 40S ribosomal subunit can bind to the mRNA and, as normally occurs, it moves to the AUG translation initiation codon (see Section 3.3). However, a silencing complex, bound within the 3′

untranslated region, prevents binding of the 60S ribosomal subunit (Fig. 5.35a). Hence, translation only occurs later in erythroid differentiation when this block is relieved and the 60S subunit can bind and initiate translation (Fig. 5.35b).

A completely different mechanism involving sequences in the 3' untranslated region regulating translation is seen in the case of *Xenopus* oocytes (for a review see Richter and Theurkauf, 2001). In this case, a number of mRNAs are maintained in a non-translated state in dormant oocytes which are arrested early in meiosis. When the oocyte is stimulated to undergo meiosis, the poly (A) tail of these mRNAs is lengthened and the mRNA is translated. Hence, in this case a connection appears to exist between the extension of the poly (A) tail to its full length and the onset of translation.

In the case of one of these mRNAs, encoding cyclin B, this has been shown to occur because in the dormant oocytes the polyadenylation factor CPSF (see Section 3.3) is not bound to the AAUAAA sequence which is involved in polyadenylation. However, another factor CPEB is bound to an adjacent region of the RNA together with a further protein known as maskin (Fig. 5.36a). Maskin, in turn, appears to interact with the translation initiation factor eIF-4E and prevent binding of eIF-4G. Hence, maskin

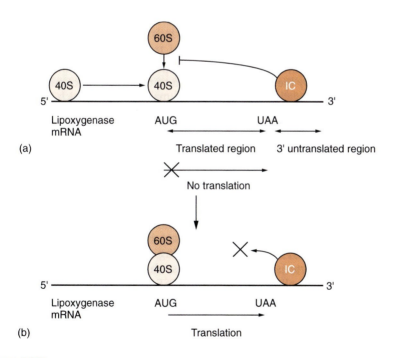

(a)

(b)

Figure 5.35

During erythroid differentiation, translation of the lipoxygenase mRNA is regulated by an inhibitory complex (IC) which binds to the 3' untranslated region of the mRNA. Early in erythroid differentiation (a), the 40S ribosomal subunit can bind to the mRNA and move to the AUG initiation codon. However, the inhibitory complex prevents the 60S ribosomal subunit from binding and so translation is inhibited. Later in the differentiation process (b), the negative effect of the inhibitory complex is blocked and translation occurs.

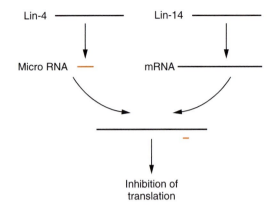

Figure 5.37

In the nematode, the lin-4 gene produces a micro RNA of 22 nucleotides which binds to the lin-14 mRNA and blocks its translation.

expression in non-expressing progenitor cells results in an increase in the number of B lymphocytes, indicating its critical role (Fig. 5.38) (Chen *et al.*, 2004). Similarly, a novel micro RNA has been shown to be produced specifically in the pancreas and to play a role in regulating insulin secretion (Poy *et al.*, 2004). Moreover, micro RNAs have been found to regulate the synthesis of numerous transcription factors in plants (see Chen, 2004), indicating the range of species which exhibit this phenomenon and reinforcing our earlier conclusion (Section 5.6) that the genes encoding transcription factors are often regulated post-transcriptionally.

Although the lin-4 RNA inhibits lin-14 by blocking translation of the mRNA, this is not the only mechanism by which micro RNAs can act. Thus, other micro RNAs have been shown to bind to complementary

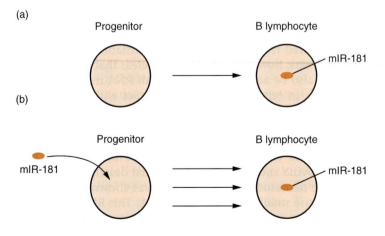

Figure 5.38

In the mouse, the miR-181 micro RNA is expressed specifically in B lymphocytes (panel a). Its artificial expression in progenitor cells induces enhanced production of B lymphocytes indicating its functional importance (panel b).

sequences in their mRNA targets and stimulate degradation of the mRNA. Interestingly, in the plant *Arabidopsis*, this process has been shown to be triggered by the loss of the cap at the 5′ end of the RNA (see Section 3.3). Following decapping, a small inhibiting RNA is produced by copying a portion of the decapped RNA. In turn, this partially double-stranded RNA is recognized as a target by enzymes which degrade the RNA (Fig. 5.39) (Gazzani *et al.*, 2004).

Hence, different small RNAs can inhibit gene expression by two distinct post-transcriptional mechanisms (Fig. 5.40 a,b) (for a review see Meister and Tuschl, 2004). Indeed, it appears that other micro RNAs can also act at the level of transcription by inducing an inactive chromatin structure in their target gene (Fig. 5.40 c). This appears to involve the small RNA inducing the methylation of histones which play a key role in regulating chromatin structure (see Section 6.6).

In recent years, many laboratories have taken advantage of the phenomenon of RNA interference (RNAi) to probe the function of specific genes, by introducing into cells small RNAs targeted against a particular gene product. Although this is a valuable experimental tool which has attracted enormous attention, it is clear that the natural phenomenon on which it is based plays a critical role in regulating gene expression at a variety of levels in a wide range of organisms.

5.8 Conclusions

A wide variety of cases exist in which gene expression can be regulated at levels other than transcription. In some lower organisms such post-transcriptional regulation may constitute the predominant form of gene control. In mammals, however, it appears to represent an adaptation to

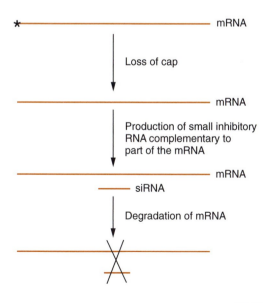

Figure 5.39

Loss of the 5′ cap at the end of an mRNA (asterisk) can promote its being copied to produce a complementary small inhibitory RNA. In turn, this partially double-stranded RNA is a target for degradation.

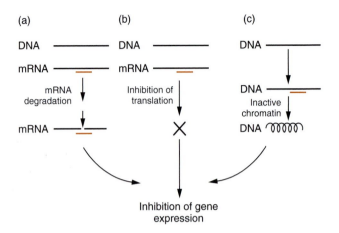

(a) (b) (c)

Figure 5.40

Micro RNAs can act to inhibit gene expression by inducing mRNA degradation (panel a), blocking mRNA translation (panel b) or inducing the reorganization of their target gene into an inactive chromatin structure (panel c).

particular situations, for example, the need to respond rapidly to the withdrawal of hormonal stimulation or stress by the regulation of RNA stability or translation. Similarly, post-transcriptional regulation may be the predominant one for the proteins which themselves regulate the transcription of other genes. Moreover, whole genome studies using both gene arrays (see Section 1.3) and the very large amount of DNA sequence data which are now available, have confirmed early studies on individual genes and shown the importance of post-transcriptional processes in generating multiple related proteins from one gene. Thus, up to 75% of human genes with multiple exons are subject to alternative splicing (Johnson *et al.*, 2003), whilst multiple new targets for RNA editing have been identified in this way (Hoopengardner *et al.*, 2003).

References

Aisen, P., Enns, C. and Wessling-Resnick, M. (2001). Chemistry and biology of eukaryotic iron metabolism. *International Journal of Biochemistry and Cell Biology* **33**, 940–959.

Baker, B.S. (1989). Sex in flies: the splice of life. *Nature* **340**, 521–524.

Bandziulis, R.J., Swanson, M.S. and Dreyfuss, G. (1989). RNA-binding proteins as developmental regulators. *Genes and Development* **3**, 431–437.

Bartel, D.P. (2004). MicroRNAs: genomics, biogenesis, mechanism, and function. *Cell* **116**, 281–297.

Bashirullah, A., Cooperstock, R.L. and Lipshitz, H.D. (2001). Spatial and temporal control of RNA stability. *Proceedings of the National Academy of Sciences of the USA* **98**, 7025–7028.

Black, D.L. (2000). Protein diversity from alternative splicing: A challenge for bioinformatics and post-genome biology. *Cell* **103**, 367–370.

Blanc, V. and Davidson, N.O. (2003). C-to-U RNA editing: mechanisms leading to genetic diversity. *Journal of Biological Chemistry* **278**, 1395–1398.

Blencowe, B.J. (2000). Exonic splicing enhancers: mechanism of action, diversity and role in human genetic diseases. *Trends in Biochemical Sciences* **25**, 106–110.

Bousquet-Antonelli, C., Presutti, C. and Tollervey, D. (2000). Identification of a regulated pathway for nuclear pre-mRNA turnover. *Cell* 102, 765–775.

Breitbart, R.E., Andreadis, A. and Nadal-Ginard, B. (1987). Alternative splicing: a ubiquitous mechanism for the generation of multiple protein isoforms from different genes. *Annual Review of Biochemistry* 56, 467–95.

Burd, C.G. and Dreyfuss, G. (1994). Conserved structure and functions of RNA binding proteins. *Science* 265, 615–621.

Chen, C.Z., Li, L., Lodish, H.F. and Bartel, D.P. (2004). MicroRNAs modulate hematopoietic lineage differentiation. *Science* 303, 83–86.

Chen, X. (2004). A microRNA as a translational repressor of APETALA2 in Arabidopsis flower development. *Science* 303, 2022–2025.

Cullen, B.R. (1991). Regulation of HIV-1 gene expression. *FASEB Journal* 5, 2361–2368.

Cullen, B.R. and Malim, H. (1991). The HIV-1 Rev protein: prototype of a novel class of eukaryotic transcriptional regulators. *Trends in Biochemical Sciences* 16, 341–348.

Curtis, D., Lehmann, R. and Zamore, P.D. (1995). Translational regulation in development. *Cell* 81, 171–178.

Danner, D. and Leder, P. (1985). Role of an RNA cleavage/poly (A) addition site in the production of membrane bound and secreted IgM mRNA. *Proceedings of the National Academy of Sciences of the USA* 82, 8658–8662.

Gazzani, S., Lawrenson, T., Woodward, C., Headon, D. and Sablowski, R. (2004). A link between mRNA turnover and RNA interference in *Arabidopsis*. *Science* 306, 1046–1048.

Gerace, L. (1995). Nuclear export signals and the fast track to the cytoplasm. *Cell* 82, 341–344.

Gingras, A.-C., Raught, B. and Sonenberg, N. (2001). Regulation of translation initiation by FRAP/mTOR. *Genes and Development* 15, 807–826.

Grabowski, P.J. (1998). Splicing regulation in neurons: tinkering with cell-specific control. *Cell* 92, 709–712.

Graveley, B.R. (2001). Alternative splicing: increasing diversity in the proteomic world. *Trends in Genetics* 17, 100–107.

Grigull, J., Mnaimneh, S., Pootoolal, J., Robinson, M.D. and Hughes, T.R. (2004). Genome-wide analysis of mRNA stability using transcription inhibitors and microarrays reveals posttranscriptional control of ribosome biogenesis factors. *Molecular and Cellular Biology* 24, 5534–5547.

Guyette, W.A., Matusik, R.A. and Rosen, J.M. (1979). Prolactin-mediated transcriptional and post-transcriptional control of casein gene expression. *Cell* 17, 1013–1023.

Hachet, O. and Ephrussi, A. (2004). Splicing of oskar RNA in the nucleus is coupled to its cytoplasmic localization. *Nature* 428, 959–963.

Handa, N., Nureki, O., Kurimoto, K., Kim, I., Sakamoto, H., Shimura, Y., Muto, Y. and Yokoyama, S. (1999). Structural basis for recognition of the *tra* mRNA precursor by the Sex-lethal protein. *Nature* 398, 579–585.

Hazelrigg, T. (1998). The destinies and destinations of RNAs. *Cell* 95, 451–460.

Hellen, C.U.T. and Sarnow, P. (2001). Internal ribosome entry sites in eukaryotic mRNA molecules. *Genes and Development* 15, 1593–1612.

Hentze, M.W. (1995). Translational regulation: versatile mechanisms for metabolic and developmental control. *Current Opinion in Cell Biology* 7, 393–398.

Hentze, M.W. and Kuhn, L.C. (1996). Molecular control of vertebrate iron metabolism : mRNA based regulatory circuits operated by iron, nitric oxide and oxidative stress. *Proceedings of the National Academy of Sciences of the USA* 93, 8175–8182.

Higuchi, M., Maas, S., Single, F.N., Hartner, J., Rozov, A., Burnashev, N., *et al.* (2000). Point mutation in an AMPA receptor gene rescues lethality in mice deficient in the RNA-editing enzyme ADAR2. *Nature* 406, 78–81.

Hoopengardner, B., Bhalla, T., Staber, C. and Reenan, R. (2003). Nervous system targets of RNA editing identified by comparative genomics. *Science* **301**, 832–836.

Jackson, R.J. and Wickens, M. (1997). Translational controls impinging on the 5' untranslated region and initiation factor proteins. *Current Opinion in Genetics and Development* **7**, 233–241.

Johnson, J.M., Castle, J., Garrett-Engele, P., Kan, Z., Loerch, P.M., Armour, C.D., *et al.* (2003). Genome-wide survey of human alternative pre-mRNA splicing with exon junction microarrays. *Science* **302**, 2141–2144.

Khodursky, A.B. and Bernstein, J.A. (2003). Life after transcription – revisiting the fate of messenger RNA. *Trends in Genetics* **19**, 113–115.

Kriventseva, E.V., Koch, I., Apweiler, R., Vingron, M., Bork, P., Gelfand, M.S. and Sunyaev, S. (2003). Increase of functional diversity by alternative splicing. *Trends in Genetics* **19**, 124–128.

Landry, J.R., Mager, D.L. and Wilhelm, B.T. (2003). Complex controls: the role of alternative promoters in mammalian genomes. *Trends in Genetics* **19**, 640–648.

Leff, S.E., Rosenfeld., M.G. and Evans, R.M. (1986). Complex transcriptional units: diversity in gene expression by alternative RNA processing. *Annual Review of Biochemistry* **55**, 1091–117.

Lynch, K.W. and Maniatis, T. (1995). Synergistic interactions between two distinct elements of a regulated splicing enhancer. *Genes and Development* **9**, 284–293.

Maas, S., Rich, A. and Nishikura, K. (2003). A-to-I RNA editing: recent news and residual mysteries. *Journal of Biological Chemistry* **278**, 1391–1394.

Manley, J.L. and Tacke, R. (1996). SR proteins and splicing control *Genes and Development* **10**, 1569–1579.

Marzluff, W.F. (1992). Histone 3' ends: essential and regulatory function. *Gene Expression* **2**, 93–97.

Meister, G. and Tuschl, T. (2004). Mechanisms of gene silencing by double-stranded RNA. *Nature* **431**, 343–349.

Mello, C.C. and Conte, D., Jr. (2004). Revealing the world of RNA interference. *Nature* **431**, 338–342.

Mohr, E. and Richter, D. (2001). Messenger RNA on the move: implications for cell polarity. *International Journal of Biochemistry and Cell Biology* **33**, 669–679.

Morris, D.R. and Geballe, A.P. (2000). Upstream open reading frames as regulators in mRNA translation. *Molecular and Cellular Biology* **20**, 8635–8642.

Nielsen, F.C., Ostergard, L., Neilsen, J. and Christiansen, J. (1995). Growth-dependent translation of IGFII mRNA by a rapamycin sensitive pathway. *Nature* **377**, 358–362.

Novina, C.D. and Sharp, P.A. (2004). The RNAi revolution. *Nature* **430**, 161–164.

O'Donovan, K.J. and Darnell, R.B. (2001). Neuronal signalling through alternative splicing: some CaRRE... Science's STKE http://stke.sciencemag.org/cgi/content/full/OC_sigtrans;2001/94/pe2.

Ostareck, D.H., Ostareck-Lederer, A., Shatsky, I.N. and Hentze, M.W. (2001). Lipoxygenase mRNA silencing in erythroid differentiation: The 3'UTR regulatory complex controls 60S ribosomal subunit joining. *Cell* **104**, 281–290.

Pachter, J.S., Yen, T.J. and Cleveland, D.W. (1987). Auto-regulation of tubulin expression is achieved through degradation of polysomal tubulin mRNAs. *Cell* **51**, 283–292.

Pain, V.M. (1996). Initiation of protein synthesis in eukaryotic cells. *European Journal of Biochemistry* **236**,747–771.

Palacios, I.M., Gatfield, D., St Johnston, D. and Izaurralde, E. (2004). An eIF4AIII-containing complex required for mRNA localization and nonsense-mediated mRNA decay. *Nature* **427**, 753–757.

Parker, R. and Song, H. (2004). The enzymes and control of eukaryotic mRNA turnover. *Nature Structural and Molecular Biology* 11, 121–127.

Powell, D.J., Freidman, J.M., Oulette, A.J., Krauter, K.S. and Darnell, J.E. (1984). Transcriptional and post-transcriptional control of specific messenger RNAs in adult and embryonic liver. *Journal of Molecular Biology* 179, 21–35.

Poy, M.N., Eliasson, L., Krutzfeldt, J., Kuwajima, S., Ma, X., Macdonald, P.E., et al. (2004). A pancreatic islet-specific microRNA regulates insulin secretion. *Nature* 432, 226–230.

Raitskin, O., Cho, D.-S.C., Sperling, J., Nishikura, K. and Sperling, R. (2001). RNA editing activity is associated with splicing factors in InRNP particles: the nuclear pre-mRNA processing machinery. *Proceedings of the National Academy of Sciences of the USA* 98, 6571–6576.

Reenan, R.A. (2001). The RNA world meets behaviour: AfiI pre-mRNA editing in animals. *Trends in Genetics* 17, 53–56.

Richter, J.D. and Theurkauf, W.E. (2001). The message is in the translation. *Science* 293, 60–62.

Roegiers, F. and Jan, Y.N. (2000). Staufen: a common component of mRNA transport in oocytes and neurones? *Trends in Cell Biology* 10, 220–224.

Ross, J. (1996). Control of messenger RNA stability in higher eukaryotes. *Trends in Genetics* 12, 171–175.

Rueter, S.M., Dawson, T.R. and Emeson, R.B. (1999). Regulation of alternative splicing by RNA editing. *Nature* 399, 75–79.

Sachs, A.B. (1993). Messenger RNA degradation in eukaryotes. *Cell* 74, 413–421.

Samuel, C.E. (2003). RNA editing minireview series. *Journal of Biological Chemistry* 278, 1389–1390.

Sandri-Goldin, R.M. (2004). Viral regulation of mRNA export. *Journal of Virology* 78, 4389–4396.

Shin, C., Feng, Y. and Manley, J.L. (2004). Dephosphorylated SRp38 acts as a splicing repressor in response to heat shock. *Nature* 427, 553–558.

Silverman, R.H. and Williams, B.R.G. (1999). Translational control perks up. *Nature* 397, 208–209.

Smith, C.W.J. and Valcárcel, J. (2000). Alternative pre-mRNA splicing: the logic of combinatorial control. *Trends in Biochemical Sciences* 25, 381–388.

Stein G.S. Stein, J.L., van Wijnen, A.J. and Liam, J.B. (1992). Regulation of histone gene expression. *Current Opinion in Cell Biology* 4, 166–173.

Stoneley, M. and Willis, A.E. (2004). Cellular internal ribosome entry segments: structures, *trans*-acting factors and regulation of gene expression. *Oncogene* 23, 3200–3207.

Takagaki, Y., Selpet, R.L., Peterson, M.L. and Manley, J.L. (1996). The polyadenylation factor CstF-64 regulates alternative processing of IgM heavy chain pre mRNA during B cell differentiation. *Cell* 87, 941–952.

Valcarcel, J., Singh, R., Zamore, P.D. and Green, M.R. (1993). The protein sex-lethal antagonises the splicing factor U2AF to regulate splicing of transformer pre-mRNA. *Nature* 362, 171–175.

Wang, J. and Manley, J.L. (1997). Regulation of pre-mRNA splicing in metazoa. *Current Opinion in Genetics and Development* 7, 205–211.

Wedekind, J.E., Dance, G.S., Sowden, M.P. and Smith, H.C. (2003). Messenger RNA editing in mammals: new members of the APOBEC family seeking roles in the family business. *Trends in Genetics* 19, 207–216.

Wickens, M. and Goldstrohm, A. (2003). A place to die, a place to sleep. *Science* 300, 753–755.

Wickens, M., Anderson, P. and Jackson, R.J. (1997). Life and death in the cytoplasm: messages from the 5′ end. *Current Opinion in Genetics and Development* 7, 220–232.

Wilusz, C.J. and Wilusz, J. (2004). Bringing the role of mRNA decay in the control of gene expression into focus. *Trends in Genetics* 20, 491–497.

Wold, B.J., Klein, W.H., Hough-Evans, B.R., Britten, R.J. and Davidson, E.H. (1978). Sea urchin embryo mRNA sequences expressed in the nuclear RNA of adult tissues. *Cell* **14**, 941–950.

Xie, J. and Black, D.L. (2001). A CaMK IV responsive RNA element mediates depolarisation-induced alternative splicing of ion channels. *Nature* **410**, 936–939.

Yaman, I., Fernandez, J., Liu, H., Caprara, M., Komar, A.A., Koromilas, A.E., *et al.* (2003). The zipper model of translational control: a small upstream ORF is the switch that controls structural remodeling of an mRNA leader. *Cell* **113**, 519–531.

Yen, T. J., Machlin, P.S. and Cleveland, D.W. (1988). Autoregulated instability of β-tubulin by recognition of the nascent amino-acids of β-tubulin. *Nature* **334**, 580–585.

Transcriptional control – chromatin structure

6

SUMMARY

- In eukaryotes gene regulation involves long-term changes which allow a cell to become and remain committed to a particular pattern of gene expression.
- These changes occur prior to a gene becoming active and involve an alteration in the chromatin structure of the whole gene from the tightly-packed solenoid structure to the more open "beads on a string" structure.
- These changes result in the chromatin of active or potentially active genes exhibiting:
  (a) Enhanced sensitivity to digestion with DNaseI
  (b) Undermethylation on specific C residues in the DNA
  (c) Changes in the post-translational modifications of the histones associated with the DNA
- At specific regulatory sites within the gene, chromatin remodeling complexes produce greater changes in the chromatin structure resulting in the appearance of sites which are hypersensitive to DNaseI digestion.
- As well as its role in global gene regulation, alterations in chromatin structure are also involved in other biological processes such as X-chromosome inactivation and genomic imprinting.

6.1 Introduction

Having established that the primary control of eukaryotic gene expression lies at the level of transcription, it is necessary to investigate the mechanisms responsible for this effect. The fact that regulation at transcription is also responsible for the control of gene expression in bacteria suggests that insights into these procedures obtained in these much simpler organisms (for a review see Travers, 1993) may be applicable to higher organisms. It is possible therefore that regulation of transcription in eukaryotes might occur by means of a protein, present in all tissues, which binds to the promoter region of a particular gene and prevents its expression. In one particular cell type, or in response to a particular signal such as heat shock, this protein would be inactivated either directly (Fig. 6.1a) or by binding of another factor (Fig. 6.1b), and would no longer bind to the gene. Hence transcription would occur only in the one cell type or in response to the signal. This mechanism is based on that regulating the

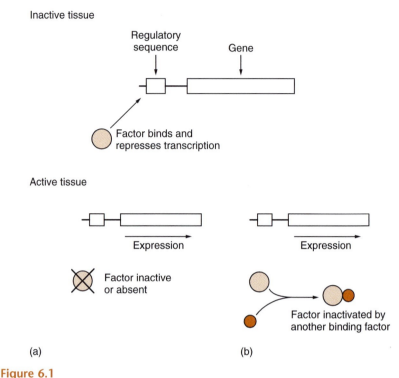

Inactive tissue

Regulatory
sequence   Gene

Factor binds and
represses transcription

Active tissue

Expression   Expression

Factor inactive
or absent

Factor inactivated by
another binding factor

(a)   (b)

Figure 6.1

Model for the activation of gene expression in a particular tissue by inactivation
(a) or binding out (b) of a repressor present in all tissues.

expression in the bacterium *E. coli* of the *lac* operon containing the genes
encoding proteins required for the metabolism of lactose. This operon is
normally repressed by the *lac* repressor protein; binding of lactose to this
protein, however, results in its inactivation and allows transcription of the
operon (for a review see Muller-Hill, 1996).

Alternatively, the fact that most eukaryotic genes are inactive in most
tissues, and become active only in one tissue or in response to a particular signal, suggests that it may be more economical to have a system in
which the gene is constitutively inactive in most tissues, without any
repressor being required. Activation of the gene would then require a
particular factor binding to its promoter. The specific expression pattern
of the gene would be controlled by the presence of this factor only in the
expressing cell type (Fig. 6.2a) or, alternatively, by the factor's requirement
for a co-factor such as a steroid hormone to convert it to an active form
(Fig. 6.2b). This mechanism is based on the regulation of the *ara* operon
in *E. coli* in which binding of the substrate arabinose to the regulatory
araC protein allows it to induce transcription of the genes required for the
metabolism of arabinose (for a review see Raibaud and Schwartz, 1984).

Hence, based on the known mechanisms of gene regulation in bacteria,
it is possible to produce models of how gene regulation might operate in
eukaryotes. Indeed, a considerable amount of evidence indicates that
many cases of gene regulation do use the activation type mechanisms seen
in the arabinose operon. Thus, as will be discussed in Chapters 7 and 8,

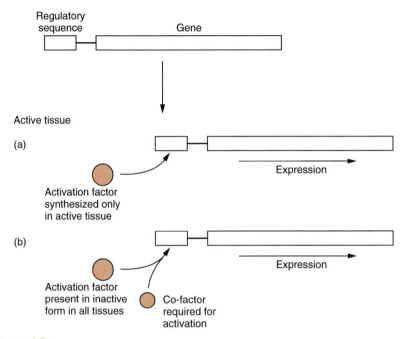

Inactive tissue   Gene is constitutively inactive. Activating
                   factor is absent or inactive

Regulatory
sequence   Gene

Active tissue

(a)

                                         Expression

Activation factor
synthesized only
in active tissue

(b)

                                         Expression

Activation factor
present in inactive   Co-factor
form in all tissues   required for
                           activation

Figure 6.2

Model for the activation of gene expression in a particular tissue by an activator
present only in that tissue (a) or activated by a co-factor (b) only in that tissue.

the effects of glucocorticoid and other steroid hormones on gene expression are mediated by the binding of the steroid to a receptor protein. This
activated complex then binds to particular sequences upstream of steroid-
responsive genes and activates their transcription.

Even in the case of steroid hormones, however, such mechanisms
cannot account entirely for the regulation of gene expression. In the
chicken, administration of the steroid hormone estrogen results in the
transcriptional activation in oviduct tissue of the gene encoding ovalbumin, as discussed in Chapter 4. In the liver of the same organism, however,
estrogen treatment has no effect on the ovalbumin gene but instead results
in the activation of a completely different gene, encoding the protein
vitellogenin. Such tissue-specific differences in the response to a particular treatment are, of course, entirely absent in single-celled bacteria and
cannot be explained simply on the basis of the activation of a single DNA-
binding protein by the hormone.

Similarly, models of this type cannot explain the data obtained by
Becker *et al.* (1987) in their studies of the regulation of the rat tyrosine
amino-transferase gene. All the protein factors binding to the regulatory
regions of this gene were detectable in both the liver, where the gene is
expressed, and in other tissues, where no expression is observed.
Moreover, the proteins isolated from these tissues were equally active in
binding to the appropriate region of the gene in deproteinized DNA. Only

in the liver, however, was actual binding of these factors to the DNA of the gene within its normal chromosomal structure detectable. Once again, these data cannot be explained solely on the basis of models in which proteins stimulate or inhibit gene expression by binding to appropriate DNA sequences. Rather, an understanding of eukaryotic gene regulation will require a knowledge of the ways in which the relatively short-term regulatory processes mediated by the binding of proteins to specific DNA sequences in this manner interact with the much longer-term regulatory processes, which establish and maintain the differences between particular tissues and also control their response to treatment with effectors such as steroids. This baseline regulation of chromatin structure thus renders eukaryotic gene regulation fundamentally different from prokaryotic gene regulation where this does not occur (for a review see Struhl, 1999). The long-term control processes which regulate chromatin structure will be discussed in this chapter, and the DNA sequences and proteins which actually regulate transcription in response to particular signals will be discussed in Chapters 7 and 8.

6.2 Commitment to the differentiated state and its stability

It is evident that the existence of different tissues and cell types in higher eukaryotes requires mechanisms that establish and maintain such differences, and that these mechanisms must be stable in the long term. Thus, despite some exceptions (see Section 2.2), tissues or cells of one type do not, in general, change spontaneously into another cell type. Inspection of mammalian brain tissue does not reveal the presence of cells typical of liver or kidney, and antibody-producing B cells do not spontaneously change into muscle cells when cultured in the laboratory. Hence, cells must be capable of maintaining their differentiated state, either within a tissue or during prolonged periods in culture.

Indeed, the long-term control processes that achieve this effect must regulate not only the ability to maintain the differentiated state more or less indefinitely but also the observed ability of cells to remember their particular cell type, even under conditions when they cannot express the characteristic features of that cell type. Thus in the experiments of Coon (1966) it was possible to regulate the behavior of cartilage-producing cells in culture according to the medium in which they were placed. In one medium the cells were capable of expressing their differentiated phenotype and produced cartilage-forming colonies synthesizing an extracellular matrix containing chondroitin sulfate. By contrast, in a medium favoring rapid division the cells did not form such colonies and, instead, divided rapidly and lost all the specific characteristics of cartilage cells, becoming indistinguishable from undifferentiated fibroblast-like cells in appearance. Nonetheless, even after 20 generations in this rapid growth medium the cells could resume the appearance of cartilage cells and synthesis of chondroitin sulfate if returned to the appropriate medium. This process did not occur if other cell types or undifferentiated fibroblast cells were placed in an identical medium. Hence the cartilage cells were capable not only of maintaining their differentiated cell type in a particular medium supplying appropriate signals, but also of remembering it in the absence of such signals and returning to the correct differentiated state when placed in the appropriate medium.

These experiments lead to the idea that cells belong to a particular lineage and that mechanisms exist to maintain the commitment of cells to a particular lineage, even in the absence of the characteristics of the differentiated phenotype. A variety of evidence exists to suggest that cells become committed to a particular differentiated stage or lineage well before they express any features characteristic of that lineage, and that such commitment can be maintained through many cell generations.

Perhaps the most dramatic example of this effect occurs in the fruit fly, *Drosophila melanogaster*. In this organism, the larva contains many disks consisting of undifferentiated cells located at intervals along the length of the body and indistinguishable from each other in appearance. Eventually, these imaginal disks (Hadorn, 1968) will form the structures of the adult, the most anterior pair producing the antennae and others producing the wings, legs, etc. In order to do this, however, the disks must pass through the intermediate pupal state where they receive the appropriate signals to differentiate into the adult structures. If they are removed from the larva and placed directly in the body cavity of the adult, they will remain in an undifferentiated state, since the signals inducing differentiation will not be received (Fig. 6.3).

This process can be continued for many generations. Thus the disk removed from the adult can be split in two, one half being used to propagate the cells by being placed directly in another adult and the other half being tested to see what it will produce when placed in a larva that is allowed to proceed to an adult through the pupal state. When this is done, it is found that a disk which would have given rise to an antenna can still do so when placed in a larva after many generations of passage through adult organisms in an undifferentiated state. This is the case even if the disk is placed at a position within the larva very different from that normally occupied by the imaginal disk producing the antenna. The cells of the imaginal disk within the normal larva must therefore have undergone a commitment event to the eventual production of a particular differentiated state, which they can maintain for many generations, prior to having ever expressed any features characteristic of that differentiated state (Fig. 6.3).

These examples of the stability of the differentiated state and commitment to it imply the existence of long-term regulatory processes capable of maintaining such stability. However, the existence of cases where such stability breaks down, either artificially following nuclear transplantation or more naturally as in lens regeneration (see Section 2.2), indicates that such processes, although stable, cannot be irreversible. Indeed, even in the imaginal disks of *Drosophila* the stability of the committed state can be broken down by culture for long periods in adults, the disks giving rise eventually to tissues other than the one intended originally when placed in the larva. This breakdown of the committed state is not a random process but proceeds in a highly characteristic manner. Wing cells are the first abnormal cells produced by a disk that should produce an eye, with other cell types being produced only subsequently. Similarly, a disk that should produce a leg never produces genitalia, although it can produce other cell types. The various transitions that can occur in this system are summarized in Fig. 6.4. Most interestingly, these various transitions are precisely those observed in the homeotic mutations in *Drosophila* (see Section 8.2) which convert one adult body part into another. Hence each

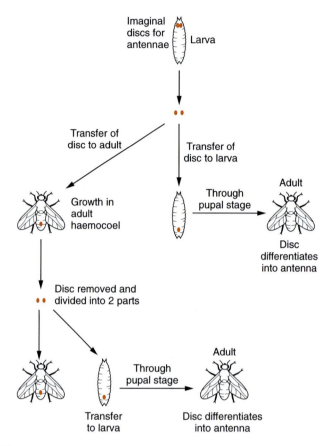

Figure 6.3

The use of larval imaginal discs to demonstrate the stability of the committed state. The disc cells maintain their commitment to produce a specific adult structure even after prolonged growth in an undifferentiated state in successive adults. Redrawn from J.B. Gurdon, *Control of Gene Expression in Animal Development* (1974), Oxford University Press, Oxford, by permission of Professor J.B. Gurdon and Oxford University Press.

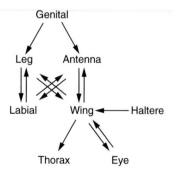

Figure 6.4

Structures that arise from specific imaginal disks of *Drosophila* following the breakdown of commitment. Redrawn from D.J. Cove, *Genetics* (1971), Cambridge University Press, by permission of Professor D.J. Cove and Cambridge University Press.

of these transitions is likely to be controlled by a specific gene whose activity changes when commitment breaks down.

These examples and the other experiments discussed in Section 2.2 eliminate irreversible mechanisms such as DNA loss as a means of explaining the process of commitment to the differentiated state. Rather, the semi-stability of this process and its propagation through many cell generations is due to the establishment of a particular pattern of association of the DNA with specific proteins. The structure formed by DNA and its associated proteins is known as chromatin. An understanding of how long-term gene regulation is achieved therefore requires a knowledge of the structure of chromatin.

6.3 Chromatin structure

If the DNA in a single human individual were to exist as an extended linear molecule, it would have a length of 5×10^{10} km and would extend 100 times the distance from the Earth to the Sun. Clearly this is not the case. Rather, the DNA is compacted by folding in a complex with specific nuclear proteins into the structure known as chromatin (for a review see Wolffe, 1995). Of central importance in this process are the five types of histone proteins (Table 6.1) whose high proportion of positively charged amino acids neutralizes the net negative change on the DNA and allows folding to occur. The basic unit of this folded structure is the nucleosome (for reviews see Kornberg and Lorch, 1999; Bradbury, 2002; Khorasanizadeh, 2004), in which approximately 200 base pairs of DNA are associated with a histone octamer containing two molecules of each of the four core histones, H2A, H2B, H3, and H4 (Fig. 6.5). In this structure approximately 146 base pairs of DNA make almost two full turns around the histone octamer, and the remainder serves as a linker DNA, joining one nucleosome to another.

The detailed arrangement of the histones and the DNA within the nucleosome has been confirmed by high resolution structural analysis using X-ray crystallography (Luger et al., 1997; Rhodes, 1997; Richmond and Davey, 2003) (color plate 9). This demonstrates that the histones form H2A-H2B and H3-H4 heterodimers which then associate to form the histone octamer with the DNA wrapped around the surface of the octamer. Interestingly, the amino-terminal tails of the histones extend beyond the surface of the nucleosome particle and are likely to be involved in interactions between nucleosomes. This is of particular interest since these amino-terminal histone tails are subject to specific modifications which are known to alter chromatin structure (see Section 6.6) and these modifications may thus act by affecting the nucleosome–nucleosome interactions which are mediated via these regions of the histone molecule.

Table 6.1 The histones

Histone	Type	Molecular weight	Molar ratio
H1	Lysine rich	23 000	1
H2A	Slightly lysine rich	13 960	2
H2B	Slightly lysine rich	13 744	2
H3	Arginine rich	15 342	2
H4	Arginine rich	11 282	2

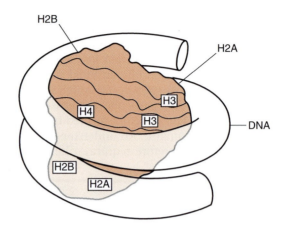

Figure 6.5

Structure of DNA and the core histones within a single nucleosome.

Within the nucleosome, the linker DNA is more accessible than the highly protected DNA tightly wrapped around the octamer and is therefore preferentially cleaved when chromatin is digested with small amounts of the DNA-digesting enzyme, micrococcal nuclease. Thus, if DNA is isolated following such mild digestion of chromatin, on gel electrophoresis it produces a ladder of DNA fragments of multiples of 200 base pairs, representing the results of cleavage in some but not all linker regions (Fig. 6.6a, track T). This can be correlated with the properties of nucleosomes isolated from the digested chromatin. Thus a partially digested chromatin preparation can be fractionated into individual nucleosomes associated with 200 bases of DNA (Fig. 6.6a, track D), dinucleosomes associated with 400 bases of DNA (Fig. 6.6a, track C), and so on (Finch *et al.*, 1975). The individual mononucleosomes, dinucleosomes, or larger complexes in each of these fractions can be observed readily in the electron microscope (Fig. 6.6b). These experiments provide direct evidence for the organization of DNA into nucleosomes within the cell.

This organization of DNA into nucleosomes, which can be directly visualized in the electron microscope as the beads on a string structure of chromatin (Fig. 6.7, page 156), constitutes the first stage in the packaging of DNA. Subsequently this structure is folded upon itself into a

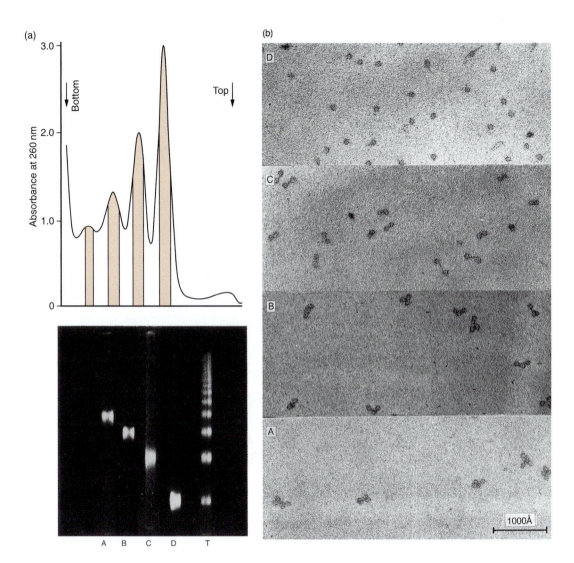

Figure 6.6

(a) Separation of mononucleosomes (D), dinucleosomes (C), trinucleosomes (B), and tetranucleo-somes (A) by sucrose gradient centrifugation. The upper panel shows the peaks of absorbance produced by the individual fractions of the gradient, the lower panel shows the DNA associated with each fraction separated by gel electrophoresis. Track T shows the DNA ladder produced from a preparation containing all the individual nucleosome fractions.

(b) Electron microscopic analysis of the fractions separated in panel (a). Mononucleosomes are clearly visible in fraction D, dinucleosomes in fraction C, and so on. Photographs kindly provided by Dr J.T. Finch, from Finch *et al.*, *PNAS* **72**, 3320–3322 (1975).

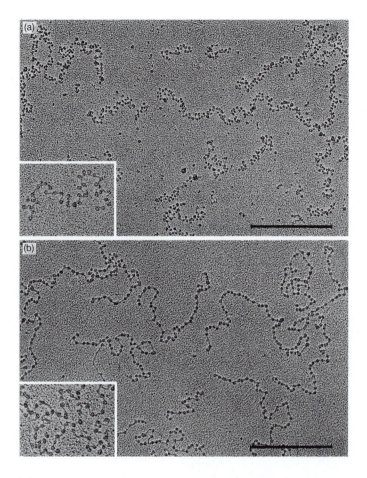

Figure 6.7

Beads on a string structure of chromatin visualized in the electron microscope in the presence (a) or absence (b) of histone H1. The bar indicates 0.5 μm. Photograph kindly provided by Dr F. Thoma, from Thoma *et al., J. Cell. Biol.* **83**, 403–427 (1979), by permission of Rockefeller Press.

much more compact structure, known as the solenoid (Fig. 6.8). In the formation of this structure histone H1 plays a critical role. As will be seen from Table 6.1, this histone is present at half the level of the other histones and is not part of the core histone octamer. Instead, one molecule of histone H1 seals the two turns which the DNA makes around the core octamer of the other histones. In the solenoid structure, these single histone H1 molecules associate with one another, resulting in tight packing of the individual nucleosomes into a 30 nm fiber (for reviews see Felsenfeld and McGhee, 1986; van Holde and Zlatanova, 1995; Mohd-Sarip and Verrijzer, 2004). This 30 nm fiber is the basic structure of chromatin in cells not undergoing division, although during cell division the DNA is further compacted by extensive looping of the 30 nm fiber to form the readily visible chromosomes. In this further compacted structure the fiber loops are linked at their bases, via specific DNA sequences known as matrix attachment regions (MAR), to a protein scaffold known as the nuclear matrix (for a review see Horn and Peterson, 2002) (Fig. 6.9, page 158).

6.4 Changes in chromatin structure in active or potentially active genes

Active DNA is organized in a nucleosomal structure

Having established that the bulk of cellular DNA is associated with histone molecules in a nucleosomal structure, an obvious question is whether genes which are either being transcribed or are about to be transcribed in a particular tissue are also organized in this manner or whether they exist as naked, nucleosome-free DNA.

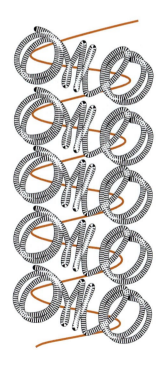

Figure 6.8

The solenoid structure of chromatin. Photograph kindly provided by Dr J. McGhee, from McGhee *et al.*, *Cell* **33**, 831–841 (1983), by permission of Cell Press.

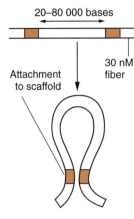

20–80 000 bases

Attachment
to scaffold

30 nM
fiber

Figure 6.9

Folding of specific regions of the 30 nM solenoid fiber to form a loop which is attached to the nuclear scaffold.

Two main lines of evidence suggest that such genes are organized into nucleosomes. Firstly, if DNA that is being transcribed is examined in the electron microscope, in most cases, the characteristic beads on a string structure (see Section 6.3) is observed, with nucleosomes visible both behind and in front of the RNA polymerase molecules transcribing the gene (Fig. 6.10; McKnight *et al.*, 1978). Thus, although this structure may break down in genes which are being extremely actively transcribed, such as occurs for the genes encoding ribosomal RNA during oogenesis, it is maintained in most transcribed genes.

Second, if DNA organized into nucleosomes is isolated as a ladder of characteristically sized fragments following mild digestion with micrococcal nuclease (see Fig. 6.6), the DNA from active genes is found in these fragments in the same proportion as in total DNA (Lacey and Axel, 1975). Similarly, no enrichment or depletion in the amount of a particular gene found in these fragments is observed when the DNA isolated from a tissue actively transcribing the gene is compared with DNA isolated from a tissue that does not transcribe it. The ovalbumin gene, for example, is found in nucleosome-sized fragments of DNA in these experiments, regardless of whether chromatin from hormonally stimulated oviduct tissue or from liver tissue is used. This is in agreement with the idea that transcribed DNA is still found in a nucleosomal structure rather than as naked DNA, which would be rapidly digested by micrococcal nuclease and hence would not appear in the ladder of nucleosome-sized DNA fragments.

Sensitivity of active chromatin to DNaseI digestion

In order to probe for other differences that may exist between transcribed and non-transcribed DNA, many workers have studied the sensitivity of these different regions to digestion with the pancreatic enzyme deoxyribonuclease I (DNaseI). Although this enzyme will eventually digest all the DNA in a cell, if it is applied to chromatin in small amounts for a short period, only a small amount of the DNA will be digested. The proportion of transcribed or non-transcribed genes present in the relatively resistant undigested DNA in a given tissue compared with the proportion in total DNA can therefore be used to detect the presence of differences between

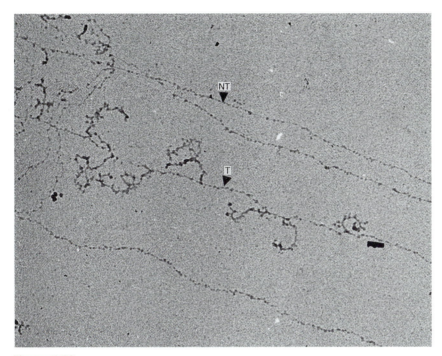

Figure 6.10

Electron micrograph of chromatin from a *Drosophila* embryo. Note the identical "beads on a string" structure of the chromatin that is not being transcribed (NT) and the chromatin that is being transcribed (T) into the readily visible ribonucleoprotein fibrils. Photograph kindly provided by Professor O.L. Miller, from McKnight *et al.*, *Cold Spring Harbor Symposium* **42**, 741 (1978), by permission of Cold Spring Harbor Laboratory.

active and inactive DNA in their sensitivity to digestion with this enzyme. The fate of an individual gene in this procedure can be followed simply by cutting the DNA surviving digestion with an appropriate restriction enzyme and carrying out the standard Southern blotting procedure (Methods Box 2.1) with a labelled probe derived from the gene of interest. The presence or absence of the appropriate band derived from the gene in the digested DNA provides a measure of the resistance of the gene to DNaseI digestion (Fig. 6.11; Methods Box 6.1).

METHODS BOX 6.1

Probing chromatin structure with DNaseI (Fig. 6.11)

- Isolate chromatin (DNA and associated histones and other proteins).
- Partially digest chromatin with DNaseI.
- Purify partially digested DNA by removing protein.
- Digest with restriction enzyme and carry out Southern blotting with probe for gene of interest (see Methods Box 2.1).
- Monitor disappearance of band for DNA of interest in samples where increasing amounts of DNaseI have been used to digest chromatin.

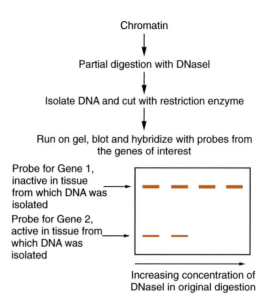

Chromatin

↓

Partial digestion with DNaseI

↓

Isolate DNA and cut with restriction enzyme

↓

Run on gel, blot and hybridize with probes from
the genes of interest

Probe for Gene 1,
inactive in tissue
from which DNA was
isolated

Probe for Gene 2,
active in tissue from
which DNA was
isolated

Increasing concentration of
DNaseI in original digestion

Figure 6.11

Preferential sensitivity of chromatin containing active genes to digestion with
DNaseI, as assayed by Southern blotting with specific probes.

This procedure has been used to demonstrate that a very wide range of
genes that are active in a particular tissue exhibit a heightened sensitivity
to DNaseI, which extends over the whole of the transcribed gene and for
some distance upstream and downstream of the transcribed region. For
example, when chromatin from chick oviduct tissue is digested, the active
ovalbumin gene is rapidly digested and its characteristic band disappears
from the Southern blot under digestion conditions which leave the DNA
from the inactive globin gene undigested. This difference between the
globin and ovalbumin genes is dependent upon their different activity in
oviduct tissue rather than any inherent difference in the resistance of the
genes themselves to digestion. Thus, if the digestion is carried out with
chromatin isolated from red blood cell precursors in which the globin gene
is active and that encoding ovalbumin is inactive, the reverse result to that
seen in the oviduct is obtained, with the globin DNA exhibiting preferential sensitivity to digestion and the ovalbumin DNA being resistant to
such digestion.

It is clear, therefore, that actively transcribed genes, though packaged
into nucleosomes, are in a more open chromatin structure than that found
for non-transcribed genes, and are hence more accessible to digestion with
DNaseI. This altered chromatin structure is not confined to the very active
genes such as globin and ovalbumin but appears to be a general characteristic of all transcribed genes, whatever their rate of transcription. Thus,
if chromatin is digested using conditions that degrade less than 10% of
the total DNA, over 90% of transcriptionally active DNA is digested, the
DNA of genes encoding rare mRNAs being as sensitive as those encoding
abundant mRNAs. Hence, the altered chromatin structure of active genes
does not appear to be dependent upon the act of transcription itself, since
the genes encoding rare mRNAs will be transcribed only very rarely.

In agreement with this idea, the altered DNaseI sensitivity of previously active genes persists even after transcription has ceased. For example, the ovalbumin gene remains preferentially sensitive to the enzyme when chromatin is isolated from the oviduct after withdrawal of estrogen when, as we have previously seen, transcription of the gene ceases (see Section 4.2). A similar preferential sensitivity to DNaseI is also observed for the genes encoding the fetal forms of globin, which are not transcribed in adult sheep cells, and for the adult globin genes in mature (14-day) chicken erythrocytes, following the switching off of transcription.

As well as being detectable after transcription has ceased permanently, such increased sensitivity can also be detected in genes about to become active, prior to the onset of transcription. As discussed previously (see Section 4.2), the transcription of the globin gene in Friend erythroleukemia cells only occurs following treatment of the cells with dimethyl sulfoxide. Increased sensitivity to DNaseI digestion is observed, however, in both the treated cells, which transcribe the gene at high levels and in the unstimulated cells which do not.

The altered, more open, chromatin structure detected by increased DNaseI sensitivity does not therefore reflect the act of transcription itself. Rather, it appears to reflect the ability to be transcribed in a particular tissue or cell type. Hence, in cells that have become committed to a particular lineage expressing particular genes, such commitment will be reflected in an altered chromatin structure which will arise prior to the onset of transcription and will persist after transcription has ceased. The ability of imaginal disk cells in *Drosophila* to maintain their commitment to give rise to a particular cell type in the absence of overt differentiation (see Section 6.2) is likely, therefore, to be due to the genes required in that cell type having already assumed an open chromatin structure. Moreover, a breakdown in this process resulting in a change in the chromatin structure of a particular regulatory gene would result in a change in the commitment of the disk cells, as illustrated in Fig. 6.4. Similarly, the altered chromatin structure of the genes required in cartilage cells would be retained in cells cultured in media not supporting expression of these genes, allowing the restoration of the differentiated cartilage phenotype when the cells are transferred to an appropriate medium (see Section 6.2).

As described in Section 6.4, active or potentially active DNA is still organized in the "beads on a string structure". However, its enhanced DNaseI sensitivity reflects the fact that it does not form the much more tightly packed solenoid structure (Section 6.3) which is formed by the bulk of DNA that is not about to be or being transcribed. Hence, commitment to a particular pattern of gene expression involves a move from the tightly packed solenoid structure to the more open "beads on a string" structure (Fig. 6.12). As the DNA in this structure is more accessible, it is therefore more readily digested by DNaseI.

Although the greater accessibility to DNaseI of active or potentially active DNA simply represents an experimental tool for dissecting its chromatin structure, the alteration of chromatin to a more open structure in committed cells is likely to be a necessary prerequisite for gene expression and hence of major biological significance. Thus, it will allow the *trans*-acting factors, which actually activate transcription of the gene, access to their appropriate DNA target sequences within it. Hence the different structure of potentially active genes can explain why a particular steroid

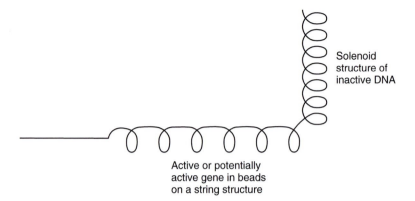

Solenoid
structure of
inactive DNA

Active or potentially
active gene in beads
on a string structure

Figure 6.12

Active or potentially active DNA is in the beads on a string chromatin structure
whereas transcriptionally inactive DNA is in the more tightly packed solenoid
structure.

hormone, such as estrogen, can induce activity of one particular gene in
one tissue and another gene in a different tissue (see Section 6.1). Thus
the ovalbumin gene in the oviduct would be in an open configuration,
allowing induction to occur, whereas the more closed configuration of the
vitellogenin gene would not allow access to the complex of hormone and
receptor, and induction would not occur. The reverse situation would
apply in liver tissue, allowing induction of the vitellogenin and not the
ovalbumin gene.

Therefore, the changes in chromatin structure detected by DNaseI digestion play an important role in establishing the commitment to express
the specific genes characteristic of a particular lineage. As well as this
change in chromatin structure a variety of biochemical changes have been
observed in DNA or its associated proteins in areas which are about to be
or are being transcribed. These biochemical changes in turn lead to the
alterations in chromatin structure, detected by DNaseI digestion and they
are discussed in the next two sections.

6.5 Alterations in DNA methylation in active or potentially active genes

Nature of DNA methylation

Although DNA consists of the four bases adenine, guanine, cytosine and
thymine, it has been known for many years that these bases can exist in
modified forms bearing additional methyl groups. The most common of
these in eukaryotic DNA is 5-methyl cytosine (Fig. 6.13), between 2 and
7% of the cytosine in mammalian DNA being modified in this way (for
reviews see Bestor and Coxon, 1993; Colot and Rossignol, 1999; Jones and
Takai, 2001).

Approximately 90% of this methylated C occurs in the dinucleotide,
CG, where the methylated C is followed on its 3′ side by a G residue.
Conveniently, this sequence forms part of the recognition sequence

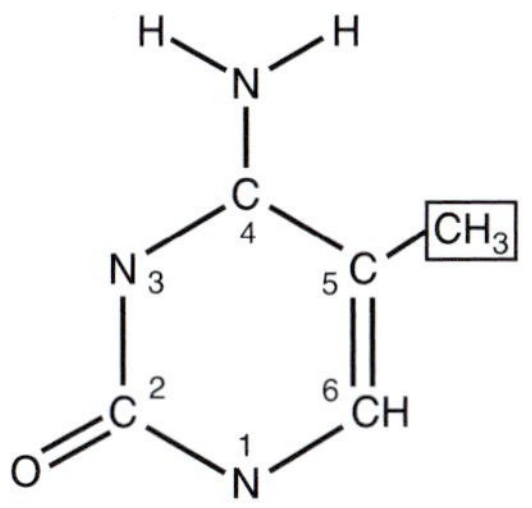

Figure 6.13

Structure of 5-methyl cytosine.

(CCGG) for two restriction enzymes, *Msp*I and *Hpa*II, which differ in their ability to cut at this sequence when the central C is methylated. Thus *Msp*I will cut whether or not the C is methylated and *Hpa*II will only do so if the C is unmethylated. This characteristic allows the use of these enzymes to probe the methylation state of the fraction of CG dinucleotides that is within cleavage sites for these enzymes. Hence, if DNA is digested with either *Hpa*II or *Msp*I, both enzymes will give the same pattern of bands only if all the C residues within the recognition sites are unmethylated. In contrast, if any sites are methylated, larger bands will be obtained in the *Hpa*II digest, reflecting the failure of the enzyme to cut at particular sites (Fig. 6.14). If this procedure is used in conjunction with Southern blot hybridization using a probe derived from a particular gene, the methylation pattern of the *Hpa*II/*Msp*I sites within the gene can be determined.

When this is done it is found that although some CG sites are always unmethylated and others are always methylated, a number of sites exhibit a tissue-specific methylation pattern, being methylated in some tissues but not in others. Such sites within a particular gene are unmethy-

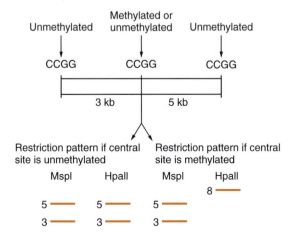

Figure 6.14

Detection of differences in DNA methylation between different tissues, using the restriction enzymes *Msp*I and *Hpa*II.

lated in tissues where the gene is active or potentially active and methylated in other tissues (for a review see Ng and Bird, 1999). For example, a particular site within the chicken globin gene is methylated in a wide variety of tissues and is therefore not digested with *Hpa*I, but is unmethylated and therefore susceptible to digestion in DNA prepared from erythrocytes (Fig. 6.15). Similarly, the tyrosine amino-transferase gene, which is expressed only in the liver (Section 6.1), is undermethylated in this tissue when compared with other tissues where it is not expressed. The possibility suggested by this finding that the *trans*-acting factors, which regulate the expression of this gene and are detectable in all tissues, may be inhibited from binding to the gene in tissues where it is methylated, is confirmed by the observation that artificial methylation of the gene prevents the binding of at least one of these factors. Interestingly, clusters of unmethylated CG sites, known as CpG islands, are also observed in constitutively expressed genes, although such sites remain unmethylated in all tissues consistent with the activity of the gene in all cell types.

In the case of tissue-specific genes which exhibit changes in methylation pattern, undermethylation, like DNaseI sensitivity, is observed prior to the onset of transcription and persists after its cessation. For example, the undermethylation of the chicken globin gene persists in mature erythrocytes, where the gene is not being transcribed but is still sensitive to DNaseI digestion. Most interestingly, the region in which unmethylated C residues are found correlates with that exhibiting heightened DNaseI sensitivity and is also depleted of histone H1. As with DNaseI sensitivity, therefore, undermethylation is a consequence of commitment to a particular pattern of gene expression, and is associated with the change in chromatin structure observed in active or potentially active genes.

Evidence that DNA methylation plays a role in regulating chromatin structure

Although undermethylation is therefore associated with the more open chromatin structure of active DNA, this association does not prove that it has a key role in creating this structure. Two lines of evidence suggest, however, that this is indeed the case and that alterations in DNA methylation can affect chromatin structure and gene expression.

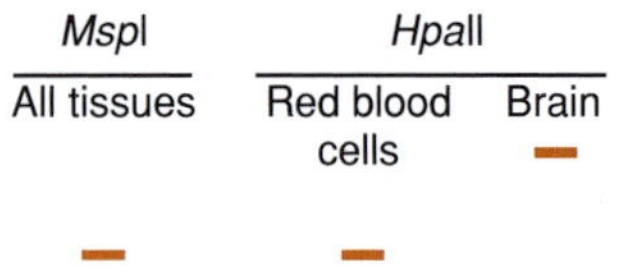

Figure 6.15

Tissue-specific methylation of *Msp*I/*Hpa*II sites in the chicken globin gene results in the methylation-sensitive enzyme, *Hpa*II, producing a band in red blood cell DNA that is identical to that produced by the methylation-insensitive *Msp*I, but producing a larger band in brain DNA. Redrawn from Weintraub *et al.*, *Cell* **24**, 333–344 (1981), by permission of Professor H. Weintraub and Cell Press.

Introduction of methylated DNA into cells

A number of experiments have shown that if DNA containing 5-methyl cytosine is introduced into cells it is not expressed, whereas the same DNA which has been demethylated is expressed. These experiments have been carried out using both eukaryotic viruses and cellular genes, such as those encoding β- and γ-globin, cloned into plasmid vectors. In these experiments the methylated DNA adopts a DNaseI-insensitive structure typical of inactive genes whereas unmethylated DNA adopts the DNaseI-sensitive structure typical of active genes (Fig. 6.16), providing direct evidence for the role of methylation differences in regulating the generation of different forms of chromatin structure.

Effect of artificially induced demethylation

If methylation differences play a crucial role in the regulation of differentiation, it should be possible to change gene expression by demethylating DNA. This has been achieved in a number of cases by treating cells with the cytidine analog, 5-azacytidine, which is incorporated into DNA but cannot be methylated, having a nitrogen atom instead of a carbon atom at position 5 of the pyrimidine ring. In the most dramatic of these cases, treatment of an undifferentiated fibroblast cell line with this compound results in the activation of key regulatory loci and the cells differentiate into multinucleate, twitching, striated muscle cells (Constantinides *et al.*, 1977) (see also Section 8.2). In other cases, although not actually producing altered gene expression, demethylation may facilitate it. Thus, if HeLa cells are treated with 5-azacytidine no dramatic effects are observed. If such cells are fused with muscle cells, however, muscle-specific genes are switched on in the treated HeLa cells, a phenomenon which is not observed when untreated HeLa cells are fused with mouse muscle cells (Chiu and Blau, 1985). Hence, treatment of the HeLa cells has altered their muscle-specific genes in such a way as to allow them to respond to *trans*-acting factors present in the mouse muscle cells. This type of regulatory process is exactly what would be predicted if methylation has a role in the alteration of chromatin structure and thereby in facilitating interactions with *trans*-acting regulatory factors.

Thus, the available evidence indicates that DNA methylation plays a central role in the regulation of gene expression, at least in mammals. This

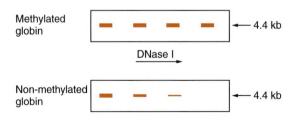

Figure 6.16

Unmethylated DNA introduced into cells adopts an open DNaseI-sensitive configuration, whereas the same DNA when methylated and then introduced into cells adopts a more tightly packed DNaseI-insensitive form. Redrawn from Keshet *et al.*, *Cell* **44**, 535–543 (1986), by permission of Professor H. Cedar and Cell Press.

conclusion is reinforced by the finding that several different DNA methyl-transferase enzymes which methylate DNA are essential for normal embryonic development, with the inactivation of any one of these result-ing in mice which are non-viable (see Okano *et al.*, 1999). Hence in mammals DNA methylation is essential for normal embryonic develop-ment. Moreover, although originally thought to be absent in *Drosophila*, 5-methyl cytosine has also been detected in this organism and shown to be regulated during embryonic development (Lyko *et al.*, 2000) further supporting a key role for this process.

Mechanism by which DNA methylation affects chromatin structure

The evidence suggesting an important role for DNA methylation in modu-lating the difference in chromatin structure between inactive and active genes raises the question of how this is achieved. It is clearly possible for the methylation differences between inactive or potentially active genes to be recognized by proteins since the proteinaceous enzyme *Hpa* II digests only unmethylated DNA (see above). Evidently, undermethylation could promote the binding of proteins which produce a more open chromatin structure (Fig. 6.17a). Alternatively, an inhibitory protein could bind to methylated DNA and promote a closed chromatin structure (Fig. 6.17b).

Although both these mechanisms may be used, current evidence supports the second mechanism. Thus, a number of different proteins able to bind specifically to methyl-CpG have been characterized and shown to

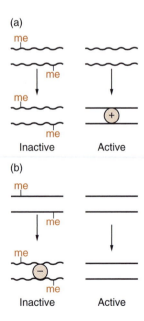

Figure 6.17

The transition from an inactive (wavy line) to an active state (straight line) of the DNA could take place via an activating protein which binds specifically to unmethylated DNA thereby activating it (panel a) or via an inhibiting protein which binds specifically to methylated DNA thereby repressing it (panel b).

play critical roles in the regulation of gene expression (for a review see Hendrich and Tweedie, 2003). For example, a specific protein MeCP2 has been shown to bind directly to methylated CG but not to unmethylated CG and its binding results in the production of a closed chromatin structure and transcriptional repression (for a review see Bird and Wolffe, 1999).

Moreover, such binding has been shown to regulate the expression of a specific gene, encoding the neurotrophic factor BDNF in neuronal cells. Prior to induction of BDNF gene expression, the BDNF gene is methylated and MeCP2 is bound to it, repressing gene expression. Activation of the gene is achieved by two mechanisms involving MeCP2. Thus, the BDNF gene becomes demethylated which decreases binding by MeCP2. Moreover, MeCP2 itself becomes modified by phosphorylation resulting in it binding less well to DNA. These two processes together result in displacement of MeCP2 from the DNA, opening the chromatin structure and allowing transcription to occur (Fig. 6.18) (for a review see Klose and Bird, 2003).

Hence, MeCP2 is able to recognize methyl CG dinucleotides and induce the formation of a closed chromatin structure, thereby regulating gene expression. The importance of this process is demonstrated by the finding that humans with mutant forms of MeCP2, which are unable to recognize meCG, exhibit Rett syndrome which is a severe developmental disorder leading to mental retardation (for a review see Tucker, 2001). Mice lacking another methyl CG binding protein, MBD1 also show defects in the nervous system, indicating that several methyl CG binding proteins

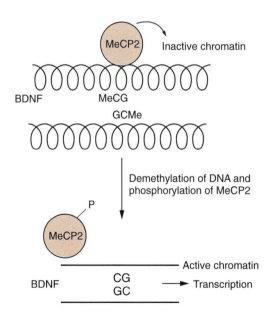

Figure 6.18

Binding of MeCP2 to methylated CG residues produces an inactive chromatin structure and prevents transcription of the BDNF gene. Activation of the gene is achieved by demethylation of the gene and phosphorylation of MeCP2, both of which reduce its ability to bind to the BDNF gene.

are required for proper functioning of the nervous system (Zhao *et al.*, 2003). Interestingly, binding of MeCP2 to meCG results in the recruitment of a multi-protein complex which includes a histone deacetylase (for a review see Bird and Wolffe, 1999) that can induce the removal of acetyl groups from histones (Fig. 6.19). As deacetylation of histones is known to be associated with transcriptionally inactive DNA, this illustrates the close link between modification of DNA and modification of histones in determining the structure of chromatin. This link is further strengthened by the observations that mice with a defective MeCP2 protein demonstrate enhanced acetylation of histone H3 (Shahbazian *et al.*, 2002). Similarly, both MeCP2 and MBD1 associate with histone methylase enzymes resulting in enhanced methylation of histone H3, indicating that DNA methylation can regulate more than one histone modification (Fuks *et al.*, 2003; Sarraf and Stancheva, 2004). Therefore, DNA methylation may exert its effects on chromatin structure by altering histone modification. The various histone modifications associated with changes in chromatin structure are discussed in the next section.

6.6   Modification of histones in the chromatin of active or potentially active genes

Given the essential role of histones in chromatin structure, it is possible that potentially active chromatin might be marked in some way by alteration of the histones within it. Interestingly, several variant forms of specific histones encoded by distinct genes have been described. Two of these, the histone H2 variant, H2A.Z and the histone H3 variant, H3.3, have been shown to be preferentially localized in active genes (for reviews see Henikoff *et al.*, 2004; van Leeuwen and Gottschling, 2003; Workman and Abmayr, 2004). Similarly, the histone H1 variant, H1b, cooperates with the Msx 1 transcription factor to inhibit transcription of muscle-specific genes (Lee *et al.*, 2004). Hence, variant forms of the histones do appear to be involved in the regulation of chromatin structure and gene expression.

However, much more extensive studies have demonstrated a critical role for modified forms of the standard histones in determining whether specific regions of chromatin are active or inactive. A number of such modifications of these proteins (involving, for example, acetylation, ubiquitination, phosphorylation and methylation) have been reported (for

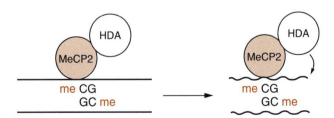

Figure 6.19

Binding of the MeCP2 protein to methyl CG leads to the recruitment of other proteins including a histone deacetylase (see Section 6.6) and leads to the chromatin forming a tightly packed inactive configuration (wavy line).

reviews see Goll and Bestor, 2002; Felsenfeld and Groudine, 2003; Jaskelioff and Peterson, 2003; Khorasanizadeh, 2004). These will be discussed in turn.

Acetylation

In the process of acetylation, the free amino group on specific lysine residues at the N-terminus of the histone molecule are modified by one of the hydrogen atoms in the amino group being substituted by an acetyl group ($COCH_3$). This modification reduces the net positive charge on the histone molecule and occurs primarily for histones H3 and H4. Hyperacetylated forms of these histones, containing several such acetyl groups, have been shown to be localized preferentially in active genes exhibiting DNaseI sensitivity whilst under-acetylation of the histones is characteristic of transcriptionally inactive regions (see, for example, Schubeler et al., 2004). Furthermore, treatment of cells with sodium butyrate, which inhibits a cellular deacetylase activity and hence increases histone acetylation, has been shown to result in DNaseI sensitivity of some regions of chromatin and to activate the expression of some previously silent cellular genes. Hence, as with DNA methylation, there is direct evidence that hyperacetylation of histones plays a role in opening the chromatin structure of active or potentially active genes.

Further evidence in favor of a role for histone acetylation in the regulation of gene expression has been obtained by the identification of several histone acetyl transferase (HAT) enzymes capable of adding acetyl groups to histone molecules (for reviews see Brown et al., 2000; Carrozza, et al., 2003). Most importantly, it has been shown that the CBP transcriptional co-activator which plays a key role in the activation of genes in response to cyclic AMP and other stimuli (see Section 8.4) has histone acetyl transferase activity directly linking this enzymatic activity with the ability to stimulate transcription. Similarly the $TAF_{II}250$ subunit of the TFIID transcription factor which is critical for basal transcription of a wide range of genes (see Section 3.2) has been shown to have histone acetyl transferase activity, as has the ATF-2 transcription factor (Kawasaki et al., 2000).

The opposite effect has been observed in the case of the nuclear receptor co-repressor which mediates the inhibiting effect of the thyroid hormone receptor on transcription when thyroid hormone is absent (see Section 8.3). Thus this factor has been shown to associate with the Sin3/RPD3 protein complex which has the ability to deacetylate histones (for reviews see Ayer, 1999; Ahringer, 2000; Ng and Bird, 2000). The inhibiting effect of the nuclear receptor co-repressor is therefore likely to involve the deacetylation of histones, thereby producing a more tightly packed chromatin structure incompatible with transcription. Hence activating factors can direct the acetylation of histones, thereby opening up the chromatin (Fig. 6.20a) whilst inhibiting factors can direct histone deacetylation thereby directing a more tightly packed chromatin structure (Fig. 6.20b).

These findings therefore link the study of transcription factors (see Chapter 8) with studies on chromatin structure and suggest that the regulation of histone acetylation/deacetylation and thereby of chromatin structure by specific factors plays a key role in the regulation of gene

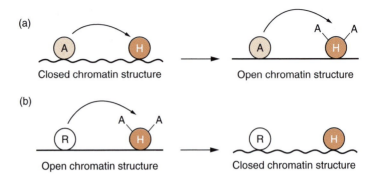

Figure 6.20

(a) An activating molecule (A) can direct the acetylation of histones (H) thereby resulting in a change in chromatin structure from a closed (wavy line) to an open (solid line) configuration. (b) An inhibiting molecule can direct the deacetylation of histones thereby having the opposite effect on chromatin structure.

expression. Similarly, as noted above (Section 6.5) a protein which binds specifically to methyl CG dinucleotides can recruit a histone deacetylase activity, thereby linking histone deacetylation to the repressive effect of DNA methylation.

Interestingly, as well as activators recruiting acetylases and repressors recruiting deacetylases, there is evidence that the acetylase/deacetylase enzymes can themselves be regulated (for a review see Stewart and Crabtree, 2000). This is seen in muscle differentiation where in myoblasts the transcriptional activator MEF2 is associated with histone deacetylases. When differentiation from myoblasts to mature myotubes occurs, the histone deacetylase is induced to move to the cytoplasm, thereby freeing MEF2 which can now activate transcription (Fig. 6.21).

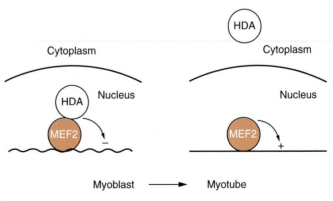

Figure 6.21

In myoblasts, the MEF2 transcriptional activator is associated with a histone deacetylase (HDA) and transcription of myotube-specific genes is repressed. When differentiation is induced, the histone deacetylase moves into the cytoplasm, allowing MEF2 to activate the transcription of myotube-specific genes.

As discussed in Section 6.3, the amino terminal ends of the histone proteins, which are modified by acetylation, project beyond the nucleosome core and could therefore potentially interact with the amino terminal ends of histones in adjacent nucleosomes or with other non-histone proteins. The effects of histone acetylation on chromatin structure may therefore operate in one of two possible ways (Fig. 6.22). Firstly, such changes might act by affecting the association of the histones with each other. In turn this would result in improved access to the DNA for factors which can stimulate transcription (Fig. 6.22a). Some evidence for this model has been provided in the case of histone acetylation where the binding to DNA of the *Xenopus* transcription factor TFIIIA was greatly enhanced when the nucleosomes contained hyperacetylated histone H3 and H4 compared with when these proteins were not acetylated.

An alternative possibility is that these modifications affect the protein–protein interaction of the histones with other regulatory molecules. This would parallel the role proposed for methylation differences in affecting the binding of positively or negatively acting factors to the DNA (see Fig. 6.17; Section 6.5). Thus, for example, acetylated histones might be recognized by a positively acting molecule leading to the destabilization of the solenoid structure and transcriptional activation. Similarly, acetylation could disrupt an association with an inhibitory molecule involved in maintaining the closed chromatin structure (Fig. 6.22b).

In agreement with this model of histone–regulatory protein interaction, it has been shown that several activating factors such as Bdf1 contain a region known as the bromodomain that binds to histones with much greater affinity when specific lysines in the histones are acetylated (Kurdistani *et al.*, 2004; for reviews see Khorasanizadeh, 2004; Loyola and

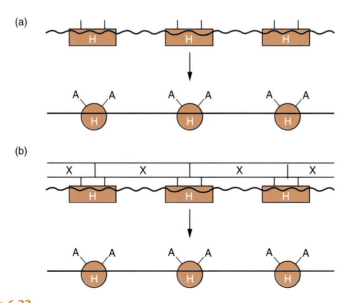

Figure 6.22

Acetylation (A) of histones (H) may activate either by directly altering the conformation of the histones themselves producing a more open chromatin structure (panel a) or indirectly by disrupting their association with an inhibiting molecule (X) (panel b).

Almouzni, 2004). Binding of these bromodomain molecules to acetylated histones therefore results in a more open chromatin structure and transcriptional activation (Fig. 6.23).

Ubiquitination and sumoylation

Ubiquitin is a small protein of only 76 amino acids, which forms a conjugate with histone H2A or H2B in which the C-terminal carboxyl group of ubiquitin is joined to the free amino group on an internal lysine residue in the histone, to form a branched molecule (Fig. 6.24; see reviews by Freiman and Tjian, 2003; Zhang, 2003). This modification reduces the net positive charge on the histone molecule, both by neutralizing the charged amino group on lysine and by introducing a number of negatively charged amino acids present in the ubiquitin molecule itself. Only a small minority of the H2A in a cell (about 5–10%) exists in this form and such ubiquitination is associated with the repression of gene expression. Thus, it has been shown that ubiquitination of histone H2A is involved in gene repression by the polycomb proteins, which promote an inactive chromatin structure (Wang *et al.*, 2004) (see Sections 7.4 and 8.2 for further discussion of polycomb proteins).

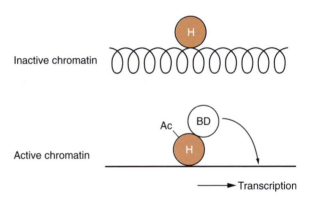

Figure 6.23

Acetylation (Ac) of histones (H) results in binding of bromodomain-containing activator proteins (BD) resulting in an open chromatin structure and subsequent transcription.

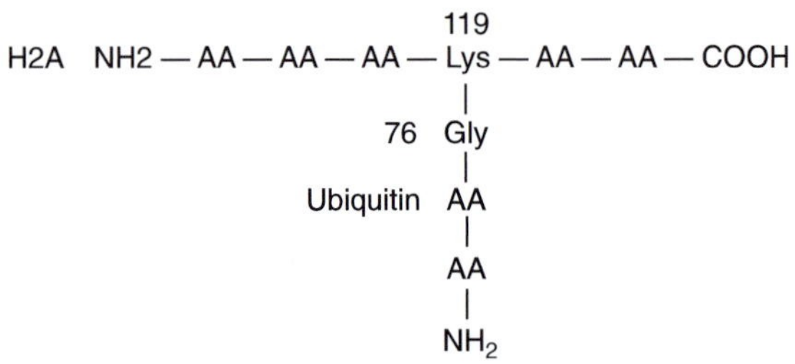

$$\text{H2A} \quad \text{NH2} - \text{AA} - \text{AA} - \text{AA} - \overset{119}{\text{Lys}} - \text{AA} - \text{AA} - \text{COOH}$$
$$|$$
$$76 \;\; \text{Gly}$$
$$|$$
$$\text{Ubiquitin} \;\; \text{AA}$$
$$|$$
$$\text{AA}$$
$$|$$
$$\text{NH}_2$$

Figure 6.24

Linkage of ubiquitin to histone H2A. The carboxyl terminal amino acid (76) of ubiquitin links to the lysine at position 119 of histone H2A. AA indicates the amino acid backbone of the molecules.

Interestingly, ubiquitination of H2B has the opposite effect to the ubiquitination of H2A since it stimulates rather than inhibits gene expression (Fig. 6.25). This is because ubiquitination of H2B promotes the methylation of histone H3 on lysines at positions 4 and 79 (Briggs *et al.*, 2002; Sun and Allis, 2002). As such, methylation promotes an open chromatin structure (see Section 6.6); this provides an explanation for the stimulating effect of H2B ubiquitination on gene expression. Moreover, it indicates that modification of one type of histone can modulate the modification of another type of histone molecule.

As well as being modified by addition of ubiquitin, lysine residues on histones can also be modified by addition of SUMO (small ubiquitin-related modifier) which, as its name suggests, is related to ubiquitin. Addition of SUMO to histones is associated with transcriptional repression (Shilo and Eisenman, 2003; for a review see Nathan *et al.*, 2003). Indeed, it has been suggested that when a gene is about to be repressed, histones are sumoylated and this results in the recruitment of enzymes which deacetylate histones, thereby resulting in a more closed chromatin structure (Fig. 6.26). (For further discussion of the effects of histone ubiquitination and sumoylation see Gill, 2004.)

Phosphorylation

Unlike acetylation or ubiquitination-sumoylation, phosphorylation targets serine amino acids in the histone molecule (for a review see Nowak and Corces, 2004). It has been shown that when cells are exposed to heat shock, phosphorylated histone H3 is concentrated at the heat-shock gene loci which are being actively transcribed and is depleted from other loci which are silenced following heat shock.

Similarly, when cells are stimulated with growth factors, histone H3 is phosphorylated on the serine residue at position 10 in the amino terminus of the protein. These phosphorylated histones are localized to genes which become transcriptionally active following growth factor stimulation such as the c-*fos* and c-*myc* genes (see Section 9.4) (Fig. 6.27). Such growth factor-induced phosphorylation of histone H3 does not occur in human patients with Coffin–Lowry syndrome who lack the phosphorylation enzyme, known as Rsk-2, which is responsible for phosphorylating histone H3 following growth factor stimulation. These patients suffer from a number of defects including mental retardation as well as facial and other abnormalities. Moreover, they show a lack of gene activation in

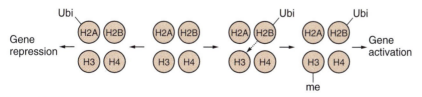

Figure 6.25

Opposite effects of ubiquitination of histones H2A and H2B. Ubiquitination of H2A represses gene expression whilst ubiquitination of H2B promotes methylation of histone H3 and therefore stimulates gene expression.

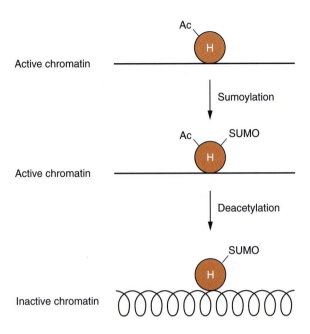

Figure 6.26

Modification of histones (H) by sumoylation stimulates their deacetylation promoting formation of inactive chromatin.

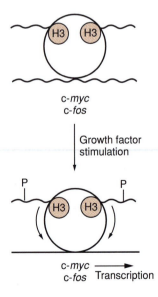

Figure 6.27

Following growth factor stimulation, histone H3 becomes phosphorylated within the nucleosomes bound to the c-*myc* and c-*fos* genes. This is associated with their moving from an inactive closed chromatin structure (wavy line) to a more open structure (solid line) allowing transcription to occur in response to the growth factor.

response to growth factor stimulation indicating the importance of Rsk-2-mediated phosphorylation of histone H3 in this process.

Interestingly, it appears that phosphorylation of serine at position 10 in histone H3 enhances the ability of histone acetyl transferases to acetylate the lysine at position 14. This has led to the idea of a "histone code" in which a particular pattern of histone modifications is characteristic of active or potentially active genes (for reviews see Paro, 2000; Jenuwein and Allis, 2001) (Fig. 6.28).

Methylation

Histones may also be modified by addition of a methyl (CH_3) group to specific lysine residues. This modification occurs particularly for histones H3 and H4 and, like the other modifications, targets amino acids at the N-terminus of the protein such as the lysine at position 9 of histone H3. Unlike the other modifications, however, methylation of the histones at particular positions such as lysine 9 or lysine 27 of histone H3 is associated with transcriptionally inactive DNA, whereas acetylation and phosphorylation are associated with active or potentially active DNA (for reviews see Zhang and Reinberg, 2001; Sims et al., 2003).

Moreover, there is evidence for an antagonistic interaction between methylation on position 9 and phosphorylation/acetylation at positions 10 and 14, all of which are found in a small region at the amino-terminus of the histone protein. Thus, for example, it appears that dephosphorylation and deacetylation of positions 10 and 14 promote methylation of the histone molecule on position 9 whilst methylation of the histone on position 9 inhibits acetylation and phosphorylation (Rea et al., 2000). Hence, the histone code described above can be extended to include methylation with the histones associated with active or potentially active DNA being acetylated on lysine 14 and phosphorylated on serine 10 whilst those associated with inactive DNA are methylated on lysine 9 (Fig. 6.29). Indeed, it has been suggested that phosphorylation on serine 10 and methylation on lysine 9 of histone H3 may act as a binary

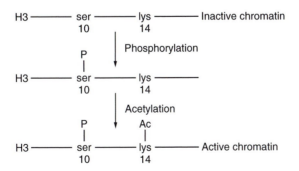

Figure 6.28

Phosphorylation of histone H3 on serine residue 10 promotes acetylation on the adjacent lysine 14 resulting in a pattern of phosphorylation/acetylation in active chromatin which is distinct from that of inactive chromatin.

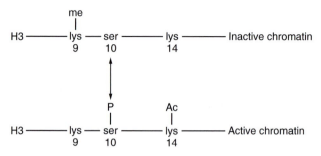

Figure 6.29

The histone H3 in inactive chromatin is modified by methylation of the lysine residue at position 9 in the amino terminus of the protein. The transition to active chromatin is associated with demethylation of this amino acid which is accompanied by phosphorylation of the adjacent serine at position 10 and acetylation of the lysine at position 14.

switch promoting gene activation or repression, respectively (for discussion see Fischle *et al.*, 2003).

It should be noted, however, that not all histone methylation is associated with inactive DNA. Thus, in contrast to methylation on position 9, the methylation of histone H4 on the arginine at position 4 promotes a more open chromatin structure and facilitates transcriptional activation by nuclear hormone receptors (Wang *et al.*, 2001) and a similar effect has also been demonstrated for lysine 4 of histone H3 (the equivalent amino acid to arginine 4 of histone H4) and ecdysone-dependent gene activation in *Drosophila* (Sedkov *et al.*, 2003). Moreover, in the region of DNA containing the mating type locus in yeast, it has been shown that transcriptionally active DNA contains histone H3 which is methylated on lysine 4, whereas adjacent transcriptionally inactive regions have histone H3 which is methylated on lysine 9 (Noma *et al.*, 2001). A similar effect has also been demonstrated in both the human (Liang *et al.*, 2004) and *Drosophila* (Schubeler *et al.*, 2004) genomes. Hence, the "histone code" is highly complex with both positive and negative effects of methylation at different positions as well as positive effects of acetylation and phosphorylation. The various modifications observed at the N-terminus of histone H3 are illustrated in Fig. 6.30.

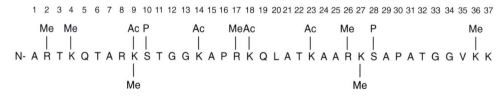

Figure 6.30

The first 37 amino acids (in the one-letter amino acid code) at the N-terminus of histone H3. The figure shows the lysine (K) and arginine (R) residues modified by methylation (Me) together with the lysine (K) residues modified by acetylation (Ac) and the serine (S) residues modified by phosphorylation (P). Note that modifications which produce transcriptional activation are shown above the line and those producing repression are shown below the line.

Evidently, this complex pattern of modifications can be readily fitted into the model shown for acetylation in Fig. 6.22. Thus, the total pattern of methylation, phosphorylation and acetylation is likely to modify chromatin structure either directly by affecting the interaction of the histones with each other to form the chromatin structure or indirectly by affecting their interaction with other regulatory proteins. In agreement with the latter possibility, it has been shown that methylation of histone H3 on lysine 9 allows it to be recognized by a specific protein HP1 which cannot recognize the unmethylated histone (Lachner *et al.*, 2001; Bannister *et al.*, 2001). HP1 is known to be able to organize the chromatin into a closed inactive structure and contains a region known as a chromodomain which is found in many proteins that inhibit transcription. Hence, recognition of histones methylated on lysine 9 by chromodomain containing proteins may be a key event in organizing the tightly packed structure of inactive chromatin (Fig. 6.31).

Interestingly, this process of histone methylation and recruitment of HP1 can be controlled by the small inhibitory RNAs which were discussed in Section 5.7. Thus, although such RNAs have predominantly been shown to inhibit gene expression at the post-transcriptional level, a recent study in *Drosophila* has shown that a small inhibitory RNA can stimulate histone H3 methylation on lysine 9 and recruitment of HP1, thereby inhibiting gene transcription by producing an inactive chromatin structure (Pal-Bhadra *et al.*, 2004). Other small inhibitory RNAs have been shown to stimulate both histone methylation and *de novo* CG methylation in species as diverse as plants (Zilberman *et al.*, 2003) and humans (Kawasaki and Taira, 2004) (for a review see Lippman and Martienssen, 2004). This further strengthens the link between the inhibitory processes of histone modification and DNA methylation which was discussed in Section 6.5. Moreover, it indicates that small inhibitory RNAs may act in different ways and in a wide range of species to inhibit transcription by

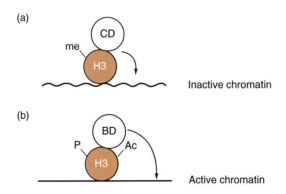

Figure 6.31

(a) Methylation of histone H3 promotes binding of chromodomain-containing proteins (CD) which can induce the closed chromatin structure characteristic of inactive DNA (wavy line). (b) In contrast, acetylation and phosphorylation promote binding of bromodomain-containing proteins (BD) which induce the more open chromatin structure characteristic of active or potentially active DNA.

stimulating the organization of an inactive chromatin structure, as well as acting at different post-transcriptional levels (see Section 5.7).

The recognition of methylated histones by inhibitory chromodomain-containing proteins contrasts with the role of acetylation in promoting recognition of the histones by proteins containing a bromodomain which was described in Section 6.6. This recognition of histones by bromodomain-containing proteins is also promoted by histone phosphorylation Thus, chromodomain-containing proteins may recognize histones methylated on specific residues and promote the tight packing of chromatin whilst recognition of phosphorylated/acetylated histones by bromodomain-containing proteins promotes the more open chromatin structure characteristic of active or potentially active DNA (Fig. 6.31).

It is likely, therefore, that histone modifications can affect chromatin structure either directly by interactions with one another or indirectly via interactions with other proteins. Regardless of the precise mechanism by which they act, however, it is clear that such modifications do play an important role in the changes in chromatin structure which are necessary for gene activation to occur.

This linkage between chromatin structure and histone modification is not confined to opening up the chromatin to allow transcriptional initiation. Thus, there is evidence that histone methylation is linked to the process of transcriptional elongation which as discussed in Sections 3.2, 3.3 and 4.3 is dependent on the phosphorylation of the C-terminal domain (CTD) of RNA polymerase II. Interestingly, the initial phosphorylation of the CTD on serine 5 results in the recruitment of a protein complex, known as Set 1, which methylates histone H3 on lysine 4. Subsequently, the CTD is phosphorylated on serine 2 and this results in the dissociation of Set 1 and the recruitment of a further complex, known as Set 2. This Set 2 complex remains associated with the RNA polymerase II as it commences transcriptional elongation and methylates histone H3 on lysine 36 in all the nucleosomes encountered by the elongating polymerase. Hence, the methylation of histone H3 is closely linked to transcriptional elongation, as well as transcriptional initiation (for reviews see Gerber and Shilatifard, 2003; Hampsey and Reinberg, 2003; Sims *et al.*, 2004) (Fig. 6.32).

6.7 Changes in chromatin structure in the regulatory regions of active or potentially active genes

DNaseI hypersensitive sites

So far in this chapter we have seen that the region of chromatin containing an active or potentially active gene has a number of distinguishing features, including undermethylation of C residues, histone modifications, and increased sensitivity to digestion with DNaseI. Such changes can extend over the entire region of the gene and some flanking sequences and, in the case of DNaseI sensitivity, result in an approximately tenfold increase in the rate at which active or potentially active genes are digested.

Following the discovery of such increased DNaseI sensitivity, many investigators studied whether, within the region of increased sensitivity, there might be particular sites which were even more sensitive to cutting with the enzyme and which would therefore be cut even before the bulk

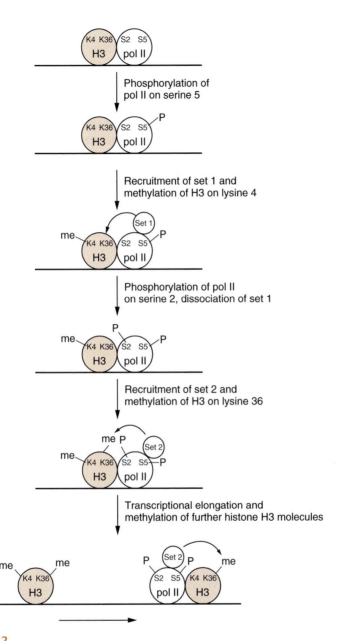

Figure 6.32

Interaction of histone methylation and transcriptional elongation. Following phosphorylation of RNA polymerase II on serine 5 of its C-terminal domain (CTD), the Set 1 complex is recruited and methylates histone H3 on lysine 4. Subsequently, following phosphorylation of polymerase II on serine 2 of the CTD, Set 1 dissociates and Set 2 is recruited. Set 2 then methylates histone H3 on lysine 36. As Set 2 remains associated with the polymerase as transcriptional elongation occurs, it will methylate histone H3 in all the nucleosomes encountered by the elongating polymerase.

of active DNA was digested. The technique used to look for such sites is based on that used to look at the overall DNaseI sensitivity of a particular region of DNA (which was described in Methods Box 6.1). Chromatin is digested with DNaseI and a restriction enzyme and then subjected to a Southern blotting procedure using a probe derived from the gene of interest. As we have seen previously, the overall sensitivity of the gene can be monitored by observing how rapidly the specific restriction enzyme fragment derived from the gene disappears with increasing amounts of the enzyme (see Fig. 6.11).

To search for hypersensitive sites, however, much lower concentrations of the enzyme are used and the appearance of discrete digested fragments derived from the gene is monitored (Fig. 6.33; see Methods Box 6.2 and compare with Methods Box 6.1). Such specific fragments have at one end the cutting site for the restriction enzyme used and, at the other, a site at which DNaseI has cut, producing a defined fragment. Since the position at which the restriction enzyme cuts in the gene is known, the position of the hypersensitive site can be mapped simply by determining the size of the fragment produced.

METHODS BOX 6.2

Detecting DNaseI hypersensitive sites (Fig. 6.33)

- Isolate chromatin (DNA and associated histones and other proteins).
- Digest with very small amounts of DNaseI.
- Purify partially digested DNA by removing protein.
- Digest with restriction enzyme and carry out Southern blotting with probe for gene of interest (see Methods Box 2.1).
- Monitor appearance of specific smaller band due to presence of DNaseI hypersensitive site within the DNA being tested.

Using this procedure a very wide variety of genes have been shown to contain such hypersensitive sites exhibiting a sensitivity to DNaseI digestion tenfold above that of the remainder of an active gene and therefore about one hundredfold above that seen in inactive DNA (for a review see Gross and Garrard, 1988). A representative list of cases in which such sites have been detected is given in Table 6.2. As with the increased sensitivity of the gene itself, many hypersensitive sites appear only in tissues where the gene is active. Thus, the increased sensitivity of globin DNA in erythrocytes to digestion is paralleled by the presence of hypersensitive sites within the gene in erythrocyte chromatin but not in that of other tissues. Similarly, the ovalbumin gene in hormonally treated chick oviduct also exhibits hypersensitive sites that are not found in other tissues, including erythrocytes (Fig. 6.34).

As with undermethylation and the sensitivity of the entire gene to digestion, DNaseI hypersensitive sites appear to be related to the potential for gene expression rather than always being associated with the act of transcription itself. Thus the hypersensitive sites near the *Drosophila*

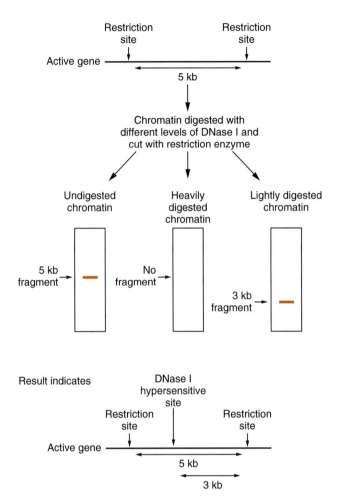

Figure 6.33

Detection of DNaseI hypersensitive sites in active genes by mild digestion of chromatin to produce a digestion product with a restriction site at one end and a DNaseI hypersensitive site at the other (right-hand panel). More extensive digestion will result in the disappearance of the band (central panel) as in the experiment illustrated in Fig. 6.11.

heat-shock genes are present in the chromatin of embryonic cells prior to any heat-induced transcription of these genes and one of the sites in the mouse α-fetoprotein gene persists in the chromatin of adult liver after the transcription of the gene (which is confined to the fetal liver) has ceased.

Hence, as with DNA methylation and generally increased sensitivity to DNaseI, the appearance of hypersensitive sites appears to be involved in gene regulation. This idea is reinforced by the location of the hypersensitive sites, which can be precisely mapped as described above. Many sites are located at the 5′ end of the genes, in positions corresponding to DNA sequences that are known to be important in regulating transcription. For example, a site present at the 5′ end of the steroid-inducible tyrosine amino-transferase gene is localized within the DNA sequence that is

Table 6.2 Examples of genes containing DNaseI hypersensitive sites

(a) Tissue-specific genes

Immune system	Immunoglobulin, complement C4
Red blood cells	α-, β- and ε-globin
Liver	α-fetoprotein, serum albumin
Nervous system	Acetylcholine receptor
Pancreas	Preproinsulin, elastase
Connective tissue	Collagen
Pituitary gland	Prolactin
Salivary gland	*Drosophila* glue proteins
Silk gland	Silk moth fibroin

(b) Inducible genes

Steroid hormones	Ovalbumin, vitellogenin, tyrosine aminotransferase
Stress	Heat-shock proteins
Viral infection	β-interferon
Amino acid starvation	Yeast HIS 3 gene
Carbon source	Yeast GAL genes, yeast ADH II gene

(c) Others

Histones, ribosomal RNA, 5S RNA, transfer RNA, cellular oncogenes c-*myc* and c-*ras*, glucose-6-phosphate dehydrogenase, dihydrofolate reductase, cysteine protease, etc.

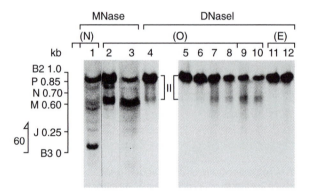

Figure 6.34

Detection of a DNaseI hypersensitive site in the ovalbumin gene in oviduct tissue (O) but not in erythrocytes (E). Track 4 shows the detection of a lower band caused by cleavage at a hypersensitive site when oviduct chromatin is digested with DNaseI. Note the progressive appearance of this band as increasing amounts of DNaseI are used to digest the oviduct chromatin (tracks 5 to 10). No cleavage is observed when similar amounts of DNaseI are used to cut erythrocyte chromatin (tracks 11 and 12). The hypersensitive site in oviduct chromatin is also cleaved, however, with micrococcal nuclease (tracks 2 to 3). Track 1 shows the pattern produced by micrococcal nuclease cleavage of naked DNA (N). Photograph kindly provided by Professor P. Chambon, from Kaye *et al.*, *EMBO Journal* **3**, 1137–1144 (1984), by permission of Oxford University Press.

responsible for the steroid inducibility of the gene (see Section 7.2). Even in cases where hypersensitive sites are located far from the site of transcriptional initiation, they appear to correspond to other regulatory sequences, such as enhancers, which can act over large distances (see Section 7.3).

In the case of the *Drosophila* gene encoding the glue protein Sgs4, the fortuitous existence of a mutant strain of fly has indicated the functional importance of hypersensitive sites. In normal flies this gene contains two hypersensitive sites, 405 and 480 bases upstream of the start of transcription. In the mutant, both sites are removed by a small DNA deletion of 100 base pairs. Despite the fact that this gene still has the start site of transcription and 350 bases of upstream sequences, no transcription occurs (Fig. 6.35), indicating the regulatory importance of the region containing the hypersensitive sites.

It is clear, therefore, that hypersensitive sites represent another marker for active or potentially active chromatin and are a feature likely to be of particular importance in gene regulation, being associated with many DNA sequences that regulate gene expression. It is therefore necessary to consider the nature and significance of these sites.

In some cases, these sites are likely to be formed where DNA is entirely free of nucleosomes and is therefore highly sensitive to DNaseI digestion This is seen, for example, in the case of the DNaseI hypersensitive sites within the enhancer element that regulates transcription of the eukaryotic virus SV40. Thus, when this virus enters cells, its DNA, which is circular and only 5000 bases in size, becomes associated with histones in a typical nucleosomal structure, which can be visualized in the electron microscope as a mini-chromosome. When this is done, however, the region containing the hypersensitive sites remains nucleosome-free and is seen as naked DNA (Fig. 6.36). A similar lack of nucleosomes in the region of hypersensitive sites is also found in the chicken β-globin gene, the 5′ hypersensitive site of this gene being excisable as a 115 bp restriction fragment lacking any associated nucleosome.

Although such cases indicate that hypersensitive sites can be produced by the complete loss of nucleosomes in a particular region, other cases exist where such sites are produced by an alteration in the structure of a nucleosome rather than its complete displacement (Fig. 6.37). It is clear, however, that whether caused by nucleosome displacement or structural alterations, the changes occurring in hypersensitive sites facilitate the entry of transcription factors or the RNA polymerase itself and hence allow the onset of transcription (Fig. 6.37). In view of the critical role played by

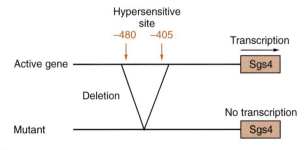

Figure 6.35

Deletion of a region containing the two hypersensitive sites upstream of the *Drosophila sgs4* gene abolishes transcription.

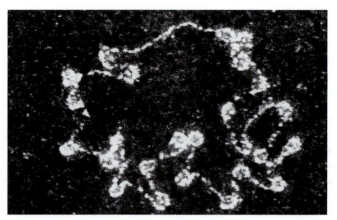

Figure 6.36

Electron micrograph of the SV40 mini-chromosome consisting of DNA and associated histones. Note the region of the enhancer and hypersensitive sites, which
appears as a thin filament of DNA free of associated proteins. Photograph kindly
provided by Professor M. Yaniv, from Saragosti *et al.*, *Cell* **20**, 65–73 (1980), by
permission of Cell Press.

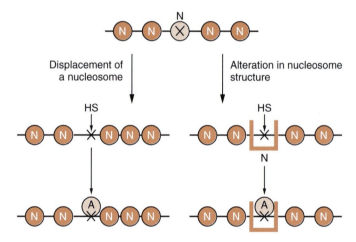

Figure 6.37

A hypersensitive site (HS) can be created either by displacement of a nucleosome or by alteration of its structure. In either case such an alteration allows a
transcriptional activator (A) to bind to its binding site (X) and activate transcription.

the changes in nucleosomes which occur at hypersensitive sites, it is necessary therefore to consider the proteins which produce such alterations in
nucleosome positioning or structure.

Chromatin remodeling by proteins capable of displacing nucleosomes or altering their structure

In a number of cases protein factors have been identified which can
displace nucleosomes or alter their structure, thereby facilitating the subse-

quent binding of molecules which stimulate transcription (for reviews see Kingston and Narlikar, 1999; Jones and Kadonaga, 2000). Thus, in the case of the heat-shock genes whose transcription is induced by elevated temperature (see Section 4.2), hypersensitive sites are produced by the binding of the GAGA protein factor to its upstream DNA binding sites in the gene promoter which results in the displacement of a nucleosome (Fig. 6.38) (for a review see Lehmann, 2004). This binding of the GAGA factor occurs in cells prior to heat treatment and hence hypersensitive sites are present prior to heat treatment. Following heat treatment, a protein factor, known as the heat-shock factor (HSF), binds to this region of DNA and transcription is stimulated (see Section 7.2). In this case, HSF is only capable of binding following heat shock and hence stimulation of transcription only occurs following such treatment (Fig. 6.39a).

In other cases, however, where the necessary transcription factors are present in all tissues, transcription may follow immediately the nucleosome-free region is generated, allowing these factors access. Thus, in the case of glucocorticoid-responsive genes, the critical regulatory event is the binding of the glucocorticoid receptor/steroid complex to a particular DNA sequence (the GRE or glucocorticoid response element), which displaces a nucleosome or alters its structure and generates a DNaseI hypersensitive site. Ubiquitous transcription factors present in all tissues, such as NFI and the TATA box binding factor TBP (see Section 3.2), immediately bind to this region and transcription begins (Fig. 6.39b) (for further details of the glucocorticoid receptor and its mode of action see Sections 7.2, 8.2 and 8.4).

Although these two situations appear different in terms of the time at which the hypersensitive site appears relative to the onset of transcription, they illustrate the basic role of hypersensitive sites, namely the displacement of nucleosomes or alteration of their structure and the generation of a site of access for regulatory proteins.

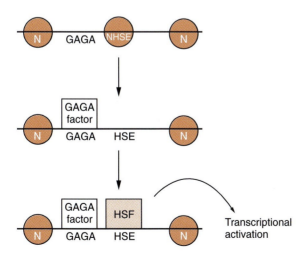

Figure 6.38

The binding of the GAGA factor to its binding site in the heat-shock genes displaces a nucleosome (N) thereby exposing the HSE binding site for the heat-shock factor (HSF). HSF then binds and activates transcription.

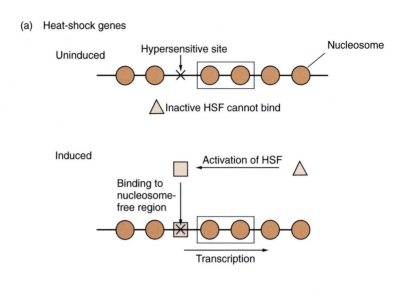

(a) Heat-shock genes

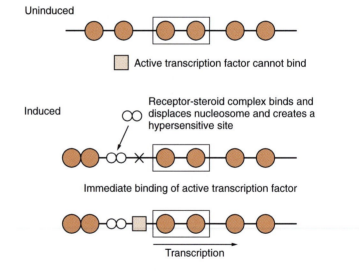

(b) Steroid-inducible genes

Figure 6.39

Two mechanisms for transcriptional activation. (a) The heat-shock transcription factor (HSF) is activated by heat and binds to a pre-existing nucleosome-free region. (b) The receptor–steroid complex displaces a nucleosome, creating a hypersensitive site, and allowing an active transcription factor to bind. X indicates the position of a hypersensitive site. Note that although the figure shows the displacement of a nucleosome, it is also possible that generation of the hypersensitive site may involve the alteration of nucleosome structure as illustrated in Fig. 6.37.

Interestingly, the glucocorticoid receptor/steroid complex does not directly alter nucleosomal structure. Rather it acts by recruiting a multiprotein complex known as the SWI/SNF complex which is able to

hydrolyze ATP and use this energy to alter nucleosomal structure, so facilitating the subsequent binding of activator molecules (for reviews see Aalfs
and Kingston, 2000; Sudarsanam and Winston, 2000).

The role of the SWI/SNF complex is not confined to steroid-responsive
genes and it appears that it can be recruited to a wide variety of different
genes by regulatory proteins which bind to these genes. The SWI/SNF
complex then acts to alter the chromatin structure of these genes, thereby
facilitating their subsequent activation by other transcription factors (Fig.
6.40). In agreement with this idea, the brahma mutation in *Drosophila*

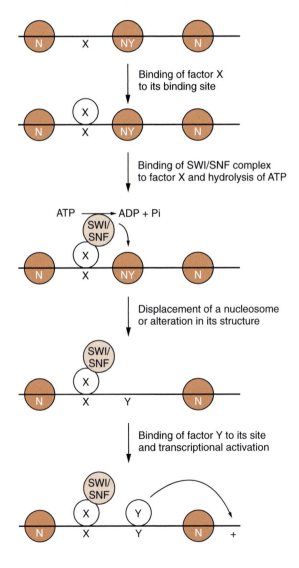

Figure 6.40

A regulatory factor (X) binds to its DNA binding site (X) and recruits the
SWI/SNF complex. This complex hydrolyzes ATP to ADP and inorganic phosphate (Pi) and uses the energy generated to alter the structure of a nucleosome
which was masking the binding site for the transcriptional activator (Y). This
allows Y to bind to its site and activate transcription.

inactivates the SWI2 component of the complex, which produces the ATP hydrolyzing activity. This results in a failure to activate the homeotic genes which play a key role in determining body pattern (see Section 8.2) (for review see Simon, 1995).

Hence the SWI/SNF complex is involved in altering the chromatin structure and thereby facilitating the transcriptional activation of genes as diverse as steroid-responsive genes and homeotic genes. Moreover, once such nucleosome disruption has been produced by SWI/SNF it can dissociate from the gene since the alteration in nucleosome structure and DNaseI hypersensitive site produced by SWI/SNF persists even after it has dissociated (Owen Hughes *et al.*, 1996).

Interestingly, the activity of the SWI/SNF complex has been shown to be modified by interaction with the linker histone H1 (Ramachandran *et al.*, 2003). Thus, in the absence of histone H1, SWI/SNF promotes the displacement of nucleosomes to the end of the DNA molecule, whereas in the presence of histone H1, SWI/SNF promotes controlled nucleosome displacement. Hence, in addition to its role in the solenoid structure of chromatin (see Section 6.3), histone H1 also regulates the displacement of nucleosomes to the correct position on the DNA (Fig. 6.41).

As with SWI/SNF, the GAGA factor described above also plays a general role in chromatin remodeling and does not act solely on the heat-shock genes. Thus, inactivation of the gene encoding the GAGA factor in *Drosophila* results in a fly with an altered body pattern known as trithorax, which is similar to the brahma mutation in that a wide range of homeotic genes are not activated (Farkas *et al.*, 1994). Moreover, the GAGA factor is also associated with a multi-protein complex, known as nucleosome remodeling factor (NURF), which resembles the SWI/SNF factor in its ability to hydrolyze ATP and alter nucleosome structure.

Hence the alteration of chromatin structure at regulatory regions is brought about by multi-protein complexes such as SWI/SNF or NURF which are active on a wide range of genes. These complexes are likely to be recruited to specific gene promoters by interaction with DNA binding proteins such as the glucocorticoid receptor or GAGA which have already

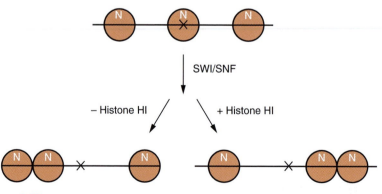

Figure 6.41

In the absence of histone H1, the SWI/SNF complex displaces nucleosomes to the end of the DNA fragment. In contrast, in the presence of histone H1 it promotes the controlled nucleosome displacement required for specific gene activation.

bound to the DNA in a gene-specific manner. In turn this binding will facilitate the subsequent binding of other activating molecules by altering nucleosome positioning or structure in an ATP-dependent manner.

In addition, the SWI/SNF complex has also been shown to be associated with the RNA polymerase II holoenzyme which, as discussed in Section 3.2, is a complex of RNA polymerase II with several basal transcription factors such as TFIIB, TFIIF and TFIIH. Moreover, it has been shown that the recruitment of the RNA polymerase II complex containing SWI/SNF factors can result in the opening up of chromatin indicating that the SWI/SNF complex can also function when brought to the DNA in this way (Gaudreau et al., 1997). In agreement with this idea, the TATA box which binds TFIID and hence recruits the RNA polymerase II complex has been shown to be of critical importance for recruiting the complexes which remodel the chromatin containing the globin gene promoter (Gui and Dean, 2003).

As well as being recruited by specific transcription factors or together with the RNA polymerase II holoenzyme, it has also been shown that SWI/SNF can be recruited to DNA by the SATB1 protein (Yasui et al., 2002). As this protein is involved in the looping of the 30 nm chromatin fiber into an even more highly compact structure (see Section 6.3), this provides a link between such looping and the chromatin remodeling/gene regulation functions carried out by SWI/SNF.

Multiple mechanisms may therefore act to recruit the SWI/SNF and NURF complexes to a particular gene. Whatever the manner in which they are recruited, however, it is clear that these ATP-dependent chromatin remodeling complexes play a key role in opening up the regulatory regions of specific genes so that trans-acting transcription factors can bind and regulate their expression. Moreover, such remodeling processes relate closely to the histone modifications discussed in Section 6.6. Thus, it has been shown that acetylation of histones facilitates recruitment of SWI/SNF to the β-interferon promoter (Agalioti et al., 2002) (see Section 7.3) and such histone modification also prevents SWI/SNF dissociating once it has bound (Hassan et al., 2001). Hence, these two processes appear to act in concert to open up the chromatin for transcription (for a review see Narlikar et al., 2002).

Although we have discussed the opening of chromatin structures by these complexes and its role in transcriptional activation, chromatin remodeling complexes can also produce a more closed chromatin structure leading to transcriptional repression (for a review see Tyler and Kadonaga, 1999). Hence, these complexes appear to play a key role in the regulation of gene expression via the alteration of chromatin structure.

6.8 Other situations in which chromatin structure is regulated

In this chapter we have discussed the role of changes in the chromatin structure of individual genes in mediating commitment to a particular differentiated state and thereby allowing the tissue-specific activation of gene expression. However, differences in chromatin structure are also involved in regulating gene expression in two other well characterized processes, namely X-chromosome inactivation and genomic imprinting. Both of these processes involve differences in expression between the two copies of a specific gene which are present on different homologous chromosomes in a diploid cell (for reviews see Lyon, 1993; Lewin, 1998).

X-chromosome inactivation

The fact that females of mammalian species have two X chromosomes whereas males have one X and one Y chromosome creates a problem of how to compensate for the difference in dosage of genes on the X-chromosome which occurs because females have two copies whereas males have only one. In mammals this problem is solved by the process of X-chromosome inactivation (for reviews see Migeon, 1994; Lee and Jaenisch, 1997; Panning and Jaenisch, 1998).

Thus, during the process of embryonic development, one of the two X chromosomes in each female cell undergoes an inactivation process so that the expression of virtually all the genes on this chromosome is inactivated whilst those on the other chromosome remain active. This results in each female cell having only one active copy of X-chromosome genes paralleling the situation in male cells which have only one X chromosome (Fig. 6.42).

This process occurs apparently randomly with cells in the early female embryo inactivating either the X chromosome inherited from the father (the paternal chromosome) or that inherited from the mother (the maternal chromosome). However, once one or other of the X chromosomes has been inactivated, this inactivation is propagated stably through cell division to all the progeny of that cell (Fig. 6.42). Such stability is dependent upon the fact that the inactive and active X chromosomes have a different chromatin structure.

Thus, within the inactive X chromosome the DNA is tightly packed into a highly condensed structure known as heterochromatin and this high density structure results in the inactive X chromosome being visible as a distinct element within the cell known as a Barr body. In turn such a condensed structure has been shown to involve decreased sensitivity to digestion with DNaseI, altered histone modification and enhanced methylation on C residues which can be observed when genes on the inactive

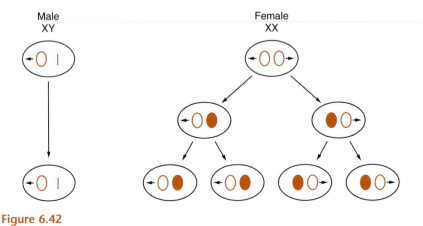

Figure 6.42

X-chromosome inactivation results in one or other of the two X chromosomes becoming inactivated (solid) in each cell whilst the other remains active (open).

X chromosome are compared with the equivalent gene on the active chromosome. Thus, for example, 60 of the 61 CpG dinucleotides in the CpG island (see Section 6.5) located around the promoter of the PGKI gene are methylated on the C residue when the inactive X chromosome is studied; whereas all these sites are unmethylated on the active X. Moreover, treatment with 5-azacytidine, leading to demethylation (see Section 6.5) can reactivate previously inactivated X-chromosome genes.

Interestingly, in *Drosophila* compensation for the reduced number of X chromosomes in male cells is achieved in the embryo, by increasing the transcriptional activity of the genes on the single male X chromosome rather than by inactivating one of the female X chromosomes. As in the case of X inactivation, however, this effect appears to involve alterations in chromatin structure since the male X chromosome has a higher level of acetylated histones than either of the female chromosomes and a histone acetyl transferase enzyme (see Section 6.6) specifically associates with the male X chromosome (for a review of comparative mechanisms of dosage compensation in mammals and *Drosophila* see Park and Kuroda, 2001).

Chromatin structure thus plays a critical role in differentially regulating the activity of genes on the X chromosomes in females and males and, in particular, in the maintenance of X-chromosome inactivation through cell division. Interestingly, the onset of X-chromosome inactivation in the embryo requires a particular region of the chromosome known as the X-inactivation center. If this region is deleted then X inactivation does not occur. A gene, known as XIST, has been mapped to the X-inactivation center and it has been shown that the inactivation of one copy of the XIST gene in mutant mice results in a failure of X-chromosome inactivation on the chromosome which lacks an active XIST gene, with the other X chromosome being preferentially inactivated (Fig. 6.43a). Hence XIST is essential for inactivation of the X chromosome on which it is located (for review see Lee and Jaenisch, 1997; Panning and Jaenisch, 1998; Kelley and Kuroda, 2000).

Interestingly, the XIST gene is transcribed only on the inactive X chromosome and not on the active chromosome, the opposite pattern to all other genes. Its critical role in X-inactivation is demonstrated by the finding that mutant mice, which show expression of XIST on both chromosomes, also show inactivation of both X chromosomes exactly as would be expected if the expression of the XIST gene does indeed inactivate genes on the X chromosome containing it (Fig. 6.43b). Moreover, if an active XIST gene is placed on a non-X chromosome, the transcription of the genes on this chromosome is inactivated (for review see Willard and Salz, 1997). Hence XIST gene transcription is sufficient to inactivate the genes to which it is linked regardless of the nature of these genes.

It has been shown that the transcription of XIST early in embryonic development on one of the two chromosomes results in the recruitment to that chromosome of a complex which can methylate histone H3 on lysines 9 and 27 (for a review see Hajkova and Surani, 2004) and therefore produce a tightly packed chromatin structure incompatible with transcription (Section 6.6). Hence, transcription of XIST on one of the two X chromosomes results in a change in chromatin structure which is propagated along the rest of that chromosome switching off all other genes (Fig. 6.44).

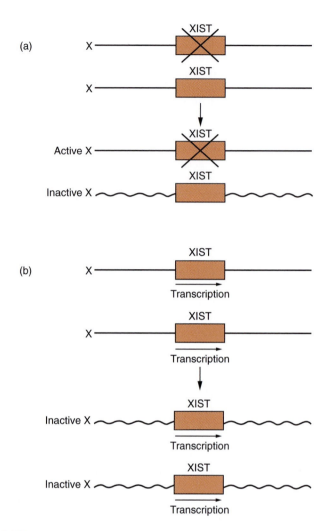

Figure 6.43

Inactivation of one copy of the XIST gene in mouse mutants results in the chromosome containing the intact XIST gene being preferentially inactivated (panel a) whilst mutant mice which transcribe the XIST gene on both X chromosomes show inactivation of both X chromosomes (panel b).

The initiation and maintenance of X-chromosome inactivation thus requires the ability to produce and stably propagate an altered chromatin structure which we have previously seen to be critical in tissue-specific gene regulation. Moreover, like the regulation of the chromatin structure of individual genes, it involves processes such as methylation of C residues in the DNA and deacetylation of histones.

Genomic imprinting

Genomic imprinting (for reviews see Ferguson-Smith and Surani, 2001; Constancia *et al.*, 2004) resembles X-chromosome inactivation in that one

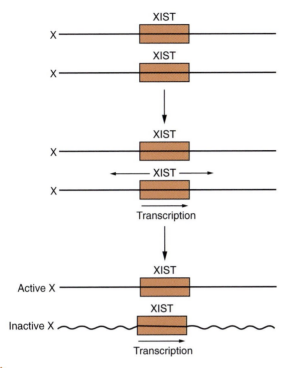

Figure 6.44

In normal mice, transcription of the XIST gene on one of the two X chromosomes is associated with its inactivation via a change in chromatin structure.

of the two copies of specific genes is inactivated whilst the other remains active. This process differs, however, in that only a few genes scattered on different autosomes (non-sex chromosomes) are inactivated and that the same copy is always inactivated in all cells and in all organisms whether male or female. Thus, it is always the maternally inherited copy of some of these, such as the genes encoding the insulin-like growth factor 2 (IGF2) protein, the SmN splicing protein and the U2AF35 related protein which are inactivated with the paternally inherited gene remaining active. Conversely the paternally inherited copies of others, such as the IGF2 receptor gene and the H19 gene, are inactivated with the maternally inherited gene remaining active (Fig. 6.45).

As with X inactivation therefore, this process results in all cells having only one functional copy of each of these genes, although in the case of genomic imprinting all cells express the same copy. This process cannot be reversed during embryonic development even when genetic crosses are used to produce embryos with two imprinted copies of the gene. Thus, embryos which inherit two maternal and no paternal copies of the Igf2 gene or two maternal but no paternal copies of the SmN gene die due to the lack of an active gene. These embryos therefore die due to the absence of a functional protein even though two copies of the gene capable of encoding this protein are present in each cell of the embryo.

Despite the lethal effects which can result when imprinting goes wrong, it is unclear what the normal function of this process is. It has been

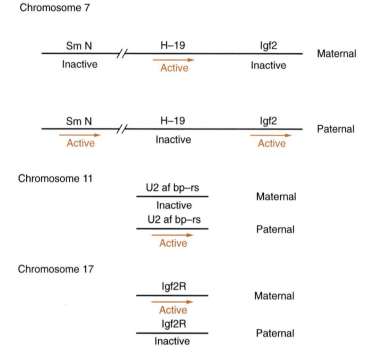

Chromosome 7

| Sm N | H–19 | Igf2 | |
| Inactive | Active | Inactive | Maternal |

| Sm N | H–19 | Igf2 | |
| Active | Inactive | Active | Paternal |

Chromosome 11

U2 af bp–rs	
Inactive	Maternal
U2 af bp–rs	
Active	Paternal

Chromosome 17

Igf2R	
Active	Maternal
Igf2R	
Inactive	Paternal

Figure 6.45

Imprinting results in the inactivation of the maternally inherited SmN, U2AFbp-rs and Igf2 genes and of the paternally inherited Igf2R and H19 genes.

suggested, for example, that imprinting may represent a means of preventing the parthenogenetic development of the unfertilized egg to produce a haploid embryo with no paternal contribution. Alternatively, it may have evolved because of the conflict between the maternal and paternal genomes in terms of the transfer of nutrients from mother to offspring. Thus, whilst it is in the paternal interest to promote the growth of the individual fetus, the maternal interest is to restrict the growth of any individual fetus so that other fetuses fathered by different males either concurrently (in multi-fetal litters) or subsequently can develop fully (for discussions see Haig, 2003; Constancia et al., 2004). Other theories range from the need for the cell to distinguish between the two copies of each chromosome which have been inherited from the mother or the father (for a review see de Villena et al., 2000) to a mechanism for producing differences between male and female offspring (for a review see Pagel, 1999).

This lack of a clear functional role for genomic imprinting is in contrast to the clear role of X-chromosome inactivation in compensating for the extra X chromosome in females compared with males. Indeed it has even been suggested that imprinting may not have a function at all but may represent a vestige of an evolutionarily ancient defence system to inactivate foreign DNA with the imprinted genes having some feature which causes this system to consider them as foreign (see Barlow, 1993 for further discussion of this idea).

Whatever its precise function (if any), it is clear that genomic imprinting resembles X-chromosome inactivation in that specific differences exist in the methylation pattern of CG dinucleotides between the active and inactive copies of the imprinted gene. Moreover, embryos which lack a functional DNA methyltransferase and which therefore cannot methylate their DNA (see Section 6.5) fail to carry out genomic imprinting (Kaneda *et al.*, 2004). Hence specific methylation of one of the two copies of the gene is critical for imprinting to occur (for a review see Feil and Khosla, 1999).

Interestingly, in some cases, the role of C-methylation in regulating imprinting can be quite complex. Thus, in the case of the H19 gene which is silenced on the paternal chromosome, a simple correlation exists with the regulatory region of the gene being methylated on the paternal chromosome but not on the maternal, as would be expected (Fig. 6.46). However, in the case of the adjacent Igf2 gene which is silenced on the maternal chromosome, no specific methylation is seen on this chromosome compared with the paternal one. Rather, the activity of the Igf2 gene is controlled by an imprinting control region (ICR) which is located between the two genes and which is methylated on the paternal chromosome where the Igf2 gene is active.

This paradox is resolved by the finding that when the ICR is methylated, a positive regulatory element (known as an enhancer: see Section 7.3) located downstream of the H19 gene acts at a distance to activate Igf2 gene expression (Fig. 6.46). In contrast, when the ICR is unmethylated on the maternal chromosome, a protein known as CTCF binds to it. This protein acts as an insulator (see Section 7.4) preventing the positive regulatory element from activating Igf2 and the gene is therefore silent. In agreement with this idea, mutations in the binding site for CTCF (Pant *et al.*, 2004) or inhibition of its synthesis (Fedoriw *et al.*, 2004) disrupt the correct pattern of expression of the H19 and Igf2 genes. Hence, in this

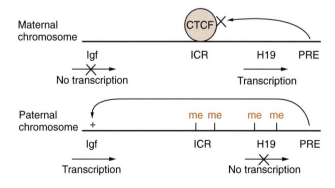

Figure 6.46

Role of C residue methylation (me) in controlling imprinting of the closely linked Igf2 and H19 genes. On the paternal chromosome, the H19 gene is methylated and is therefore only expressed from the maternal chromosome where it is unmethylated. In contrast, methylation of the imprinting control region (ICR) on the paternal chromosome prevents binding of an insulator protein (CTCF) and allows a distinct positive regulatory element (PRE) to specifically activate Igf2 expression only on the paternal chromosome.

chromosomal region, methylation of C residues on the paternal chromosome results in silencing of the H19 gene and expression of the Igf2 gene on this chromosome (for reviews see Allshire and Bickmore, 2000; Reik and Murrell, 2000; Arney, 2003).

It is clear therefore that, as with X chromosome inactivation, imprinting involves similar modifications and changes in chromatin structure to those which are used to regulate cellular commitment and gene transcription in specific cell types.

6.9 Conclusions

A variety of changes take place in the chromatin of genes during the process of commitment to a particular pathway of differentiation. Such changes involve both modification of the DNA itself by undermethylation, to the histones with which it is associated, and to the general packaging of the DNA in chromatin. In the last few years, the study of these changes has moved from the simple description of changes such as DNaseI sensitivity to a more mechanistic approach. This has identified three key processes regulating chromatin structure namely, DNA methylation (Section 6.5), histone modifications, particularly acetylation (Section 6.6) and ATP-dependent remodeling of chromatin structure by complexes such as SWI/SNF and NURF (Section 6.7). Indeed, these processes are closely linked, as described earlier, with DNA methylation stimulating histone changes such as deacetylation and methylation, whilst in turn such changes in histone modification can regulate SWI/SNF recruitment to the promoter (for review see Richards and Elgin, 2002).

The combination of all the various processes discussed in this chapter results in three levels of chromatin structure within the cell (Fig. 6.47). Thus although the bulk of inactive DNA is organized into a tightly packed solenoid structure, active or potentially active genes are organized into a more open "beads on a string" structure, and short regions within the gene are either nucleosome-free or have structurally altered nucleosomes.

The role of these changes in allowing cells to maintain a commitment to a particular differentiated state and to respond differently to inducers of gene expression is well illustrated in the case of the steroid hormones and their effect on gene expression. Thus the difference in the response

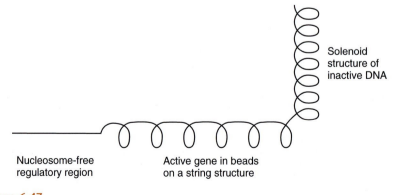

Solenoid structure of inactive DNA

Nucleosome-free regulatory region

Active gene in beads on a string structure

Figure 6.47

Levels of chromatin structure in active and inactive DNA.

of different tissues to treatment with estrogen (see Section 6.1) is likely to be due to the fact that in one tissue certain steroid-responsive genes will be inaccessible within the solenoid structure and will therefore be incapable of binding the receptor–hormone complex that is necessary for activation. In other genes, which are in the beads on a string structure and therefore more accessible, such binding of the complex to defined sequences in the gene will occur. Even in this case, however, gene activation will not occur as a direct consequence of this interaction, as might be the case in bacteria. Rather, the binding will result in the displacement of a nucleosome from the region of DNA, generating a hypersensitive site and allowing other regulatory proteins to interact with their specific recognition sequences and cause transcription to occur.

Both the action of the receptor–hormone complex and the subsequent binding of other transcription factors to nucleosome-free DNA clearly involve the interaction of regulatory proteins with specific DNA sequences. The next two chapters will discuss the DNA sequences and proteins involved in these interactions.

References

Aalfs, J.D. and Kingston, R.E. (2000). What does "chromatin remodelling" mean? *Trends in Biochemical Sciences* **25**, 548–555.

Agalioti, T., Chen, G. and Thanos, D. (2002). Deciphering the transcriptional histone acetylation code for a human gene. *Cell* **111**, 381–392.

Ahringer, J. (2000). NuRD and SIN3 histone deacetylase complexes in development. *Trends in Genetics* **16**, 351–356.

Allshire, R. and Bickmore, W. (2000). Pausing for thought on the boundaries of imprinting. *Cell* **102**, 705–708.

Arney, K.L. (2003). H19 and Igf2–enhancing the confusion? *Trends in Genetics* **19**, 17–23.

Ayer, D.E. (1999). Histone deacetylases: transcriptional repression with SINers and NuRDs. *Trends in Cell Biology* **9**, 193–198.

Bannister, A.J., Zegerman, P., Partridge, J.F., Miska, E.A., Thomas, J.O., Allshire, R.C. and Kouzarides, T. (2001). Selective recognition of methylated lysine 9 on histone H3 by the HP1 chromo domain. *Nature* **410**, 120–124.

Barlow, D.P. (1993). Methylation and imprinting: from host defence to gene regulation? *Science* **260**, 309–310.

Becker, P.B., Ruppert, S. and Schutz, G. (1987). Genomic fingerprinting reveals cell type-specific binding of ubiquitous factors. *Cell* **51**, 435–443.

Bestor, T.H. and Coxon, A. (1993). The pros and cons of DNA methylation. *Current Biology* **3**, 384–386.

Bird, A.P. and Wolffe, A.P. (1999). Methylation-induced repression – belts, braces and chromatin. *Cell* **99**, 451–454.

Bradbury, E.M. (2002). Chromatin structure and dynamics: state-of-the-art. *Molecular Cell* **10**, 13–19.

Briggs, S.D., Xiao, T., Sun, Z.-W., Caldwell, J.A., Shabanowitz, J., Hunt, D.F., et al. (2002). *Trans*-histone regulatory pathway in chromatin. *Nature* **418**, 498.

Brown, C.E., Lechner, T., Howe, L. and Workman, J.L. (2000). The many HATs of transcription coactivators. *Trends in Biochemical Sciences* **25**, 15–19.

Carrozza, M.J., Utley, R.T., Workman, J.L. and Cote, J. (2003). The diverse functions of histone acetyltransferase complexes. *Trends in Genetics* **19**, 321–329.

Chiu, C.-P. and Blau, H. (1985). 5-azacytidine permits gene activation in a previously non-inducible cell type. *Cell* **40**, 417–424.

Colot, V. and Rossignol, J.L. (1999). Eukaryotic DNA methylation as an evolutionary device. *BioEssays* **21**, 402–411.

Constancia, M., Kelsey, G. and Reik, W. (2004). Resourceful imprinting. *Nature* **432**, 53–57.

Constantinides, P.G., Jones, P.A. and Gevers, W. (1977). Functional striated muscle cells from non-myoblast precursors following 5-azacytidine treatment. *Nature* **267**, 364–366.

Coon, H.G. (1966). Clonal stability and phenotypic expression of chick cartilage cells *in vitro*. *Proceedings of the National Academy of Sciences of the USA* **55**, 66–73.

Farkas, G., Gausz, J., Gallioni, M., Reuter, G., Yurkovics, H. and Karch, F. (1994). The trithorax-1 gene encodes the *Drosophila* GAGA factor. *Nature* **371**, 806–808.

Fedoriw, A.M., Stein, P., Svoboda, P., Schultz, R.M. and Bartolomei, M.S. (2004). Transgenic RNAi reveals essential function for CTCF in H19 gene imprinting. *Science* **303**, 238–240.

Feil, R. and Khosla, S. (1999). Genomic imprinting in mammals an interplay between chromatin and DNA methylation? *Trends in Genetics* **15**, 431–435.

Felsenfeld, G. and Groudine, M. (2003). Controlling the double helix. *Nature* **421**, 448–453.

Felsenfeld, G. and McGhee, J.D. (1986). Structure of the 30 nm chromatin fibre. *Cell* **44**, 375–377.

Ferguson-Smith, A.C. and Surani, M.A. (2001). Imprinting and the epigenetic asymmetry between parental genomes. *Science* **293**, 1086–1089.

Finch, J.T., Noll, M. and Kornberg, R.D. (1975). Electron microscopy of defined lengths of chromatin. *Proceedings of the National Academy of Sciences of the USA* **72**, 3320–3322.

Fischle, W., Wang, Y. and Allis, C.D. (2003). Binary switches and modification cassettes in histone biology and beyond. *Nature* **425**, 475–479.

Freiman, R.N. and Tjian, R. (2003). Regulating the regulators: lysine modifications make their mark. *Cell* **112**, 11–17.

Fuks, F., Hurd, P.J., Wolf, D., Nan, X., Bird, A.P. and Kouzarides, T. (2003). The methyl-CpG-binding protein MeCP2 links DNA methylation to histone methylation. *Journal of Biological Chemistry* **278**, 4035–4040.

Gaudreau, L., Schmid, A., Blaschke, D., Ptashne, M. and Horz, W. (1997). RNA polymerase II holoenzyme recruitment is sufficient to remodel chromatin at the yeast PHO5 promoter. *Cell* **89**, 55–62.

Gerber, M. and Shilatifard, A. (2003). Transcriptional elongation by RNA polymerase II and histone methylation. *Journal of Biological Chemistry* **278**, 26303–26306.

Gill, G. (2004). SUMO and ubiquitin in the nucleus: different functions, similar mechanisms? *Genes and Development* **18**, 2046–2059.

Goll, M.G. and Bestor, T.H. (2002). Histone modification and replacement in chromatin activation. *Genes and Development* **16**, 1739–1742.

Gross, D.S. and Garrard, W.T. (1988). Nuclease hypersensitive sites in chromatin. *Annual Review of Biochemistry* **57**, 159–197.

Gui, C.-Y. and Dean, A. (2003). A major role for the TATA box in recruitment of chromatin modifying complexes to a globin gene promoter. *Proceedings of the National Academy of Sciences of the USA* **100**, 7009–7014.

Hadorn, E. (1968). Transdetermination in cells. *Scientific American* **219**, 110–120.

Haig, D. (2003). Family matters. *Nature* **421**, 491–492.

Hajkova, P. and Surani, M.A. (2004). Programming the X chromosome. *Science* **303**, 633–634.

Hampsey, M. and Reinberg, D. (2003). Tails of intrigue: phosphorylation of RNA polymerase II mediates histone methylation. *Cell* **113**, 429–432.

Hassan, A.H., Neely, K.E. and Workman, J.L. (2001). Histone acetyltransferase complexes stabilise SWI/SNF binding to promoter nucleosomes. *Cell* **104**, 817–827.

Hendrich, B. and Tweedie, S. (2003). The methyl-CpG binding domain and the evolving role of DNA methylation in animals. *Trends in Genetics* **19**, 269–277.

Henikoff, S., Furuyama, T. and Ahmad, K. (2004). Histone variants, nucleosome assembly and epigenetic inheritance. *Trends in Genetics* **20**, 320–326.

Horn, P.J. and Peterson, C.L. (2002). Chromatin higher order folding: wrapping up transcription. *Science* **297**, 1824–1827.

Jaskelioff, M. and Peterson, C.L. (2003). Chromatin and transcription: histones continue to make their marks. *Nature Cell Biology* **5**, 395–399.

Jenuwein, T. and Allis, C.D. (2001). Translating the histone code. *Science* **293**, 1074–1080.

Jones, K.A. and Kadonaga, J.T. (2000). Exploring the transcription–chromatin interface. *Genes and Development* **14**, 1992–1996.

Jones, P.A. and Takai, D. (2001). The role of DNA methylation in mammalian epigenetics. *Science* **293**, 1068–1070.

Kaneda, M., Okano, M., Hata, K., Sado, T., Tsujimoto, N., Li, E. and Sasaki, H. (2004). Essential role for de novo DNA methyltransferase Dnmt3a in paternal and maternal imprinting. *Nature* **429**, 900–903.

Kawasaki, H. and Taira, K. (2004). Induction of DNA methylation and gene silencing by short interfering RNAs in human cells. *Nature* **431**, 211–217.

Kawasaki, H., Schiltz, L., Chiu, R., Itakura, K., Taira, K., Nakatani, Y. and Yokoyama, K.K. (2000). ATF-2 has intrinsic histone acetyltransferase activity which is modulated by phosphorylation. *Nature* **405**, 195–200.

Kelley, R.L. and Kuroda, M.I. (2000). Noncoding RNA genes in dosage compensation and imprinting. *Cell* **103**, 9–12.

Khorasanizadeh, S. (2004). The nucleosome: from genomic organization to genomic regulation. *Cell* **116**, 259–272.

Kingston, R.E. and Narlikar, G.J. (1999). ATP-dependent remodelling and acetylation as regulators of chromatin fluidity. *Genes and Development* **13**, 2339–2352.

Klose, R. and Bird, A. (2003). MeCP2 repression goes nonglobal. *Science* **302**, 793–795.

Kornberg, R.D. and Lorch, Y. (1999). Twenty-five years of the nucleosome, fundamental particle of the eukaryote chromosome. *Cell* **98**, 285–294.

Kurdistani, S.K., Tavazoie, S. and Grunstein, M. (2004). Mapping global histone acetylation patterns to gene expression. *Cell* **117**, 721–733.

Lacey, E. and Axel, R. (1975). Analysis of DNA of isolated chromatin subunits. *Proceedings of the National Academy of Sciences of the USA* **72**, 3978–3982.

Lachner, M., O'Carroll, D., Rea, S., Mechtler, K. and Jenuwein, T. (2001). Methylation of histone H3 lysine 9 creates a binding site for HP1 proteins. *Nature* **410**, 116–120.

Lee, H., Habas, R. and Abate-Shen, C. (2004). MSX1 cooperates with histone H1b for inhibition of transcription and myogenesis. *Science* **304**, 1675–1678.

Lee, J.T. and Jaenisch, R. (1997). The (epi) genetic control of mammalian X-chromosome inactivation. *Current Opinion in Genetics and Development* **7**, 274–280.

Lehmann, M. (2004). Anything else but GAGA: a nonhistone protein complex reshapes chromatin structure. *Trends in Genetics* **20**, 15–22.

Lewin, B. (1998). The mystique of epigenetics. *Cell* **93**, 301–303.

Liang, G., Lin, J.C.Y., Wei, V., Yoo, C., Cheng, J.C., Nguyen, C.T., et al. (2004). Distinct localization of histone H3 acetylation and H3-K4 methylation to the transcription start sites in the human genome. *Proceedings of the National Academy of Sciences of the USA* **101**, 7357–7362.

Lippman, Z. and Martienssen, R. (2004). The role of RNA interference in heterochromatic silencing. *Nature* **431**, 364–370.

Loyola, A. and Almouzni, G. (2004). Bromodomains in living cells participate in deciphering the histone code. *Trends in Cell Biology* **14**, 279–281.

Luger, A, Mäder, A.W., Richmond, R.K., Sargent, D.F. and Richmond, T.J. (1997). Crystal structure of the nucleosome core particle at 2.8 Å resolution. *Nature* **389**, 251–260.

Lyko, F., Ramsahoye, B.H., and Jaenisch, R. (2000). DNA methylation in *Drosophila melanogaster*. *Nature* **408**, 538–540.

Lyon, M.F. (1993). Epigenetic inheritance in mammals. *Trends in Genetics* **9**, 123–128.

McKnight, S. L., Bustin, M. and Miller, O.L. (1978). Electron microscope analysis of chromosome metabolism in the *Drosophila melanogaster* embryo. *Cold Spring Harbor Symposium on Quantitative Biology* **42**, 741–754.

Migeon, B.R. (1994). X-chromosome inactivation: molecular mechanisms and genetic consequences. *Trends in Genetics* **10**, 230–235.

Mohd-Sarip, A. and Verrijzer, C.P. (2004). A higher order of silence. *Science* **306**, 1484–1485.

Muller-Hill, B.W. (ed.) (1996). *The lac Operon: A Short History of a Genetic Paradigm*. de Gruyter, Berlin.

Narlikar, G.J., Fan, H.-Y. and Kingston, R.E. (2002). Co-operation between complexes that regulate chromatin structure and transcription. *Cell* **108**, 475–487.

Nathan, D., Sterner, D.E. and Berger, S.L. (2003). Histone modifications: now summoning sumoylation. *Proceedings of the National Academy of Sciences of the USA* **100**, 13118–13120.

Ng, H.-H. and Bird, A. (1999). DNA methylation and chromatin modification. *Current Opinion in Genetics and Development* **9**, 158–163.

Ng, H.-H. and Bird, A. (2000). Histone deacetylases: silencers for hire. *Trends in Biochemical Sciences* **25**, 121–126.

Noma, K.-I., Allis, C.D. and Grewal, S.I.S. (2001). Transitions in distinct histone H3 methylation patterns at the heterochromatin domain boundaries. *Science* **293**, 1150–1155.

Nowak, S.J. and Corces, V.G. (2004). Phosphorylation of histone H3: a balancing act between chromosome condensation and transcriptional activation. *Trends in Genetics* **20**, 214–220.

Okano, M., Bell, D.W., Haber, D.A. and Li, E. (1999). DNA methyltransferases Dnmt3a and Dnmt3b are essential for de novo methylation and mammalian development. *Cell* **99**, 247–257.

Owen H., Utley, T., Cole, R.T., Peterson, J.C.L. and Workman, J.L. (1996). Persistent site-specific remodelling of a nucleosome array by transient action of the SWI/SNF complex. *Science* **273**, 513–516.

Pagel, M. (1999). Mother and father in surprise genetic agreement. *Nature* **397**, 19–20.

Pal-Bhadra, M., Leibovitch, B.A., Gandhi, S.G., Rao, M., Bhadra, U., Birchler, J.A. and Elgin, S.C. (2004). Heterochromatic silencing and HP1 localization in *Drosophila* are dependent on the RNAi machinery. *Science* **303**, 669–672.

Panning, B. and Jaenisch, R. (1998). RNA and the epigenetic regulation of X chromosome inactivation. *Cell* **93**, 305–308.

Pant, V., Kurukuti, S., Pugacheva, E., Shamsuddin, S., Mariano, P., Renkawitz, R., *et al.* (2004). Mutation of a single CTCF target site within the *H19* imprinting control region leads to loss of *Igf2* imprinting and complex patterns of de novo methylation upon maternal inheritance. *Molecular and Cellular Biology* **24**, 3497–3504.

Park, Y. and Kuroda, M.I. (2001). Epigenetic aspects of X-chromosome dosage compensation. *Science* **293**, 1083–1085.

Paro, R. (2000). Formatting genetic text. *Nature* **406**, 579–580.

Raibaud, O. and Schwartz, M. (1984). Positive control of transcription initiation in bacteria. *Annual Review of Genetics* **18**, 173–206.

Ramachandran, A., Omar, M., Cheslock, P. and Schnitzler, G.R. (2003). Linker histone H1 modulates nucleosome remodelling by human SWI/SNF. *Journal of Biological Chemistry* **278**, 48590–48601.

Rea, S., Eisenhaber, F., O'Carroll, D., Strahl, B.D., Sun, Z.-W., Schmid, M., et al. (2000). Regulation of chromatin structure by site-specific histone H3 methyltransferases. *Nature* 406, 593–599.

Reik, W. and Murrell, A. (2000). Silence across the border. *Nature* 405, 408–409.

Rhodes, D. (1997). The nucleosome core all wrapped up. *Nature* 389, 231–233.

Richards, E.J. and Elgin, S.C.R. (2002). Epigenetic codes for heterochromatin formation and silencing: rounding up the usual suspects. *Cell* 108, 489–500.

Richmond, T.J. and Davey, C.A. (2003). The structure of DNA in the nucleosome core. *Nature* 423, 145–150.

Sarraf, S.A. and Stancheva, I. (2004). Methyl-CpG binding protein MBD1 couples histone H3 methylation at lysine 9 by SETDB1 to DNA replication and chromatin assembly. *Molecular Cell* 15, 595–605.

Schubeler, D., MacAlpine, D.M., Scalzo, D., Wirbelauer, C., Kooperberg, C., van Leeuwen, F., et al. (2004). The histone modification pattern of active genes revealed through genome-wide chromatin analysis of a higher eukaryote. *Genes and Development* 18, 1263–1271.

Sedkov, Y., Cho, E., Petruk, S., Cherbas, L., Smith, S.T., Jones, R.S., et al. (2003). Methylation at lysine 4 of histone H3 in ecdysone-dependent development of *Drosophila*. *Nature* 426, 78–83.

Shahbazian, M.D., Young, J.I., Yuva-Paylor, L.A., Spencer, C.M., Antalffy, B.A., Noebels, J.L., et al. (2002). Mice with truncated MeCP2 recapitulate many Rett Syndrome features and display hyperacetylation of histone H3. *Neuron* 35, 243–254.

Shilo, Y. and Eisenman, R.N. (2003). Histone sumoylation is associated with transcriptional repression. *Proceedings of the National Academy of Sciences of the USA* 100, 13225–13230.

Simon, J. (1995). Locking in stable states of gene expression: transcriptional control during *Drosophila* development. *Current Opinion in Cell Biology* 7, 376–385.

Sims, R.J., III, Nishioka, K. and Reinberg, D. (2003). Histone lysine methylation: a signature for chromatin function. *Trends in Genetics* 19, 629–639.

Sims, R. J., III, Belotserkovskaya, R. and Reinberg, D. (2004). Elongation by RNA polymerase II: the short and long of it. *Genes and Development* 18, 2437–2468.

Stewart, S. and Crabtree, G.R. (2000). Regulation of the regulators. *Nature* 408, 46–47.

Struhl, K. (1999). Fundamentally different logic of gene regulation in eukaryotes and prokaryotes. *Cell* 98, 1–4.

Sudarsanam, P. and Winston, F. (2000). The Swi/Snf family nucleosome-remodelling complexes and transcriptional control. *Trends in Genetics* 16, 345–351.

Sun, Z.-W. and Allis, C.D. (2002). Ubiquitination of histone H2B regulates H3 methylation and gene silencing in yeast. *Nature* 418, 104–108.

Travers, A. (1993). *DNA–Protein Interactions*. Chapman and Hall, London.

Tucker, K.L. (2001). Methylated cytosine and the brain: a new base for neuroscience. *Neuron* 30, 649–652.

Tyler, J.K. and Kadonaga, J.T. (1999). The "dark side" of chromatin remodelling: repressive effects on transcription. *Cell* 99, 443–446.

van Holde, K. and Zlatanova, J. (1995). Chromatin higher order structure: chasing a mirage? *Journal of Biological Chemistry* 270, 8373–8376.

van Leeuwen, F. and Gottschling, D.E. (2003). The histone minority report: the variant shall not be silenced. *Cell* 112, 591–593.

de Villena, F., de la Casa-Esperón, E. and Sapienza, C. (2000). Natural selection and the function of genome imprinting: beyond the silenced minority. *Trends in Genetics* 16, 573–579.

Wang, H., Huang, Z.-Q., Xia, L., Feng, Q., Erdjument-Bromage, H., Strahl, B.D., *et al* (2001). Methylation of histone H4 at arginine 3 facilitating transcriptional activation by nuclear hormone receptor. *Science* **293**, 853–857.

Wang, H., Wang, L., Erdjument-Bromage, H., Vidal, M., Tempst, P., Jones, R.S., and Zhang, Y. (2004). Role of histone H2A ubiquitination in Polycomb silencing. *Nature* **431**, 873–878.

Willard, H.F. and Salz, H.K. (1997). Dosage compensation remodelling chromatin with RNA. *Nature* **386**, 228–229.

Wolffe, A. (1995). *Chromatin: Structure and Function.* 2nd edition. Academic Press London.

Workman, J.L. and Abmayr, S.M. (2004). Histone H3 variants and modifications on transcribed genes. *Proceedings of the National Academy of Sciences of the USA* **101**, 1429–1430.

Yasui, D., Miyano, M., Cal, S., Varga-Weisz, P. and Kohwi-Shigematsu, T. (2002). SATB1 targets chromatin remodelling to regulate genes over long distances. *Nature* **419**, 641–645.

Zhang, Y. (2003). Transcriptional regulation by histone ubiquitination and deubiquitination. *Genes and Development* **17**, 2733–2740.

Zhang, Y. and Reinberg, D. (2001). Transcription regulation by histone methylation: interplay between different covalent modifications of the core histone tails. *Genes and Development* **15**, 2343–2360.

Zhao, X., Ueba, T., Christie, B.R., Barkho, B., McConnell, M.J., Nakashima, K., *et al.* (2003). Mice lacking methyl-CpG binding protein 1 have deficits in adult neurogenesis and hippocampal function. *Proceedings of the National Academy of Sciences of the USA* **100**, 6777–6782.

Zilberman, D., Cao, X. and Jacobsen, S.E. (2003). ARGONAUTE4 control of locus-specific siRNA accumulation and DNA and histone methylation. *Science* **299**, 716–719.

Transcriptional control – DNA sequence elements

7

SUMMARY

- In eukaryotes, protein-coding genes which need to be expressed together are not closely linked in the genome.
- The expression of such genes is coordinated by short DNA sequences which are predominantly located upstream of the start site of transcription and which activate gene expression in specific situations.
- Other positively acting DNA elements are located further away from the gene and are known as enhancers or locus-control regions.
- Specific DNA elements, known as silencers and insulators, can have an inhibitory effect on gene expression.
- All these sequences act by binding regulatory proteins which either affect chromatin structure or alter the activity of the basal transcriptional complex, thereby regulating gene expression.
- In contrast, regulation of transcription by RNA polymerases I and II is much less complex than that of protein-coding genes transcribed by RNA polymerase II.

7.1 Introduction

Relationship of gene regulation in prokaryotes and eukaryotes

As discussed in Chapter 6, various alterations occur in the chromatin structure of a particular gene prior to the onset of transcription. Once such changes have occurred, the actual onset of transcription takes place through the interaction of defined proteins (transcription factors) with specific DNA sequences adjacent to the gene. This final stage of gene regulation is clearly analogous to the activation or repression of gene expression in prokaryotes, which was discussed briefly in Section 6.1. However, before the manner in which transcription factors and DNA sequences act to regulate gene expression in higher eukaryotes can be considered, it is necessary to mention one aspect of eukaryotic systems that does not exist in bacteria.

Coordinately regulated genes are not linked in eukaryotes

In prokaryotes, when several genes encoding particular proteins are expressed in response to a particular signal, the genes are found tightly linked together in an operon. In response to the activating signal, all the genes in the operon are transcribed as one single polycistronic (multi-gene) mRNA molecule, and translation of each protein from internal AUG initiation codons (see Section 3.3) results in the desired coordinate production of the proteins encoded by the individual genes. A typical example of this is seen in the case of the lactose-inducible genes located in the *lac* operon. The efficient use of lactose requires not only the enzyme β-galactosidase, which cleaves the lactose molecule, but also a permease enzyme, to facilitate uptake of lactose by the cell, and a modifying *trans*-acetylase enzyme. The genes encoding these three molecules must be activated in response to lactose and this is achieved by linking them together in an operon whose expression is dependent upon the presence of lactose (Fig. 7.1).

In higher eukaryotic organisms this does not occur. Thus operon-type structures transcribed into polycistronic RNAs do not exist in higher organisms. Rather, individual genes are transcribed by RNA polymerase II into individual monocistronic RNAs encoding single proteins (see Section 3.3). Hence, genes whose protein products are required in parallel are present in the genome as individual genes whose expression must be coordinated. Moreover, such coordinately expressed genes are not even closely associated in the genome in a manner that might be thought to facilitate their regulation, but are very often present on different chromosomes within the eukaryotic nucleus. Thus, the production of a functional antibody molecule by the mammalian B cell requires the synthesis of both the immunoglobulin heavy- and light-chain proteins, which together make up the functional antibody molecule. The genes encoding the heavy- and light-chain proteins are, however, found on separate chromosomes; the heavy-chain locus being on chromosome 14 in humans whereas light-chain genes are found on both chromosomes 2 and

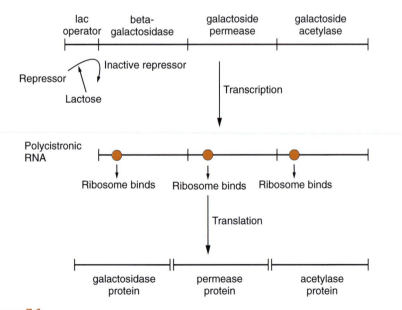

Figure 7.1

Structure of the *lac* operon of *E. coli*, in which the three genes are transcribed into one single RNA and translated into individual proteins following binding of the ribosome at three sites in the RNA molecule.

22. Similarly, the production of a functional globin molecule requires the association of an α-globin-type protein and a β-globin-type protein. Yet the genes encoding the various members of the α-globin family are on chromosome 16 in humans whereas the genes encoding the β-globins are found on chromosome 11.

   This difference between prokaryotes and eukaryotes is likely to reflect a need for greater flexibility in regulating gene expression in eukaryotes. Thus the bacterial arrangement, although providing a simple mechanism for coordinating the expression of different proteins, also means that, in general, expression of one protein will also necessitate expression of the other coordinately expressed proteins. In contrast, the eukaryotic system allows α-globin to be produced in parallel with β-globin in the adult organism but to be associated with another β-globin-like protein, namely γ-globin, in the fetus.

The Britten and Davidson model for the coordinate regulation of unlinked genes

This greater flexibility does, however, necessitate some means of coordinately regulating gene expression, allowing the production of α- and β-globin in the adult reticulocyte, γ- and β-globin in the embryonic reticulocyte, and the heavy and light chains of immunoglobulin in the antibody-producing B cell. A model of the mechanism of such coordinate regulation was put forward by Britten and Davidson (1969). They proposed that genes regulated in parallel with one another in response to a particular signal would contain a common regulatory element which would cause the activation of the gene in response to that signal (Fig. 7.2). Individual genes could contain more than one regulatory element, some of which would be shared with other genes, which in turn could possess

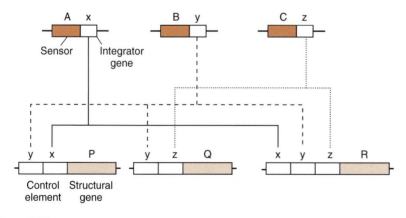

Figure 7.2

Britten and Davidson model for the coordinate expression of unlinked genes in eukaryotes. Sensor elements (A, B, C) detect changes requiring alterations in gene expression and switch on appropriate integrator genes (x, y, z) whose products activate the structural genes (P, Q, R) containing the appropriate control elements. Note the flexibility of the system whereby a particular structural gene can be activated with or without another structural gene by selecting which integrator gene is activated.

elements not present in the first gene. Specific signals causing gene activation would act by stimulating a specific integrator gene whose product would activate all the genes containing one particular sequence element. This mechanism would allow the observed activation of distinct but overlapping sets of genes in response to specific signals via the activation of particular integrator genes.

Although this model was proposed over 30 years ago, when our understanding of eukaryotic gene regulation was considerably more limited than at present, it continues to serve as a useful framework for considering gene regulation. In modern terms, the integrator gene would be considered as encoding a transcription factor (see Chapter 8) which binds to the regulatory sequences and activates expression of the corresponding genes. As discussed in Section 8.4, the activation of such a factor by a particular signal has now been shown to operate both by *de novo* synthesis of the factor (in an individual tissue or in response to a particular signal), as envisaged by Britten and Davidson, and by direct activation of a pre-existing protein, for example, by phosphorylation in response to the activating signal. The concept that such factors act by binding to DNA sequences held in common between coordinately regulated genes is now well established, and it is therefore necessary to consider the nature of these sequences. In the original model of Britten and Davidson (1969) it was considered that this role would be played by a class of repeated DNA sequences about 200–300 base pairs in length which had been shown to be present in many copies in the genome. It is now clear, however, that the target DNA binding sites for transcription factors are much shorter sequences, approximately 10–30 base pairs in length and these short sequence elements will now be discussed (for a review see Maniatis *et al.*, 1987).

7.2 Short sequence elements located within or adjacent to the gene promoter

Short regulatory elements

In prokaryotes the sequence elements that play an essential role in transcriptional regulation are located immediately upstream of the point at which transcription begins, and form part of the promoter which directs transcription of the gene. Such sequences can be divided into two classes, namely those found in all genes, which play an essential role in the process of transcription itself, and those found in one or a few genes, which mediate their response to a particular signal (Schmitz and Galas, 1979). It would be expected by analogy, therefore, that the sequences that play a role in the regulation of eukaryotic transcription would be located similarly, upstream of the transcription start site within or adjacent to the promoter. Hence, a comparison of such sequences in different genes should reveal both basic promoter elements necessary for all transcription (which would be present in all genes) and those necessary for a particular pattern of regulation (which would be present only in genes exhibiting a specific pattern of regulation). The role of such sequences can be confirmed either by destroying them by deletion or mutation or by transferring them to other genes in an attempt to confer the specific pattern of regulation of the donor gene upon the recipient.

The 70 kDa heat-shock protein gene

To illustrate this method of analysis, we will focus upon the gene encoding the 70 kDa molecular weight heat-shock protein (hsp70) and compare it with other genes showing different patterns of regulation. As discussed in Section 4.2, exposure of a very wide variety of cells to elevated temperature results in the increased synthesis of a few heat-shock proteins, of which hsp70 is the most abundant. Such increased synthesis is mediated in part by increased transcription of the corresponding gene, which can be visualized as a puff within the polytene chromosome of *Drosophila* (for a review see Ashburner and Bonner, 1979). Hence, examination of the sequences located upstream of the start site for transcription in this gene should identify potential sequences involved in its induction by temperature elevation, as well as those involved in the general mechanism of transcription. The sequences present in this region of the *hsp70* gene which are also found in other genes are listed in Table 7.1, and their arrangement is illustrated in Fig. 7.3 (Williams *et al.*, 1989; for further information on these sequences see Davidson *et al.*, 1983, Jones *et al.*, 1988 and references therein).

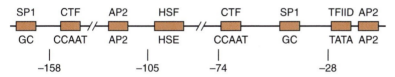

Figure 7.3

Transcriptional control elements in the human *hsp70* gene promoter. The protein binding to a particular site is indicated above the line and the corresponding DNA element below the line. These elements are described more fully in Table 7.1.

Table 7.1 Sequences present in the upstream region of the *hsp70* gene which are also found in other genes

Name	Consensus	Other genes containing sequences
TATA box	TATA A/T A A/T	Very many genes
CCAAT box	TGTGGCTNNNAGCCAA	α- and β-globin, herpes simplex virus thymidine kinase, cellular oncogenes c-*ras*, c-*myc*, albumin, etc.
Sp1 box	GGGCGG	Metallothionein IIA, type II procollagen, dihydrofolate reductase, etc.
CRE	T/G T/A CGTCA	Somatostatin, fibronectin, α-gonadotrophin, c-*fos*, etc.
AP2 box	CCCCAGGC	Collagenase, class 1 antigen H-2K^b, metallothionein IIA
Heat-shock consensus	CTNGAATNTTCTAGA	Heat-inducible genes *hsp83*, *hsp27*, etc.

A comparison of this type reveals a number of sequence motifs shared by the *hsp70* gene and other non-heat-inducible genes as well as one which is unique to heat-inducible genes. These include the TATA box which was discussed in Section 3.2 and several others which will be considered in turn.

The CCAAT and Spl boxes

The basal transcriptional complex containing TBP, RNA polymerase II and other associated factors can produce only a low rate of transcription. This rate is enhanced by the binding of other transcription factors to sites upstream of the TATA box which enhances either the stability or the activity of the basal complex. The binding sites for several of these factors are present in a wide variety of different genes with different patterns of activity. These sites act as targets for the binding of specific factors which are active in all cell types and their binding therefore results in increased transcription in all tissues.

An example of this is the Spl box, two copies of which are present in the hsp70 gene promoter (Fig. 7.3). This GC-rich DNA sequence binds a transcription factor known as Spl which is present in all cell types. Similarly, the CCAAT box, which is located upstream of the start site of transcription of a wide variety of genes (including the *hsp70* gene) which are regulated in different ways, is also believed to play an important role in allowing transcription of the genes containing it by binding constitutively expressed transcription factors.

The heat-shock element

In contrast to these very widespread sequence motifs, another sequence element in the *hsp70* gene is shared only with other genes whose transcription is increased in response to elevated temperature. This sequence is found 62 bases upstream of the start site for transcription of the *Drosophila hsp70* gene and at a similar position in other heat-inducible genes (Davidson *et al.*, 1983). This heat-shock element (HSE) is therefore believed to play a critical role in mediating the observed heat inducibility of transcription of these genes. In order to confirm that this is the case, it is necessary to transfer this sequence from the *hsp70* gene to another gene, which is not normally heat inducible, and show that the recipient gene now becomes inducible. This was achieved by Pelham (1982) who transferred the HSE onto the non-heat-inducible thymidine kinase (*tk*) gene taken from the eukaryotic virus, herpes simplex. When the hybrid gene was introduced into cells and the temperature subsequently raised, increased thymidine kinase production was detected, showing that the HSE had rendered the *tk* gene inducible by elevated temperature (Fig. 7.4).

This experiment therefore proves that the common sequence element found in the heat-inducible genes is responsible directly for their heat inducibility. The manner in which these experiments were carried out also permits a further conclusion with regard to the way in which this sequence acts. Thus, the HSE used by Pelham was taken from the *Drosophila hsp70* gene and, in this cold-blooded organism, would be activated normally by the thermally stressful temperature of 37°C. The cells into which the hybrid gene was introduced, however, were mammalian cells which grow

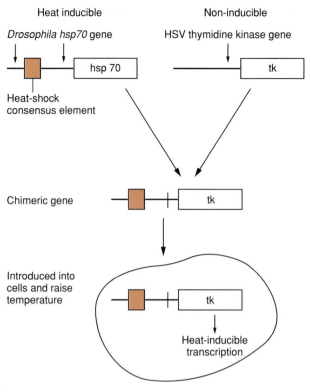

Heat inducible

Drosophila hsp70 gene

Non-inducible

HSV thymidine kinase gene

hsp 70

tk

Heat-shock
consensus element

Chimeric gene

tk

Introduced into
cells and raise
temperature

tk

Heat-inducible
transcription

Figure 7.4

Demonstration that the heat-shock consensus element mediates heat inducibility. Transfer of this sequence to a gene (thymidine kinase) which is not normally inducible renders this gene heat inducible.

normally at 37°C and only express the heat-shock genes at the higher temperature of 42°C.

In these experiments the hybrid gene was induced only at 42°C, the heat-shock temperature characteristic of the cell into which it was introduced, and not at 37°C, the temperature characteristic of the species from which the DNA sequence came. This means that the HSE does not possess some form of inherent temperature sensor or thermostat which is set to go off at a particular temperature, since in this case the *Drosophila* sequence would activate transcription at 37°C, even in mammalian cells. Rather, it must act by being recognized by a cellular protein which is activated in response to elevated temperature and, by binding to the heat-shock element, produces increased transcription. Evidently, although the elements of this response are conserved sufficiently to allow the mammalian protein to recognize the *Drosophila* sequence, the mammalian protein will, of course, only be activated at the mammalian heat-shock temperature and hence induction will only occur at 42°C.

Hence these experiments not only provide evidence for the importance of the HSE in causing heat-inducible transcription, but also indicate that it acts by binding a protein. This indirect evidence that the HSE acts by binding a protein can be confirmed directly by using a variety of techniques which allow the proteins binding to a specific DNA sequence to be analyzed (see Section 7.2 for a description of some of these methods). When an analysis of this type is carried out on the upstream regions of the heat-shock genes, it can be shown that in non-heat-shocked cells, TBP is bound to the TATA box and the GAGA factor is bound to upstream sequences (Fig. 7.5a). The binding of the GAGA factor is believed to displace a nucleosome and create the DNaseI hypersensitive sites observed in these genes in non-heat-shocked cells (see Section 6.7). By contrast, in heat-shocked cells, where high level transcription of the gene is occurring, an additional protein, which is bound to the HSE, is detectable on the upstream region (Fig. 7.5b).

Hence the induction of the heat-shock genes is indeed accompanied by the binding of a protein, known as the heat-shock factor (HSF), to the HSE, as suggested by the experiments of Pelham (1982). The binding of this factor to a gene whose chromatin structure has already been altered to render it potentially activatable, results in stimulation of transcription exactly as discussed in Chapter 6 and illustrated in Fig. 6.39. In agreement with this, the purified heat-shock factor can bind to the HSE and stimulate the transcription of the *hsp70* gene in a cell-free nuclear extract, while having no effect on the transcription of the non-heat-inducible actin gene (Parker and Topol, 1984).

The activation of these heat-inducible genes can therefore be fitted very readily into the Britten and Davidson scheme for the regulation of gene expression. The HSE represents the common sequence present in the similarly regulated genes, whereas the heat-shock factor would represent the product of the integrator gene which regulates their expression. Unlike the original model, however, it is clear that the heat-shock factor is not synthesized *de novo* in response to thermal stress. Rather it is present in unstressed cells (Parker and Topol, 1984) and can activate the heat-shock genes following exposure to elevated temperature even in the presence of protein synthesis inhibitors preventing its *de novo* synthesis (Zimarino and Wu, 1987). It is clear, therefore, that upon heat shock a previously inac-

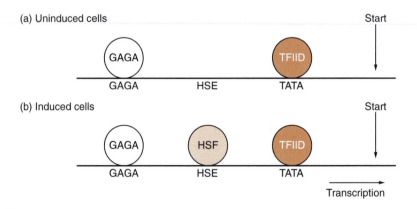

Figure 7.5

Proteins binding to the promoter of the *hsp70* gene before (a) and after (b) heat shock.

tive factor is activated to a DNA-binding form by a post-translational modification involving an alteration of the protein itself. This activation has been produced in an isolated cell-free nuclear extract by elevated temperature, and appears to involve a temperature-dependent change in the protein from a monomer to a trimer which allows it to bind to the heat-shock element and activate transcription (for a review see Morimoto, 1998). (For further discussion of the mechanism of HSF activation see Section 8.4.)

Other response elements

Although some modification of the Britten and Davidson model, with regard to the activation of the transcription factor itself, is therefore necessary, the heat-shock system does provide a clear example of the role of short common sequences within the promoters of particular genes in mediating their common response to a specific stimulus. A number of similar elements, which are found in the promoters of genes activated by other signals, have now been identified and have been shown to be capable of transferring the specific response to another marker gene (reviewed by Davidson *et al.*, 1983). A selection of such sequences is listed in Table 7.2.

As indicated in Table 7.2, these sequences act by binding specific proteins which are synthesized or activated in response to the inducing signal. Such transcription factors are discussed further in Chapter 8. It is noteworthy, however, that many of the sequences in Table 7.2 exhibit

Table 7.2 Sequences that confer response to a particular stimulus

Consensus sequences	Response to	Protein factor	Gene containing sequences
CTNGAATNTT CTAGA	Heat	Heat-shock transcription factor	*hsp70, hsp83, hsp27*, etc.
T/G T/A CGTCA	Cyclic AMP	CREB/ATF	Somatostatin, fibronectin, α-gonadotrophin *c-fos, hsp70*
TGAGTCAG	Phorbol esters	AP1	Metallothionein IIA, α-antitrypsin, collagenase
CC(A/T)$_6$GG	Growth factor in serum	Serum response factor	*c-fos, Xenopus* γ-actin
RGRACANNN TGTYCY	Glucocorticoid	Glucocorticoid receptor	Metallothionein IIA, tryptophan oxygenase, uteroglobin, lysozyme
RGGTCANNN TGACCY	Estrogen	Estrogen receptor	Ovalbumin, conalbumin, vitellogenin
RGGTCATGA CCY	Thyroid hormone	Thyroid hormone receptor	Growth hormone, myosin heavy chain
TGCGCCCGCC AGTTTCNN binding TTTCNC/T	Heavy metals Interferon-α	Mep-1 Stat-1 Stat-2	Metallothionein genes Oligo A synthetase, guanylate-binding protein
TTNCNNNAA	Interferon-γ	Stat-1	Guanylate-binding protein, Fc γ receptor

N, any base; R, purine; Y, pyrimidine.

dyad symmetry, a similar sequence being found in the 5' to 3' direction on each strand. The estrogen response element for example, has the sequence:

5' AGGTCANNNTGACCT 3'
3' TCCAGTNNNACTGGA 5'

the two halves of the ten-base palindrome being separated by three random bases. Such symmetry in the binding sites for these transcription factors indicates that they bind to the site in a dimeric form consisting of two protein molecules.

Various sequences that confer response to several different signals have therefore been identified. Exactly as suggested by Britten and Davidson, one gene can possess more than one such element, allowing multiple patterns of regulation. Thus comparison of the sequences listed in Table 7.2 with those contained in the *hsp70* gene, listed in Table 7.1, reveals that, in addition to the HSE, this gene also contains the cyclic AMP response element (CRE) which mediates the induction of a number of genes, such as that encoding somatostatin, in response to treatment with cyclic AMP. Similarly, while genes may share particular elements, flexibility is provided by the presence of other elements in one gene and not another, allowing the induction of a particular gene in response to a given stimulus which has no effect on another gene. For example, although the *hsp70* gene and the metallothionein IIA gene share a binding site for the transcription factor AP2, only the metallothionein gene has a binding site for the glucocorticoid receptor, which confers responsivity to glucocorticoid hormone induction, and hence only this gene is inducible by hormone treatment.

In some cases, the sequence elements that confer response to a particular stimulus can be shown to be related to one another. Thus the sequence mediating response to glucocorticoid treatment is similar to that which mediates response to another steroid, namely estrogen. Similarly, both the estrogen- and thyroid hormone-responsive elements contain identical sequences showing dyad symmetry forming a palindromic repeat of the sequence GGTCA. In the estrogen-responsive element, however, the two halves of this dyad symmetry are separated by three bases which vary between different genes, whereas in the thyroid hormone-responsive element the two halves are contiguous (Table 7.3a; see Gronemeyer and Laudet, 1995; Khorasanizadeh and Rastinejad, 2001).

In addition, as well as being arranged as palindromic repeats, the GGTCA core sequence can also be arranged as two direct repeats (Table 7.3b). In this arrangement, the spacing between the two repeats can again regulate which hormone produces a response, with a spacing of four bases forming an alternative thyroid hormone response element whilst a spacing of one base confers responsivity to 9-cis retinoic acid, two or five bases to all trans-retinoic acid and a spacing of four bases to vitamin D (Table 7.3b, see Gronemeyer and Moras, 1995; Khorasanizadeh and Rastinejad, 2001).

Such similarities in these different hormone response elements are paralleled by a similarity in the individual cytoplasmic receptor proteins which form a complex with each of these hormones and then bind to the corresponding DNA sequence. All of these receptors can be shown to be

Table 7.3   Relationship of various hormone response elements

(a) Palindromic repeats

Glucocorticoid	RGRACANNNTGTYCY
Estrogen	RGGTCANNNTGACCY
Thyroid	RRGTCA- - - TGACCY

(b) Direct repeats

9-*cis* retinoic acid	AGGTCAN$_1$AGGTCA
All-*trans* retinoic acid	AGGTCAN$_2$AGGTCA
	AGGTCAN$_4$AGGTCA
Vitamin D3	AGGTCAN$_5$AGGTCA
Thyroid hormone	AGGTCAN$_3$AGGTCA

N indicates that any base can be present at that position; R indicates a purine, i.e. A or G;
Y indicates a pyrimidine, i.e. C or T. A dash indicates that no base is present, the gap
having been introduced to align the sequence with the other sequences.

members of a large family of related DNA binding proteins whose
hormone and DNA binding specificities differ from one another. Hence a
particular DNA sequence confers a response to a particular hormone
because it binds the appropriate receptor which also binds that hormone.
The exchange of particular regions of these receptor proteins with those
of other family members has provided considerable information on the
manner in which sequence-specific binding to DNA occurs, and this is
discussed in Section 8.2.

   Although the sequence elements shown in Table 7.2 are all involved in
the response to particular inducers of gene expression, it is clear that other
short sequence elements or combinations of elements are involved in
controlling the tissue-specific patterns of expression exhibited by eukaryotic genes. Thus the octamer motif (ATGCAAAT), which is found in both
the immunoglobulin heavy- and light-chain promoters, can confer B-cell-
specific expression when linked to a non-regulated promoter (Wirth *et al.*,
1987). Similarly, short DNA sequences that bind liver-specific transcription factors have been identified in the region of the rat albumin
promoter, known to be involved in mediating the liver-specific expression
of this gene.

DNA binding by short sequence elements

In discussing short sequence elements which confer a particular pattern
of gene regulation we have assumed that they act by binding specific regulatory proteins and indeed the work of Pelham (1982) (see above) provided
indirect evidence that this is the case for the HSE. However, it is necessary
to prove directly that this is so for the HSE and other short DNA sequences.
A variety of methods have therefore been devised for demonstrating that
regulatory protein(s) bind to a particular DNA sequence and for characterizing these factors. Two of the most important of these will be described
(for further discussion of these and other methods, see Latchman, 2004
and for practical details of how to carry them out see Latchman, 1999).

The DNA mobility shift assay

When a particular DNA sequence is first identified as being involved in a specific pattern of gene regulation, the next step is usually to carry out a DNA mobility shift assay in which the DNA sequence and cell extract are mixed and protein binding to the DNA is detected by the DNA moving more slowly in the gel. Hence, protein binding to DNA results in the appearance of a retarded band, giving the technique its alternative names of band shift or gel retardation assay (Fig. 7.6, track A) (Methods Box 7.1).

Once such a band has been detected, one can for example, carry out the assay using extracts of different cell types or of cells treated in different ways to see if the presence or absence of the DNA binding activity correlates with the pattern of gene activity conferred by the DNA sequence

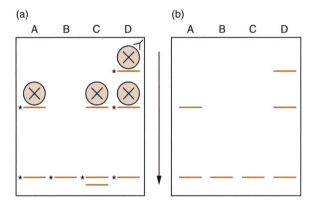

Figure 7.6

DNA mobility shift assay. Panel (a) shows schematically the factors binding to a radioactively labeled oligonucleotide (*) whilst panel b shows what will be seen when the gel is visualized by autoradiography. In track A, a protein (X) in the extract used has bound to the DNA resulting in a retarded radioactive complex of lower mobility. In track B, the cell extract used does not have the DNA binding protein so no retarded complex forms. In track C, a large excess of unlabeled competitor DNA has been added. It binds protein X but no retarded band is seen on autoradiography since the competitor DNA is not radioactive. In track D, an antibody to protein X has been added so that a super-shifted complex of even lower mobility is formed.

METHODS BOX 7.1

DNA mobility shift assay (Fig. 7.6)

- Radioactively label DNA fragment or oligonucleotide containing the DNA sequence of interest.
- Mix with whole cell or nuclear extract and incubate.
- Run the mixture on a non-denaturing gel and observe the position of the radioactive bands by autoradiography.
- Detect protein binding to the DNA by the appearance of a retarded band in the autoradiograph (Fig. 7.6).

(Fig. 7.6, tracks A and B). Similarly, it is possible to characterize the DNA binding specificity of the binding factor by adding a large excess of unlabeled DNA sequences as well as the radioactively labeled sequence. If the unlabeled "competitor" sequence can also bind the factor, it will do so and since it is in excess it will prevent the labeled sequence from binding the factor. However, since the competitor DNA is unlabeled, no radioactive band will be detected (Fig. 7.6, track C). This method can therefore be used to determine whether a novel DNA binding activity has a similar binding specificity to a known protein and is therefore likely to be identical or closely related to it. This relationship between the binding activity and known proteins can also be probed if an antibody to the known protein is available. Thus, the antibody can be added to the assay to see whether it binds to the DNA binding factor forming a complex of even lower mobility, which is known as a "super-shifted" complex (Fig. 7.6, track D).

DNaseI footprinting assay

The DNA mobility shift assay is therefore of great value in characterizing the proteins binding to a specific DNA sequence. However, it does not provide any information as to where the protein binds within the DNA sequence or its position relative to other proteins binding to adjacent DNA sequences in the regulatory region of a particular gene. This is achieved by the DNaseI footprinting assay in which binding of a protein protects its binding site in the DNA from digestion (Fig. 7.7; Methods Box 7.2).

Although technically more difficult to carry out, this method thus offers advantages over the DNA mobility shift assay. Firstly, it delineates the bases which are actually bound by the binding protein. Second, it is possible to visualize multiple footprints on a single DNA molecule, thereby elucidating the pattern of proteins bound to the promoter or regulatory region of a particular gene.

As with the DNA mobility shift assay, it is also possible to determine whether a particular binding activity/footprint is observed with extracts prepared from cells exposed to different treatments or from different cell

---

METHODS BOX 7.2

DNaseI footprinting assay (Fig. 7.7)

- Radioactively label DNA fragment or oligonucleotide containing the DNA sequence of interest *at one end only*.
- Mix with whole cell or nuclear extract and incubate.
- Digest the mixture with DNaseI to produce a series of DNA fragments each differing by only a single base.
- Run the DNA fragments on a *denaturing* polyacrylamide gel capable of resolving single base differences and detect radioactive bands by autoradiography.
- Visualize area where protein was bound as a gap in the ladder of DNA fragments due to the protein protecting this region of DNA from digestion (Fig. 7.7).

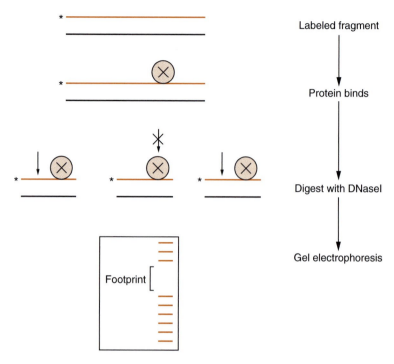

Figure 7.7

DNaseI footprinting assay. A protein (X) that binds to the test sequence, which is radioactively labeled at one end protects it from digestion by DNaseI. Hence, the DNA fragments corresponding to digestion in this region will not be formed and will appear as a "footprint" in the ladder of bands formed by DNaseI digestion at other, unprotected, parts of the DNA. Hence, protein binding to specific sequences can be detected and localized as a footprint.

types and relate this to the pattern of expression of the gene being studied. Similarly, the DNA sequence specificity of any binding activity can be studied by adding an excess of an unlabeled oligonucleotide containing a sequence related to that of a particular footprint. Thus, if the DNA binding activity producing the footprint can bind to the unlabeled sequence, it will do so and hence the footprint will disappear since that region of DNA will no longer be protected from digestion. DNaseI footprinting is therefore a valuable technique for examining the interaction of proteins with particular DNA sequences.

Chromatin immunoprecipitation

Both DNA mobility shift assays and DNaseI footprinting are carried out using purified DNA fragments and cell extracts. They therefore demonstrate what proteins can bind to a DNA sequence, rather than determine which proteins actually do so in the intact cell. The technique of chromatin immunoprecipitation (ChIP) overcomes this problem by using an antibody to a particular transcription factor to immunoprecipitate and purify the DNA fragments to which it is bound within the normal chromatin structure (Fig. 7.8; Methods Box 7.3).

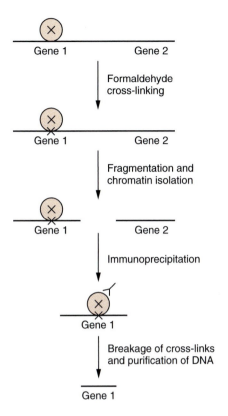

Figure 7.8

Chromatin immunoprecipitation assay in which a DNA fragment (gene 1) is purified on the basis that it binds a particular transcription factor (X) in the chromatin structure of the intact cell.

METHODS BOX 7.3

Chromatin immunoprecipitation (Fig. 7.8)

- Fix living cells with formaldehyde stably to cross-link transcription factors to their DNA binding sites.
- Fragment chromatin into small fragments and purify.
- Immunoprecipitate the transcription factor of interest and its target DNA, using an antibody to the transcription factor.
- Break the DNA–protein cross-links and isolate DNA.
- Characterize the isolated DNA.

Using this method, it is possible to identify whether a particular gene of interest is present in the immunoprecipitate under particular conditions, thereby testing whether the transcription factor is bound to the gene in intact cells under those conditions (Fig. 7.9a). Hence, the effect of specific treatments such as heat shock, steroids, etc., on the binding of a particular factor to a specific DNA binding-site can be investigated under different conditions within the natural chromatin structure of the gene (for a review see Orlando, 2000).

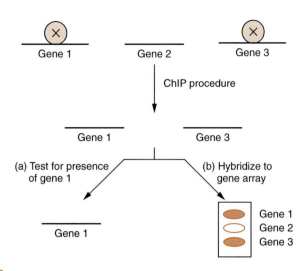

Figure 7.9

The DNA isolated in the ChIP procedure (Fig. 7.8) can be tested for the presence of a specific gene (gene 1, panel a) or hybridized to a gene array which will detect all the genes which bind the factor under test (genes 1 and 3, panel b).

In addition, however, the ChIP technique can be used in conjunction with DNA arrays or gene chips (see Section 1.3) in which all the DNA in a particular genome is arrayed on a filter. By labeling the immuno-precipitated DNA and hybridizing it to such a filter, all the target genes in the cell for a particular factor can be characterized (Ren *et al.*, 2000) (Fig. 7.9b).

This genome-wide location analysis, which combines ChIP assays and DNA arrays, will become of increasing importance as the genomes of more and more organisms are sequenced. Thus, in yeast a number of putative regulatory DNA sequences have been identified on the basis of their sequence and their conservation between different yeast strains (see Clifton *et al.*, 2003; Kellis *et al.*, 2003). By using genome-wide location analysis, such sequences can be tested for their ability to bind specific tran-scription factors, under different conditions. Hence, all the binding sites for a particular factor in the yeast genome can be defined and the effect on such binding of specific treatments can be assessed (see Lee *et al.*, 2002; Harbison *et al.*, 2004).

In this way, it is possible to begin to characterize the complex regula-tory networks which are characteristic of eukaryotes in which transcription factors evidently bind at large numbers of different sites across the genome. Indeed, a recent study using genome-wide location analysis showed that the yeast Ste12 transcription factor binds to distinct binding sites and therefore regulates distinct sets of genes during mating compared with filamentous growth (Zeitlinger *et al.*, 2003). A similar approach aimed at identifying all the regulatory elements in the human genome and defining the factors which bind to them is now being piloted (Encode Project Consortium, 2004) although such an analysis is obvi-ously much more complex in humans compared with the less complex yeast.

Mechanism of action of promoter regulatory elements

It is clear, therefore, that short DNA sequence elements located near the start site of transcription play an important role in regulating gene expression in eukaryotes. As indicated in Table 7.2 and discussed above, such sequences mediate transcriptional activation by binding a specific protein. This binding may give rise to gene activity in one of two ways (Fig. 7.10). Firstly, as discussed in Section 6.7 and illustrated by the glucocorticoid receptor, binding of a specific protein may result in displacement of a nucleosome and generation of a DNaseI hypersensitive site, allowing easy access to the gene for other transcription factors. The direct activation of

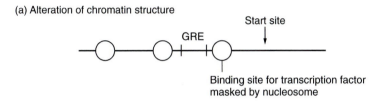

(a) Alteration of chromatin structure

Start site

GRE

Binding site for transcription factor masked by nucleosome

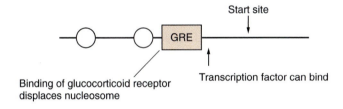

Start site

GRE

Binding of glucocorticoid receptor displaces nucleosome

Transcription factor can bind

(b) Interaction with other proteins

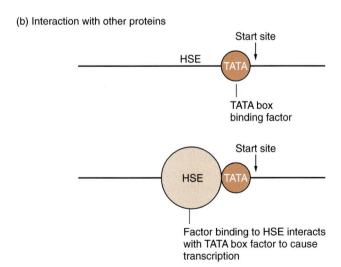

Start site

HSE   TATA

TATA box binding factor

Start site

HSE   TATA

Factor binding to HSE interacts with TATA box factor to cause transcription

Figure 7.10

Roles of short sequence elements in gene activation. These elements can either bind a factor that displaces a nucleosome and unmasks a binding site for another factor (a) or can bind a factor that directly activates transcription (b).

transcription by such factors constitutes the second mechanism of gene induction, and is illustrated both by the binding of other non-regulated factors to glucocorticoid-regulated genes following binding of the receptor and by the binding of the heat-shock factor to its consensus sequence in the heat-inducible genes. These factors are likely to act by interacting with proteins necessary for transcription, such as TBP or RNA polymerase itself. This interaction facilitates the formation of a stable transcription complex, which may enhance the binding of RNA polymerase to the DNA or alter its structure in a manner which increases its activity (for further discussion see Section 8.3; Latchman, 2004).

It should be noted that these two mechanisms of action are not exclusive. Thus the glucocorticoid receptor, whose binding to a specific DNA sequence displaces a nucleosome, also contains an activation domain capable of interacting with other bound transcription factors (see Section 8.3). Hence, following binding of the receptor, transcription is increased by inter-factor interactions between the receptor and other bound factors. Similarly, the heat-shock factor can induce the acetylation of histone H4 (Thomson et al., 2004) which results in a more open chromatin structure (see Section 6.6), indicating that it can modulate chromatin structure as well as stimulating the basal transcriptional complex. Hence, the binding of transcription factors to short DNA sequence elements can activate transcription by two distinct mechanisms with at least some factors utilizing both of these mechanisms.

Short DNA sequence elements act in a similar manner to gene regulatory elements in prokaryotes, and are located at a similar position in the promoter region close to the start site of transcription. Unlike the situation in prokaryotes, however, important elements involved in the regulation of eukaryotic gene expression are also found at very large distances from the site of initiation of transcription. The nature of such regulatory elements will now be discussed.

7.3 Enhancers

Regulatory sequences that act at a distance

The first indication that sequences located at a distance from the start site of transcription might influence gene expression in eukaryotes came with the demonstration that sequences over 100 bases upstream of the transcriptional start site of the histone H2A gene were essential for its high-level transcription (Grosschedl and Birnsteil, 1980). Moreover, although this sequence was unable to act as a promoter and direct transcription, it could increase initiation from an adjacent promoter element up to one hundred-fold when located in either orientation relative to the start site of transcription.

Subsequently, a vast range of similar elements have been described in both cellular genes and those of eukaryotic viruses, and have been called enhancers because, although they lack promoter activity and are unable to direct transcription themselves, they can dramatically enhance the activity of promoters (for reviews see Hatzopoulos et al., 1988; Muller et al., 1988; Thompson and McKnight, 1992; Pennisi, 2004). Hence, if an enhancer element is linked to a promoter, such as that derived from the β-globin gene, the activity of the promoter can be increased several hundred-fold.

Variation in the position and orientation at which an enhancer element was placed relative to the promoter has led to three conclusions with regard to the action of enhancers. These are: (a) an enhancer element can activate a promoter when placed up to several thousand bases from the promoter; (b) an enhancer can activate a promoter when placed in either orientation relative to the promoter; or (c) an enhancer can activate a promoter when placed upstream or downstream of the transcribed region, or within an intervening sequence which is removed from the RNA by splicing (see Section 5.2). These characteristics constitute the definition of an enhancer and are summarized in Fig. 7.11.

Tissue-specific activity of enhancers

Following the discovery of enhancers, it was rapidly shown that many genes expressed in specific tissues also contained enhancers. Such enhancers frequently exhibited a tissue-specific activity, being able to enhance the activity of other promoters only in the tissue in which the

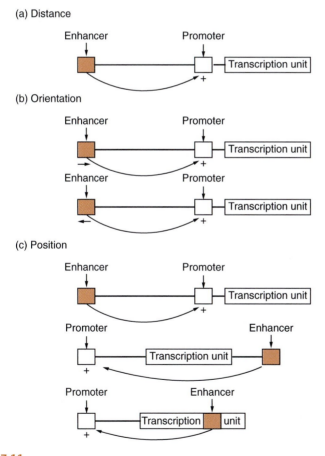

Figure 7.11

Characteristics of an enhancer element which can activate a promoter at a distance (a), in either orientation relative to the promoter (b), and when positioned upstream, downstream, or within a transcription unit (c).

gene from which they were derived is normally active and not in other tissues (Fig. 7.12). The tissue-specific activity of such enhancers acting on their normal promoters is likely, therefore, to play a critical role in mediating the observed pattern of gene regulation.

Thus, as previously discussed (Section 2.4), the genes encoding the heavy and light chains of the antibody molecule contain an enhancer located within the large intervening region separating the regions encoding the joining and constant regions of these molecules. When this element is linked to another promoter, such as that of the β-globin gene, it increases its activity dramatically when the hybrid gene is introduced into B cells. In contrast, however, no effect of the enhancer on promoter activity is observed in other cell types, such as fibroblasts, indicating that the activity of the enhancer is tissue-specific.

Similar tissue-specific enhancers have also been detected in genes expressed specifically in the liver (α-fetoprotein, albumin, α-1-antitrypsin), the endocrine and exocrine cells of the pancreas (insulin, elastase, amylase), the pituitary gland (prolactin, growth hormone), and many other tissues. The tissue-specific activity of these enhancer elements is likely to play a crucial role in the observed tissue-specific pattern of expression of the corresponding gene. Thus in the case of the insulin gene, early experiments involving linkage of different upstream regions of the gene to a marker gene, and subsequent introduction into different cell types, identified a region approximately 250 bases upstream of the transcriptional start site as being of crucial importance in producing high-level

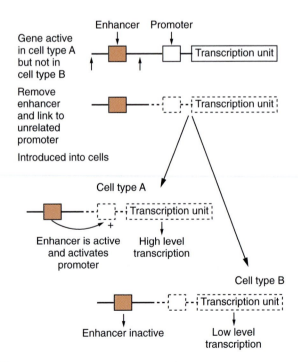

Figure 7.12

A tissue-specific enhancer can activate the promoter of its own or another gene only in one particular tissue and not in others.

expression in pancreatic endocrine cells. This position corresponds exactly to the position of the tissue-specific enhancer, indicating the importance of this element in gene regulation. Similarly, mutation of conserved sequences within the tissue-specific enhancers of genes expressed in the exocrine cells of the pancreas, such as elastase and chymotrypsin, abolishes the tissue-specific pattern of expression of these genes.

In the case of the insulin gene, the importance of the enhancer element in producing tissue-specific gene expression was further demonstrated by experiments in which this enhancer (together with its adjacent promoter) was linked to the gene encoding the large T antigen of the eukaryotic virus SV40, whose production can be measured readily using a specific antibody. The resulting construct was introduced into a fertilized mouse egg and the expression of large T antigen analyzed in all tissues of the transgenic mouse that developed following the return of the egg to the oviduct. Expression of large T antigen was detectable only in the pancreas and not in any other tissue (Fig. 7.13) and was observed specifically in the β cells of the pancreatic islets which produce insulin (Fig. 7.14; Hanahan 1985). The enhancer of the insulin gene is therefore capable of conferring the specific pattern of insulin gene expression on an unrelated gene *in vivo*.

Hence, enhancer elements constitute another type of DNA sequence which is involved in the activation of genes in a particular tissue or in response to a particular stimulus, as suggested by the Britten and Davidson model. As envisaged by the model, the binding of cell type-specific

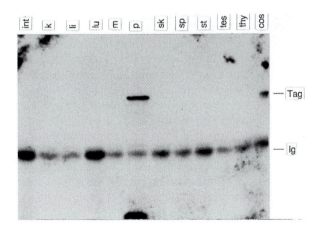

Figure 7.13

Assay for expression of a hybrid gene in which the SV40 T-antigen protein-coding sequence is linked to the insulin gene enhancer and promoter. The gene was introduced into a fertilized egg and a transgenic mouse containing the gene in every cell of its body isolated. Expression of the T antigen is assayed by immunoprecipitation of protein from each tissue with an antibody specific for T antigen. Note that expression of the T antigen is detectable only in the pancreas (p) and not in other tissues, indicating the tissue-specific activity of the insulin gene enhancer. The track labeled "cos" contains protein isolated from a control cell line expressing T antigen. The Ig band in all tracks is derived from the immunoglobulin antibody used to precipitate the T antigen. Photograph kindly provided by Professor D. Hanahan, from Hanahan, *Nature* **315**, 115–122 (1985), by permission of Macmillan Magazines Ltd.

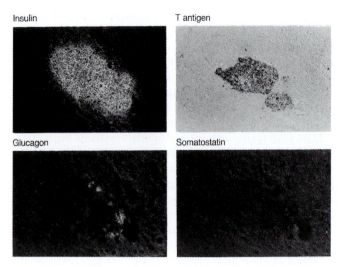

Figure 7.14

Immunofluorescence assay of pancreas preparations from the transgenic mice described in the legend to Fig. 7.13 with antibodies to the indicated proteins. Note that the distribution of T antigen parallels that of insulin and not that of the other pancreatic proteins. Photograph kindly provided by Professor D. Hanahan, from Hanahan, *Nature* **315**, 115–122 (1985), by permission of Macmillan Magazines Ltd.

proteins to the enhancer element has been demonstrated for many different enhancers, including those in the immunoglobulin and insulin genes. This has been achieved using the techniques described in Section 7.2 which can be used to determine the protein(s) binding to any specific DNA sequence regardless of whether it is located adjacent to the promoter or within an enhancer.

Therefore, in many cases the tissue-specific expression of a gene will be determined both by the enhancer element and sequences adjacent to the promoter. In the liver-specific pre-albumin gene, for example, gene activity is controlled both by the promoter itself, which is active only in liver cells, and by an upstream enhancer element, which activates any promoter approximately tenfold in liver cells and not at all in other cell types. Similarly, in the immunoglobulin genes, when the enhancer and the promoter itself are separated, both exhibit B-cell-specific activity in isolation but the maximal expression of the gene is observed only when the two elements are brought together.

The importance of enhancer elements in the regulation of gene expression therefore necessitates consideration of the mechanism by which they act.

Mechanism of action of enhancers

In considering the nature of enhancers, we have drawn a distinction between these elements, which act at a distance, and the sequences discussed in Section 7.2, which are located immediately adjacent to the start site of transcription. In fact, however, closer inspection of the sequences within enhancers indicates that they are often composed of the same sequences found adjacent to promoters. For example, the immunoglobulin heavy-chain enhancers contain the octamer motif (ATGCAAAT) which, as discussed previously, is also found in the immunoglobulin promoters. Use of DNA mobility shift and DNaseI foot-printing assays (see Section 7.2) has shown that these promoter and enhancer elements bind the identical B-cell-specific transcription factor (as well as a related protein found in all cell types) and play an important role in the B-cell-specific expression of the immunoglobulin heavy-chain gene. Interestingly, within the enhancer, the octamer motif is found within a modular structure containing binding sites for several different transcription factors, which act together to activate gene expression (Fig. 7.15).

The close relationship of enhancer and promoter elements is further illustrated by the *Xenopus hsp70* gene, in which multiple copies of the HSE are located at positions far upstream of the start site and function as a heat-inducible enhancer element when transferred to another gene (Bienz and Pelham, 1986). Similarly, the heat-shock element of the *Drosophila hsp70* gene, which we have used as the basic example of a promoter motif, has been shown to function as an enhancer when multiple copies are placed at a position well upstream of the transcriptional start site (Bienz and Pelham, 1986).

Enhancers therefore appear to consist of sequence motifs, which are also present in similarly regulated promoters and may be present within the enhancer associated with other control elements or in multiple copies. In many cases, an enhancer will therefore consist of an array of different sequence motifs which function together to produce strong activation of transcription. The array of different regulatory proteins which assemble on such an enhancer has been called an enhanceosome (for a review see Merika and Thanos, 2001).

It seems likely, therefore, that enhancers may activate gene expression by either or both of the mechanisms described previously for promoter elements, namely a change in chromatin structure leading to nucleosome displacement or by direct interaction with the proteins of the transcriptional apparatus. Indeed, the enhanceosome of the interferon β promoter has been shown to function by both these mechanisms. Thus, the enhanceosome complex of regulatory proteins which assembles on the enhancer first recruits a histone acetylase complex, stimulating the subsequent recruitment of the chromatin remodeling complex SWI/SNF. As described in Sections 6.6 and 6.7, this will open up the chromatin and then allow the binding of the complex of RNA polymerase II and associated proteins which will direct enhanced transcription (for reviews see Struhl, 2001; Cosma, 2002; Fry and Peterson, 2002) (Fig. 7.16).

In the case of chromatin structure changes, it is readily apparent that such changes caused by a protein binding to the enhancer could be prop-

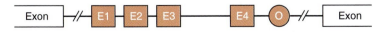

Figure 7.15

Protein-binding sites in the immunoglobulin heavy-chain gene enhancer. O indicates the octamer motif.

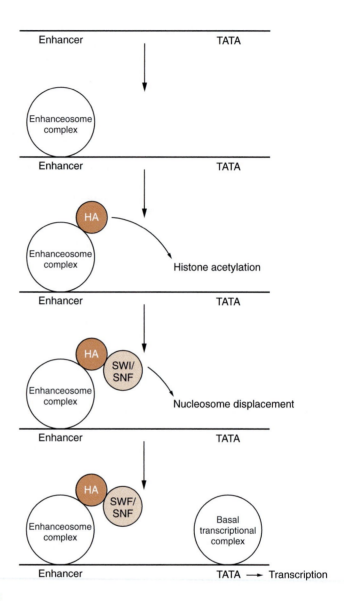

Figure 7.16

Activation of the β-interferon promoter involves binding of the enhanceosome complex to the enhancer. This facilitates the recruitment of a histone acetylase (HA) which in turn recruits the SWI/SNF complex, leading to nucleosome displacement and binding of the basal transcriptional complex.

agated over large distances in both directions, causing the observed distance, position and orientation independence of the enhancer. In agreement with this possibility, DNaseI hypersensitive sites have been mapped within a number of enhancer elements, including the immunoglobulin enhancer, and the nucleosome-free gap in the DNA of the eukaryotic virus SV40 (see Section 6.7) is located at the position of the enhancer.

At first sight, models involving the binding of protein factors to the enhancer followed by direct interaction with proteins of the transcriptional apparatus are more difficult to reconcile with the action at a distance characteristic of enhancers. Nonetheless, the binding to enhancers of very many proteins crucial for transcriptional activation (for a review see Hatzopoulos et al., 1988) suggests that enhancers can indeed function in this manner. Models to explain this postulate that the enhancer serves as a site of entry for a regulatory factor. The factor would then make contact with the promoter-bound transcriptional apparatus either by sliding along the DNA, or via a continuous scaffold of other proteins, or by the looping out of the intervening DNA (for a review of the mechanisms of long distance gene activation see Bulger and Groudine, 1999) (Fig. 7.17).

Of these possibilities, both the sliding model and the continuous scaffold model are difficult to reconcile with the observed large distances over which enhancers act. Similarly, these models cannot explain the observations of Atchison and Perry (1986) who found that the immunoglobulin enhancer activates equally two promoters placed 1.7 kb and 7.7 kb away on the same DNA molecule, since they would postulate that sliding or scaffolded molecules would stop at the first promoter (Fig. 7.18). Moreover, it has been shown that an enhancer can act on a promoter

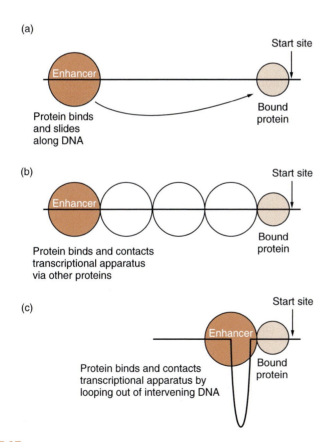

Figure 7.17

Possible models for the action of enhancers located at a distance from the activated promoter.

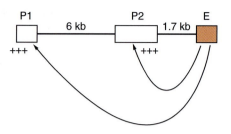

Figure 7.18

An enhancer activates an adjacent promoter and a more distant one equally well.

when the two are located on two separate DNA molecules linked only by a protein bridge, which would disrupt sliding or scaffolded molecules (Muller *et al.*, 1989).

Such observations are explicable, however, via a model in which proteins bound at the promoter and enhancer proteins have an affinity for one another and make contact via looping out of the intervening DNA. Moreover, such a model can explain readily the critical importance of DNA structure on the action of enhancers. Thus it has been shown that removal of precise multiples of ten bases (one helical turn) from the region between the SV40 enhancer and its promoter has no effect on its activity, but deletion of DNA corresponding to half a helical turn disrupts enhancer function severely.

Interestingly, some proteins which bind to enhancers actually bend the DNA so that interactions can occur between regulatory proteins bound at distant sites on the DNA (Fig. 7.19) (for a review see Werner and Burley, 1997). This is seen in the case of the T-cell receptor α chain gene enhancer where the LEF-1 factor binds to a site at the center of the enhancer and bends the DNA so that other regulatory transcription factors can interact with one another. Indeed, such effects are not confined to enhancers since binding of TBP to the TATA box in gene promoters (see Section 3.2) also

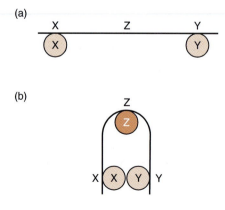

Figure 7.19

Binding of a factor (Z) which bends the DNA can produce interactions between other regulatory proteins (X and Y) which bind at distant sites on the DNA.

bends the DNA, suggesting that this may be a widespread mechanism for producing interactions between regulatory proteins.

7.4 Negatively acting sequence elements

Silencers

Thus far we have assumed that enhancers act in an entirely positive manner. In a tissue containing an active enhancer-binding protein, the enhancer will activate a promoter, whereas in other tissues where the protein is absent or inactive, the enhancer will have no effect. Such a mechanism does indeed appear to operate for the majority of enhancers which, when linked to promoters, activate gene expression in one or a few cell types and have no effect in other cell types. In contrast, however, some sequences appear to act in an entirely negative manner, inhibiting the expression of genes which contain them. Following the initial identification of such an element in the cellular oncogene c-*myc*, similar elements, which are referred to as silencers, have been defined in genes encoding proteins as diverse as collagen type II, growth hormone and glutathione transferase P. Like enhancers, silencers can act on distant promoters when present in either orientation but have an inhibitory rather than a stimulatory effect on the level of gene expression.

As with enhancer elements it is likely that these silencers act either at the level of chromatin structure by recruiting factors which direct the tight packing of adjacent DNA or by binding a protein which then directly inhibits transcription by interacting with RNA polymerase and its associated factors. Examples of silencers which appear to act in each of these ways have been observed (Fig. 7.20). Thus the silencer element located approximately 2 kb from the promoters of the repressed mating type loci in yeast plays a crucial role in organizing this region into the tightly packed structure characteristic of non-transcribed DNA (for review see Dillin and Rine, 1995).

Interestingly the silencer element appears to represent a site for attachment of the DNA to the nuclear matrix. Hence, it may act by promoting

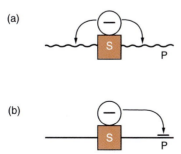

Figure 7.20

A silencer element (S) can inhibit activity of the promoter (P) either (a) by binding a protein which organizes the DNA into a tightly packed chromatin structure or (b) by binding an inhibitory transcription factor which represses promoter activity.

the further condensation of the chromatin solenoid to form a loop of DNA attached to the nuclear matrix which is the most condensed structure of chromatin (see Section 6.3 and Fig. 6.9). This could be achieved by recruiting protein factors which direct tight packing of the adjacent DNA, in the same way as the SWI/SNF complex and the GAGA factor direct the unpacking of the adjacent DNA (see Section 6.7). Indeed specific factors such as polycomb have been identified which have exactly this effect, inducing the ubiquitination of histone H2A and the methylation of histone H3 on lysine 27 (see Section 6.6), thereby promoting the tight packing of chromatin (Kirmizis *et al.*, 2004) (Fig. 7.21). Moreover, inactivation of polycomb by mutation results in aberrant activation of homeotic genes exactly as would be expected for a protein which promotes an inactive chromatin structure. Hence chromatin structure is controlled by the antagonistic effects of proteins such as polycomb which direct tight packing of the chromatin and others such as GAGA/trithorax and brahma (see Section 6.7) which direct its opening (for reviews see Orlando, 2003; Levine *et al.*, 2004).

As well as acting at the level of chromatin structure, it is also possible for silencers to act by binding negatively acting transcription factors which interact with the basal transcriptional complex of RNA polymerase II and associated factors (Fig. 7.20b). Thus, the silencer in the gene encoding lysozyme appears to act, at least in part, by binding the thyroid hormone receptor which in the absence of thyroid hormone has a directly inhibitory effect on gene activity (see Section 8.3, for further discussion of the mechanisms of gene repression).

Hence, just as enhancer elements can act at a distance to enhance gene expression by opening up the chromatin or by directly stimulating transcription, silencers can inhibit gene expression either by promoting a more tightly packed chromatin structure or by directly repressing transcription.

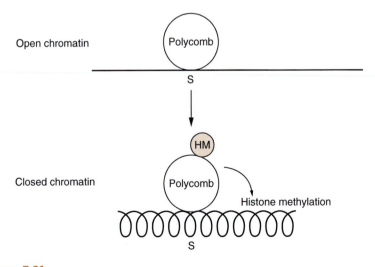

Figure 7.21

Binding of polycomb to a silencer sequence can recruit a histone methylase (HM) which produces a tightly-packed chromatin structure by methylating histones.

Insulators

The existence of elements such as enhancers or silencers which can act over large distances to alter gene expression leads to the question of how the action of such elements is confined to the appropriate gene or genes. Thus, some mechanism must exist to prevent the effect of enhancers or silencers spreading to other genes in adjacent regions of the chromosome and producing an inappropriate pattern of gene expression.

This problem is solved by the existence of insulator sequences which, as their name implies, block the spread of enhancer or silencer activity so that it is confined to the appropriate gene or genes and other adjacent genes are "insulated" from these effects (Fig. 7.22) (for reviews see Bell *et al.*, 2001; Labrador and Corces, 2002; West *et al.*, 2002). Such insulators are seen, for example, in the region of human chromosome 11 which contains several closely linked genes encoding β-globin-type proteins (see Section 7.1). Thus, as discussed in Section 7.5, this region contains a strong enhancer-like element, known as a locus control region (LCR), which stimulates the expression of the β-globin genes in erythroid cells. As shown in Fig. 7.23, the region containing the LCR and the β-globin genes is flanked by two insulator sequences which prevent the LCR inappropriately activating the flanking genes encoding the folate receptor and the odorant receptor gene (Fig. 7.23).

Interestingly, the activity of insulator elements can be regulated so that they are active in one situation and inactive in another. An example of

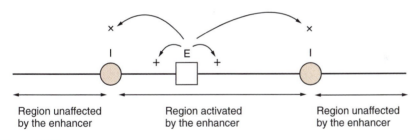

Region unaffected by the enhancer   **Region activated by the enhancer**   **Region unaffected by the enhancer**

Figure 7.22

An insulator element (I) limits the action of an enhancer (E) so that it only activates gene expression in a specific region and not in adjacent regions.

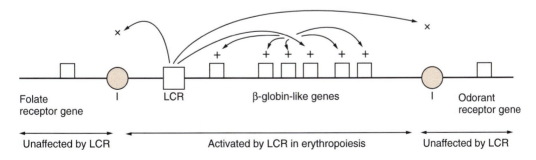

Folate receptor gene   I   LCR   β-globin-like genes   I   Odorant receptor gene

Unaffected by LCR   Activated by LCR in erythropoiesis   Unaffected by LCR

Figure 7.23

Insulator elements (I) flanking the β-globin gene cluster prevent the locus control region (LCR) from stimulating the expression of adjacent genes and confine its effect to the β-globin genes.

this was discussed in Chapter 6 (Section 6.8) where the imprinting control region prevents the enhancer of one H19 gene from activating the Igf2 gene on the maternal chromosome but allows it to do so on the paternal chromosome. Thus, the imprinting control region is acting as an insulator element blocking the ability of an enhancer to exert its stimulatory effect on an adjacent region on the maternal chromosome but having no effect on the paternal chromosome (see Fig. 6.46).

As discussed in Section 6.8, the imprinting control region/insulator is only active when it binds a protein, known as CTCF, which is essential for its insulator activity. Hence, as with the other DNA sequences discussed in this chapter, it appears that insulator sequences achieve their effect by binding specific proteins. Evidently, the manner in which such protein binding allows the insulator to block the action of an enhancer or silencer will depend on the mechanism by which the enhancer or silencer itself is actually acting to alter gene expression (see Sections 7.3 and 7.4).

Thus, for example, if an enhancer or a silencer acts to induce changes in chromatin structure which are propagated over large distances, then an insulator with a protein bound to it would presumably act by preventing this change in chromatin structure from spreading past the point where the protein was bound. In agreement with this idea, it has been shown that the polycomb and trithorax proteins which are known to act by altering chromatin structure (see Section 6.7) bind to specific insulator elements in *Drosophila* (Gerasimova and Corces, 1998).

Similarly, if looping of the DNA takes place so as to bring proteins bound to the promoter and to the enhancer into contact, the silencer might change this looping pattern so that an enhancer and a promoter which were separated by the insulator were not brought into contact. This could involve, for example, the DNA being thrown into smaller loops so that the promoter and the distant enhancer were on separate loops which would not allow contact between the bound proteins (Fig. 7.24) (for discussion see Bell *et al.*, 2001).

It is clear therefore that insulators play a key role in gene expression and its regulation by acting to ensure that the long distance effects of enhancer and silencer sequences are confined to the appropriate DNA region.

7.5 Locus control regions

The locus control region

When specific genes are introduced into fertilized mouse eggs and used to create transgenic mice, the introduced gene is frequently expressed at very low levels and this expression is not increased when multiple copies of the genes are introduced. Similarly different copies of the introduced gene which integrate into the host chromosomes at different positions are expressed at very different levels, suggesting that gene activity is being influenced by adjacent chromosomal regions. This effect occurs even when the gene is introduced together with its adjacent promoter and enhancer elements. This has led to the concept that some sequences, which are necessary for high level gene expression independent of the position of the gene in the genome, are absent from the introduced gene.

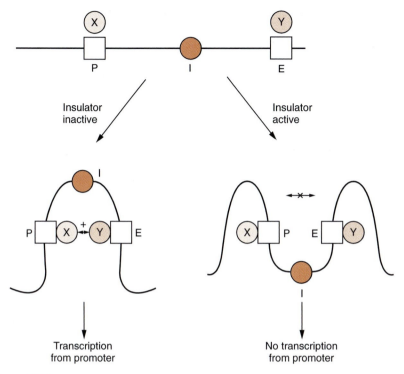

Figure 7.24

An insulator (I) may affect the looping pattern of the DNA so that when it is active proteins (X) bound at a promoter (P) cannot interact with proteins (Y) bound at an enhancer so preventing the enhancer from stimulating the promoter. In contrast, when the insulator is inactive, the looping pattern allows such interactions to occur and the enhancer can activate the promoter (compare Fig. 7.19).

This idea has been supported by studies in the β-globin gene cluster (see Section 7.1) which contains four functional β-globin-like genes and a non-functional pseudogene (Fig. 7.25). Thus a region located 10–20 kb upstream of the β-globin genes has been shown to confer high level, position independent expression when linked to a single globin gene and introduced into transgenic mice. Moreover, this element acts in a tissue-specific manner since its presence allows the globin gene to be expressed in the correct pattern with high level expression in erythroid cells and not in other cell types.

Most importantly, this element is not only required for expression of the β-globin genes in transgenic animals but also plays a role in the natural expression of the globin genes. Thus its deletion in human individuals leads to a lack of expression of any of the genes in the cluster. This occurs even in cases where all the genes, together with their promoter and enhancer regions, remain intact leading to a lethal disease known as Hispanic thalassemia in which no functional hemoglobin is produced (Fig. 7.25).

The role of this element in stimulating activity of all the genes in the β-globin cluster has led to its being called a locus control region (LCR: for reviews see Bulger and Groudine, 1999; Li *et al.*, 1999). As with enhancers,

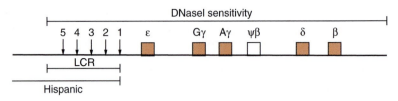

Figure 7.25

Organization of the β-globin gene locus showing the position of the locus control region (LCR) and the five DNaseI hypersensitive sites within it (arrows). The region showing enhanced DNaseI sensitivity in the reticulocyte lineage is indicated together with the functional genes (solid boxes) encoding epsilon globin (ε), the two forms of gamma globin (Gγ and Aγ), delta globin (δ) and beta globin (β) and the non-functional β-globin-like pseudogene (open box). The extent of the deletion in patients with Hispanic thalassemia is indicated by the line. Note that this deletion removes the LCR but leaves the genes themselves intact.

following its original definition in the β-globin cluster, similar LCR elements have been identified in other genes including the α-globin cluster, the major histocompatibility locus and the CD2 and lysozyme genes. It is clear therefore that LCRs constitute another important element essential for the correct regulation of gene expression.

Mechanism of action of LCRs

DNA sequence analysis of LCRs has indicated that they are rich in sequence motifs which are also found in promoter and enhancer elements and which could bind transcription factors. As with promoters and enhancers therefore it is in principle possible that the LCR could function either by direct interaction with the transcriptional apparatus or by affecting chromatin structure. In practice, however, the second of these is likely to be correct. Thus in many cases regions with LCR activity do not affect gene activity when introduced transiently into cells under conditions where the exogenous DNA is not packaged into chromatin but can do so when the same gene construct integrates into the host chromosome.

It is likely therefore that the LCR functions by affecting chromatin structure of the adjacent genes so that they can be transcribed. In agreement with this the region defined as the LCR on the basis of its functional activity contains five DNaseI hypersensitive sites (see Section 6.7) which appear early in erythroid development prior to the expression of any of the genes in the β-globin cluster (Fig. 7.25). Moreover, the LCR is able to induce the adjacent region containing the β-globin genes to assume the enhanced DNaseI sensitivity which is characteristic of active or potentially active genes (see Section 6.4) and the LCR marks the boundary of the region of the genome which shows this preferential sensitivity in erythroid cells (Fig. 7.25).

Thus, the presence of this region would render the adjacent DNA capable of being expressed in a position-independent manner, allowing high level, tissue-specific expression of the gene in transgenic mice, regardless of the position in the genome into which it integrated. In contrast a gene lacking this sequence would lack this stimulatory action and would also be subject to the influence of adjacent regulatory elements which might inhibit its expression (Fig. 7.26). In agreement with this idea, a transgene containing the CD2 locus control region is not inactivated when inserted into a tightly packed region of the DNA known as heterochromatin, whereas this does occur when the same gene is inserted into this region without the LCR.

As with other DNaseI hypersensitive sites (see Section 6.7) the sites in the LCR are likely to be free of nucleosomes and may be in a highly super-coiled form subject to torsional stress. This may allow them to propagate a wave of altered chromatin structure to the adjacent region causing the formation of the beads on a string structure characteristic of active or potentially active DNA throughout the adjacent region (see Section 6.4).

Interestingly, several LCRs contain MAR (matrix attachment region) sequences which are involved in the attachment of large domains or loops of chromatin to the nuclear matrix or scaffold (see Section 6.3). Thus a region controlled by a single LCR, such as that containing the β-globin genes, may constitute a single large domain or loop which is attached to the nuclear scaffold and whose chromatin structure is regulated as a single unit. In agreement with this idea, the MAR sequences contained within the LCR of the immunoglobulin gene have been shown to enhance the accessibility of the adjacent chromatin in B lymphocytes but not in other cell types (Jenuwein et al., 1997). Similarly, in T lymphocytes, the SATB1 protein which is involved in the looping of chromatin (see Sections 6.3 and 6.7) promotes histone acetylation and regulates gene expression over a 10 kb region of DNA (Cai et al., 2003).

Hence the LCR controls the structure of a specific region of DNA ensuring an appropriate pattern of gene regulation. Once an entire region of chromatin has been opened up in this way by the LCR, the promoter and enhancer elements of each individual gene in the region would direct its specific expression pattern. This is well illustrated in the β-globin gene cluster in which the DNaseI hypersensitive sites within the LCR and the

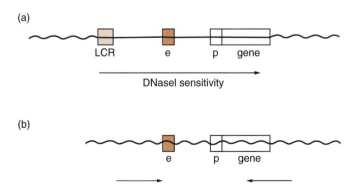

Figure 7.26

An inserted gene containing an LCR will be organized into an open chromatin conformation characterized by enhanced DNaseI sensitivity compared with flanking DNA (panel a). In contrast an inserted gene lacking the LCR will be subject to repression by adjacent regions which direct its organization into a closed chromatin organization (panel b).

overall DNaseI sensitivity appear very early in erythroid development prior to the expression of any of the β-globin genes. Subsequently the ε-globin, γ-globin and β-globin genes are expressed successively in embryonic, fetal and adult erythroid cells, respectively, with expression being preceded in each case by the appearance of DNaseI hypersensitive sites in the promoter/enhancer regions of the individual gene (Fig. 7.27). The LCR thus constitutes a control element which directs the chromatin structure of a large region of DNA, allowing the subsequent activation of individual genes within the cluster via the formation of DNaseI hypersensitive sites and transcription factor binding at their individual promoter and enhancer regions.

7.6 Regulation of transcription by RNA polymerases I and III

The very wide variety of patterns of expression in genes transcribed by RNA polymerase II is thus paralleled by a wide variety of DNA sequences which can produce such expression patterns by binding specific transcription factors which will be discussed in detail in Chapter 8. The much simpler patterns of expression of genes transcribed by RNA polymerases I and III (see Section 3.2) is paralleled by their simpler regulation which is generally achieved by altering the level of activity of basal transcription factors which are involved in the expression of all the genes transcribed by a particular polymerase (for reviews see White, 1994; Grummt, 2003).

Thus, for example, during mitosis, the transcription of the ribosomal genes is inhibited by regulating the activity of the basal transcription factors UBF and SL1/TIF-1B which are essential for transcription by RNA polymerase I (see Section 3.2). Upon entry into mitosis, both factors are phosphorylated, which decreases their activity and hence blocks tran-

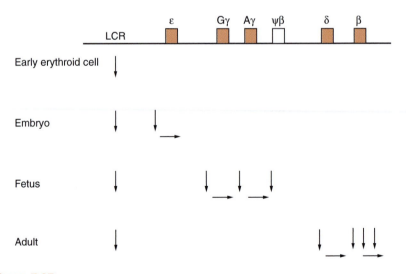

Figure 7.27

The appearance of multiple DNaseI hypersensitive sites within the LCR (shown as a single arrow for simplicity) in early erythroid cells, precedes the appearance of other hypersensitive sites adjacent to each individual gene which occurs later in development as these genes are sequentially expressed.

scription of the ribosomal genes by RNA polymerase I. After the cell completes mitosis, both factors are dephosphorylated and this activates SL1. However, ribosomal gene transcription does not occur until entry into the G1 phase of the next cell cycle, since it also requires phosphorylation of UBF on a different region of the protein, which is necessary for its activation (Fig. 7.28).

Interestingly, the fact that the TBP factor is involved in transcription by all three RNA polymerases (see Section 3.2) appears to be used to regulate the balance between transcription by the different RNA polymerases. Thus, the Dr1 inhibitory factor binds to TBP and inhibits its ability to recruit other transcription factors to RNA polymerase II and III gene promoters. It has no effect, however, on the activity of TBP within the RNA polymerase I transcriptional complex (White *et al.*, 1994). Hence Dr1 may regulate the balance between the transcription of the ribosomal genes which are the only genes transcribed by RNA polymerase I and all the other genes in the cell (Fig. 7.29).

The activity of RNA polymerase III is also altered in specific situations by altering the level of its specific basal transcription factors (see Section 3.2). Thus the activation of genes transcribed by RNA polymerase III following treatment of cells with serum is due to enhanced levels of active TFIIIC whereas the down-regulation of polymerase III transcription during the differentiation of embryonal carcinoma cells is due to a decrease in TFIIIB levels (for a review see White, 1994).

As with RNA polymerase II, these regulatory processes involve factors which exert their effects by binding at the DNA level. In addition, however, in the case of RNA polymerase III the presence of transcriptional control elements within the transcribed region (see Section 3.2), and hence in both the DNA and the RNA product is exploited in a unique regulatory mecha-

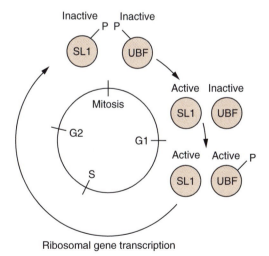

Figure 7.28

Regulation of transcription of the ribosomal RNA genes by RNA polymerase I. During mitosis, transcription is blocked by phosphorylation of the UBF and SL1 basal transcription factors. Following mitosis both factors are dephosphorylated which activates SL1. Subsequently, UBF is phosphorylated on a different region of the protein which activates it and allows transcription to occur during the cell cycle.

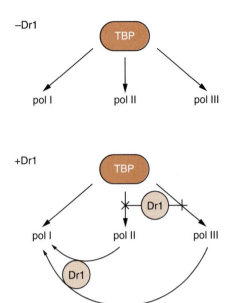

Figure 7.29

The Dr1 factor inhibits the activity of TBP within the basal transcription complex of genes transcribed by RNA polymerase II and III but not in that of the ribosomal genes which are transcribed by RNA polymerase I. Hence Dr1 may regulate the balance between the transcription of ribosomal genes by RNA polymerase I and that of all other genes in the cell by RNA polymerases II and III.

nism involving the basal transcription factor TFIIIA which binds to the internal regulatory sequence of the 5S ribosomal RNA genes.

Thus, the toad, *Xenopus laevis*, contains two types of 5S genes: the oocyte genes, which are transcribed only in the developing oocyte before fertilization, and the somatic genes, which are transcribed in cells of the embryo and adult. The internal control region of both these types of genes bind TFIIIA, whose binding is necessary for their transcription (see Section 3.2). However, sequence differences between the two types of gene (Fig. 7.30) result in a higher affinity of the TFIIIA factor for the somatic compared with the oocyte genes. Hence, the oocyte genes are only transcribed in the oocyte, where there are abundant levels of TFIIIA, and not in other cells, where the levels are only sufficient for activity of the somatic genes. In the developing oocyte TFIIIA is synthesized at high levels and transcription of the oocyte genes begins. As more and more 5S RNA molecules containing the TFIIIA binding site accumulate in the maturing oocyte, they bind the transcription factor. This factor is thus sequestered with the 5S RNA into storage particles and is unavailable for transcription of the 5S rRNA genes. Hence the level of free transcription factor falls below that necessary for the transcription of the oocyte genes and their transcription ceases, while transcription of the higher affinity somatic genes is unaffected (Fig. 7.31).

A simple mechanism involving differential binding of a transcription factor to two related genes is thus used in conjunction with the binding of the same factor to its RNA product to modulate expression of the 5S genes.

+46   +98

TCGGAAGCCAAGCAGGGTCGGGCCTGGTTAGTACTTGGATGGGAGACCGCCTGG
        G T

Figure 7.30

Sequence of the internal control element of the *Xenopus* somatic 5S genes. The two base changes in the oocyte 5S genes which result in a lower affinity for TFIIIA are indicated below the somatic sequence.

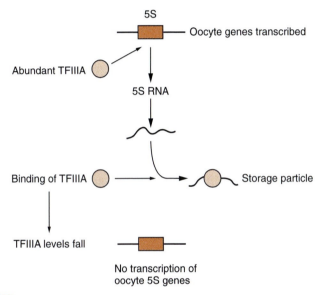

Figure 7.31

Sequestration of the TFIIIA factor by binding to 5S RNA made in the oocyte results in a fall in the level of the factor, switching off transcription of the oocyte-specific 5S genes. The somatic 5S genes, which have a higher affinity binding site for TFIIIA, continue to be transcribed.

7.7   Conclusions

In this chapter we have discussed a number of sequence elements, which can confer a particular pattern of gene regulation causing increased expression in response to a particular signal or in a particular tissue. Such sequences are present in a variety of locations close to or distant from the start site of transcription, in intervening sequences or, in the case of polymerase III transcription units, within the region encoding the RNA product itself. The interaction of the different positive and negative elements within these various regions controls the pattern of expression of a particular gene (for a review see Bonifer, 2000). Indeed, it has been argued that the complex pattern of sequences regulating a single gene (promoters, enhancers, silencers, etc.) is one of the defining features of a eukaryote and is necessary to produce the complex patterns of gene expression required in the adult organism and in development (for a review see Levine and Tjian, 2003). In all cases, however, such regions act by binding a protein which recognizes specific sequences within the control region. The activity of these proteins is crucial to the increased transcription mediated by the sequences to which they bind. The nature and function of these proteins is discussed in the next chapter.

References

Ashburner, M. and Bonner, J.J. (1979). The induction of gene activity in *Drosophila* by heat shock. *Cell* **17**, 241–254.

Atchison, M.L. and Perry, R.P. (1986). Tandem kappa immunoglobulin promoters are equally active in the presence of the kappa enhancer: implications for models of enhancer function. *Cell* **46**, 253–262.

Bell, A.C., West, A.G. and Felsenfeld, G. (2001). Insulators and boundaries: versatile regulatory elements in the eukaryotic genome. *Science* **291**, 447–450.

Bienz, M. and Pelham, H.R.B. (1986). Heat shock regulatory elements function as an inducible enhancer when linked to a heterologous promoter. *Cell* **45**, 753–760.

Bonifer, C. (2000). Developmental regulation of eukaryotic gene loci which *cis*-regulatory information is required? *Trends in Genetics* **16**, 310–315.

Britten, R.J. and Davidson, E.H. (1969). Gene regulation for higher cells: a theory. *Science* **165**, 349–358.

Bulger, M. and Groudine, M. (1999). Looping versus linking: toward a model for long-distance gene activation. *Genes and Development* **13**, 2465–2477.

Cai, S., Han, H.-J., and Kohwi-Shigematsu, T. (2003). Tissue-specific nuclear architecture and gene expression regulated by SATB1. *Nature Genetics* **34**, 42–51.

Clifton, P., Sudarsanam, P., Desikan, A., Fulton, L., Fulton, B., Majors, *et al.* (2003). Finding functional features in *Saccharomyces* genomes by phylogenetic footprinting. *Science* **301**, 71–76.

Cosma, M.P. (2002). Ordered recruitment: gene-specific mechanism of transcription activation. *Molecular Cell* **10**, 227–236.

Davidson, E.H., Jacobs, H.T. and Britten, R.J. (1983). Very short repeats and co-ordinate induction of genes. *Nature* **301**, 468–470.

Dillin, A. and Rine, J. (1995). On the origin of a silencer. *Trends in Biochemical Sciences* **20**, 231–235.

The ENCODE project consortium (2004). The ENCODE (ENCyclopedia Of DNA Elements) Project. *Science* **306**, 636–640.

Fry, C.J. and Peterson, C.L. (2002). Unlocking the gates to gene expression. *Science* **295**, 1847–1848.

Gerasimova, T.I. and Corces, V.G. (1998). Polycomb and trithorax group proteins mediate the function of a chromatin insulator. *Cell* **92**, 511–521.

Gronemeyer, H. and Laudet, V. (1995). Nuclear receptors. *Protein Profile* **2**, 1173–1308.

Gronemeyer, H. and Moras, D. (1995). Nuclear receptors: how to finger DNA. *Nature* **375**, 190–191.

Grosschedl, R. and Birnsteil, M.L. (1980). Spacer DNA sequences up-stream of the TATAAATA sequence are essential for promotion of histone H2A transcription *in vivo*. *Proceedings of the National Academy of Sciences of the USA* **77**, 7102–7106.

Grummt, I. (2003). Life on a planet of its own: regulation of RNA polymerase I transcription in the nucleolus. *Genes and Development* **17**, 1691–1702.

Hanahan, D. (1985). Heritable formation of pancreatic beta-cell tumours in transgenic mice expressing recombinant insulin/simian virus 40 oncogenes. *Nature* **315**, 115–122.

Harbison, C.T., Gordon, D.B., Lee, T.I., Rinaldi, N.J., Macisaac, K.D., Danford, T.W., *et al.* (2004). Transcriptional regulatory code of a eukaryotic genome. *Nature* **431**, 99–104.

Hatzopoulos, A.K., Schlokat, U. and Gruss, P. (1988). Enhancers and other cisacting sequences. In: *Transcription and Splicing* (eds B.D. Hames and D.M. Glover), IRL Press, Oxford, pp 43–96.

Jenuwein, T., Forrester, W.C., Fernandez-Herrero, L.A. Laible, G., Dull, M. and Grosschedl, R. (1997). Extension of chromatin accessibility by nuclear matrix attachment regions. *Nature* **385**, 269–272.

Jones, N.C., Rigby, P.W.J. and Ziff, E.B. (1988). *Trans*-acting protein factors and the regulation of eukaryotic transcription. *Genes and Development* **2**, 267–281.

Kellis, M., Patterson, N., Endrizzi, M., Birren, B. and Lander, E.S. (2003). Sequencing and comparison of yeast species to identify genes and regulatory elements. *Nature* **423**, 241–254.

Khorasanizadeh, S. and Rastinejad, F. (2001). Nuclear–receptor interactions on DNA-response elements. *Trends in Biochemical Sciences* **26**, 384–390.

Kirmizis, A., Bartley, S.M., Kuzmichev, A., Margueron, R., Reinberg, D., Green, R., and Farnham, P.J. (2004). Silencing of human polycomb target genes is associated with methylation of histone H3 Lys 27. *Genes and Development* **18**, 1592–1605.

Labrador, M. and Corces, V.G. (2002). Setting the boundaries of chromatin domains and nuclear organisation. *Cell* **111**, 151–154.

Latchman, D.S. (ed.). (1999). *Transcription Factors: a Practical Approach*. Oxford University Press, Oxford, New York.

Latchman, D.S. (2004). *Eukaryotic Transcription Factors*. Fourth Edition. Elsevier/Academic Press, pp 360.

Lee, T.I., Rinaldi, N.J., Robert, F., Odom, D.T., Bar-Joseph, Z., Gerber, G.K., et al. (2002). Transcriptional regulatory networks in *Saccharomyces cerevisiae*. *Science* **298**, 799–804.

Levine, M. and Tjian, R. (2003). Transcription regulation and animal diversity. *Nature* **424**, 147–151.

Levine, S.S., King, I.F.G. and Kingston, R.E. (2004). Division of labour in Polycomb group repression. *Trends in Biochemical Sciences* **29**, 478–485.

Li, Q., Harju, S. and Peterson, K.R. (1999). Locus control regions coming of age at a decade plus. *Trends in Genetics* **15**, 403–408.

Maniatis, T., Goodbourn, S. and Fischer, J.A. (1987). Regulation of inducible and tissue-specific gene expression. *Science* **236**, 1237–1245.

Merika, M. and Thanos, D. (2001). Enhanceosomes. *Current Opinion in Genetics and Development* **11**, 205–208.

Morimoto, R.I. (1998). Regulation of the heat shock transcriptional response: cross talk between a family of heat shock factors, molecular chaperones, and negative regulators. *Genes and Development* **12**, 3788–3796.

Muller, H.-P., Soga, J.W. and Schaffner, W. (1989). An enhancer stimulates transcription in *trans* when linked to the promoter via a protein bridge. *Cell* **58**, 767–777.

Muller, M.M., Gerster, T. and Schaffner, W. (1988). Enhancer sequences and the regulation of gene transcription. *European Journal of Biochemistry* **176**, 485–495.

Orlando, V. (2000). Mapping chromosomal proteins *in vivo* by formaldehyde-crosslinked-chromatin immunoprecipitation. *Trends in Biochemical Sciences* **25**, 99–104.

Orlando, V. (2003). Polycomb, epigenomes and control of cell identity. *Cell* **112**, 599–606.

Parker, C.S. and Topol, J. (1984). A *Drosophila* RNA polymerase II transcription factor binds to the regulatory site of an *hsp70* gene. *Cell* **37**, 273–283.

Pelham, H.R.B. (1982). A regulatory upstream promoter element in the *Drosophila hsp70* heat-shock gene. *Cell* **30**, 517–528.

Pennisi, E. (2004). Searching for the genome's second code. *Science* **306**, 632–634.

Ren, B., Robert, F., Wyrick, J.J., Aparicio, O., Jennings, E.G., Simon, I., *et al.* (2000). Genome-wide location and function of DNA binding proteins. *Science* 290, 2306–2309.

Schmitz, A. and Galas, D. (1979). The interaction of RNA polymerase and *lac* repressor with the *lac* control region. *Nucleic Acids Research* 6, 111–137.

Struhl, K. (2001). A paradigm for precision. *Science* 293, 1054–1055.

Thomson, S., Hollis, A., Hazzalin, C.A. and Mahadevan, L.C. (2004). Distinct stimulus-specific histone modifications at hsp70 chromatin targeted by the transcription factor heat shock factor-1. *Molecular Cell* 15, 585–594.

Thompson, C.C. and McKnight, S.L. (1992). Anatomy of an enhancer. *Trends in Genetics* 8, 232–236.

Werner, M.H. and Burley, S.K. (1997). Architectural transcription factors: proteins that remodel DNA. *Cell* 88, 733–736.

West, A.G., Gaszner, M., and Felsenfeld, G. (2002). Insulators: many functions, many mechanisms. *Genes and Development* 16, 271–288.

White, R.J. (1994). *RNA Polymerase III Transcription*. R.G Landes Company. Austen. TX, p 147.

White, R.J., Khoo, B.C.-E., Inostroza, J.A., Reinberg, D. and Jackson, S.P. (1994). Differential regulation of RNA polymerase I, II and III by the TBP repressor Dr1. *Science* 266, 448–450.

Williams, G.T., McClanahan T.K. and Morimoto, R.I. (1989). E1a transactivation of the human *hsp70* promoter is mediated through the basal transcriptional complex. *Molecular and Cellular Biology* 9, 2574–2587.

Wirth, T., Staudt, L. and Baltimore, D. (1987). An octamer oligonucleotide upstream of a TATA motif is sufficient for lymphoid specific promoter activity. *Nature* 329, 174–178.

Zeitlinger, J., Simon, I., Harbison, C.T., Hannett, N.M., Volkert, T.L., Fink, G.R. and Young, R.A. (2003). Program-specific distribution of a transcription factor dependent on partner transcription factor and MAPK signalling. *Cell* 113, 395–404.

Zimarino, V. and Wu, C. (1987). Induction of sequence-specific binding of *Drosophila* heat-shock activator protein without protein synthesis. *Nature* 327, 727–730.

Transcriptional control – transcription factors

8

SUMMARY

- DNA sequences which affect the rate of transcription act by binding regulatory proteins, known as transcription factors.
- Transcription factors have a modular structure in which different domains of the protein are responsible for different functions such as DNA binding or transcriptional activation.
- A number of different DNA binding domains have been identified and used to classify transcription factors into families with related DNA binding domains.
- Following DNA binding, transcription factors can either activate or repress transcription.
- Specific activation domains in transcription factors activate transcription by stimulating the assembly/activity of the basal transcription complex either directly or indirectly via co-activators.
- Inhibitory transcription factors can repress transcription either indirectly by blocking the effect of a positively acting factor or directly via the basal transcriptional complex.
- Transcription factors can themselves be regulated by controlling their synthesis or their activity so allowing them to produce a particular pattern of gene expression either in a specific cell type or in response to a specific signal.

8.1   Introduction

As discussed in Chapter 7, the expression of specific genes in particular cell types or tissues is regulated by DNA sequence motifs present within promoter or enhancer elements which control the alteration in chromatin structure of the gene that occurs in a particular lineage, or the subsequent induction of gene transcription. It was assumed for many years that such sequences would act by binding a regulatory protein which was only synthesized in a particular tissue or was present in an active form only in that tissue. In turn, the binding of this protein would result in the observed effect on gene expression. Indeed, as described in Section 7.2 cell extracts can be used in DNA mobility shift or DNase I footprinting assays to show that they contain protein(s) able to bind to a specific sequence.

The isolation and characterization of such factors proved difficult, however, principally because they were present in very small amounts.

Hence, even if they could be purified, the amounts obtained were too small to provide much information as to the properties of the protein.

This obstacle was overcome by the cloning of the genes encoding a number of different transcription factors. Two general approaches were used to achieve this (for reviews, see Latchman, 1999; Kadonaga, 2004). In one approach (Fig. 8.1), exemplified by the work of Kadonaga and Tjian (1986), the transcription factor Spl was purified by virtue of its ability to bind to its specific DNA binding site. The partial amino acid sequence of the protein was then obtained from the small amount of material isolated, and was used in conjunction with the genetic code to predict a set of DNA oligonucleotides, one of which would encode this region of the protein. The oligonucleotides were then hybridized to a complementary DNA library prepared from Spl-containing HeLa cell mRNA. A cDNA clone derived from the Spl mRNA must contain the sequence capable of encod-

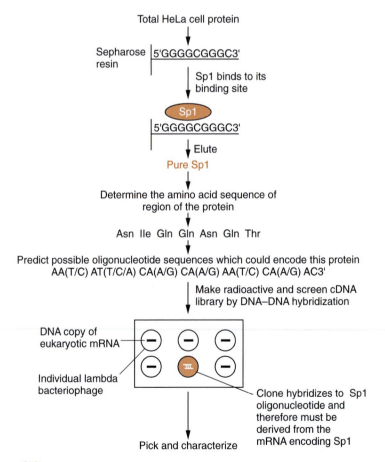

Figure 8.1

Isolation of cDNA clones for the Spl transcription factor by screening with short oligonucleotides predicted from the protein sequence of Spl. Because several different triplets of bases can code for any given amino acid, multiple oligonu-cleotides that contain every possible coding sequences are made. Positions at which these oligonucleotides differ from one another are indicated by the brack-ets containing more than one base.

ing the protein and hence will hybridize to the probe. In this experiment one single clone derived from the Spl mRNA was isolated by screening a library of 1 million recombinants prepared from the whole population of HeLa cell mRNAs (Kadonaga *et al.*, 1987).

An alternative, more direct, approach to the cloning of transcription factors is exemplified by the work of Singh *et al.* (1988) on the NFκB protein, which is involved in regulating the expression of the immunoglobulin genes in B cells (Fig. 8.2). As in the previous method, a cDNA library was constructed containing copies of all the mRNAs in a specific cell type. However, the library was constructed in such a way that the sequences within it would be translated into their corresponding proteins. This was achieved by inserting the cDNA into the coding region of the bacteriophage β-galactosidase gene, resulting in the translation of the eukaryotic insert as part of the bacteriophage protein. Most interestingly, these fusion proteins were capable of binding DNA with the same specificity as the original transcription factor encoded by the cloned mRNA. Hence the library could be screened directly with the radiolabeled DNA binding site for a particular transcription factor. A clone containing the mRNA for this factor, and hence expressing it as a fusion protein, bound the labeled DNA and could be identified readily and isolated.

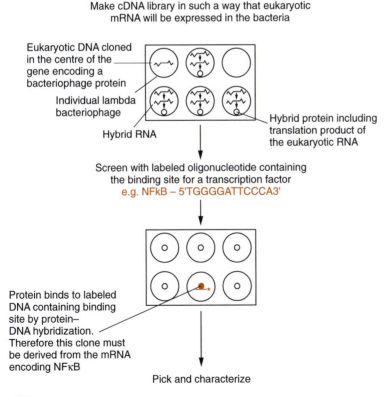

Figure 8.2

Isolation of cDNA clones for the NFκB transcription factor by screening an expression library with a DNA probe containing the binding site for the factor.

Unlike the previous method, this procedure involves DNA–protein rather than DNA–DNA binding and can be used without prior purification of the transcription factor, provided its binding site is known. Since most factors are identified on the basis of their binding to a particular site, this is not a significant problem and the use of these two methods has resulted in the isolation of the genes encoding a wide variety of transcription factors.

In turn, this has resulted in an explosion of information on these factors (for general reviews see Kadonaga, 2004; Latchman, 2004). Thus, once the gene for a factor has been cloned, Southern blotting (Section 2.2) can be carried out to study the structure of the gene, Northern blotting (Section 1.3) can be used to search for RNA transcripts derived from it in different cell types and related genes expressed in other tissues or other species can be identified.

More importantly, considerable information can be obtained from the cloned gene about the corresponding protein and its activity. Thus, not only can the DNA sequence of the gene be used to predict the amino acid sequence of the corresponding protein, but the existence of functional domains within the protein with particular activities can also be defined. As described above, if the gene encoding a transcription factor is expressed in bacteria, it continues to bind DNA in a sequence-specific manner. Hence if the gene is broken up into small pieces and each of these is expressed in bacteria (Fig. 8.3), the abilities of each portion to bind to DNA, to other proteins, or to a potential regulatory molecule can be assessed using, for example, DNA mobility shift assays, exactly as with cellular extracts (see Section 7.2). This mapping can also be achieved by transcribing and translating pieces of the DNA into protein fragments in the test tube and testing their activity in the same way.

Each of the domains identified in this way can be altered by mutagenesis of the DNA and subsequent expression of the mutant protein as before. The testing of the effect of these mutations on the activity mediated by the particular domain of the protein will thus allow the identification of the amino acids that are critical for each of the observed properties of the protein.

In this way large amounts of information have accumulated on individual transcription factors. Rather than attempt to consider each factor individually, we will focus on the properties necessary for such a factor and illustrate our discussion by referring to the manner in which these are achieved in individual cases.

It should be evident from the foregoing discussion that the first property such a factor requires is the ability to bind to DNA in a sequence-specific manner, and this is discussed in Section 8.2. Subsequently, the bound factor must influence transcription either positively or negatively by interacting with other transcription factors or with the RNA polymerase itself. Section 8.3 therefore considers the means by which DNA-bound transcription factors actually regulate transcription. Finally, in the case of transcription factors which activate a particular gene in one tissue only, some means must be found to ensure that the transcription factor is active only in that tissue. Section 8.4 discusses how this is achieved, either by the expression of the gene encoding the factor only in one particular tissue or by a tissue-specific modification which results in the activation of a factor present in all cell types.

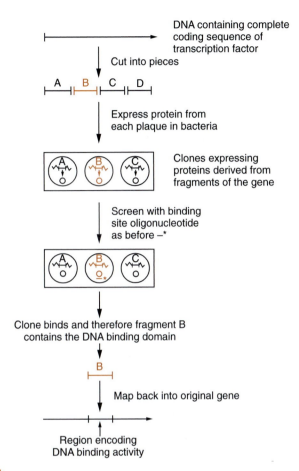

Figure 8.3

Mapping of the DNA binding region of a transcription factor by testing the ability of different regions to bind to the appropriate DNA sequence when expressed in bacteria.

8.2   DNA binding by transcription factors

Introduction

Extensive studies of eukaryotic transcription factors have identified several structural elements, which either bind directly to DNA or which facilitate DNA binding by adjacent regions of the protein (for reviews see Garvie and Wolberger, 2001; Latchman, 2004). These motifs will be discussed in turn, using transcription factors that contain them to illustrate their properties.

The helix-turn-helix motif

The homeobox

The small size and rapid generation time of the fruit fly *Drosophila melanogaster* has led to it being one of the best-characterized organisms genetically and a number of mutations which affect various properties of the organism have been described. These include mutations which affect the development of the fly, resulting, for example, in the production of additional legs in the position of the antennae (Fig. 8.4). Genes of this type are likely to play a crucial role in the development of the fly and, in particular, in determining the body plan, and are known as homeotic genes (for a review see Lawrence and Morata, 1994).

The critical role for the products of these genes, identified genetically, suggested that they would encode regulatory proteins which would act at particular times in development to activate or repress the activity of other genes encoding proteins required for the production of particular structures. This idea was confirmed when the genes encoding these proteins were cloned. Thus, these proteins were shown to be able to bind to DNA in a sequence-specific manner and to be able to induce increased transcription of genes which contained this binding site. Thus in the case of the homeotic gene *fushi tarazu* (ftz), mutation of which produces a fly with only half the normal number of segments, the protein has been shown to bind specifically to the sequence TCAATTAAATGA. When the gene encod-

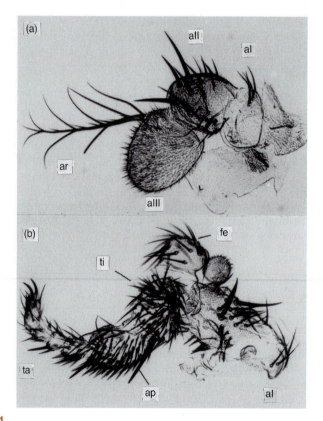

Figure 8.4

Effect of a homeotic mutation, which produces a middle leg (b) in the region that would contain the antenna of a normal fly (a). al, all and alll: first, second and third antennal segments; ar, arista; ta, tarsus; ti, tibia; fe, femur; ap, apical bristle. Photograph kindly provided by Professor W.J. Gehring, from Schneuwly, S., Klemenz, R., Gehring, W.J., *Nature* 1, 816–818 (1987), by permission of *Nature*.

ing this protein is introduced into *Drosophila* cells with a marker gene containing this sequence, transcription of the marker gene is increased. This up-regulation is entirely dependent on binding of the Ftz protein to this sequence in the promoter of the marker gene, since a 1 bp change in this sequence, which abolishes binding, also abolishes the induction of transcription (Fig. 8.5).

The product of another homeotic gene, the engrailed protein, binds to the identical sequence to that bound by Ftz. Its binding does not produce increased transcription of the marker gene, however, and indeed it prevents the activation by Ftz. Hence, the expression of Ftz alone in a cell would activate particular genes, whereas Ftz expression in a cell also expressing the engrailed product would have no effect (Fig. 8.6). In this way interacting homeotic gene products expressed in particular cells could control the developmental fate of the cells by regulating the expression of specific target genes.

Interestingly, there is evidence that the homeotic genes may be necessary not only for the actual production of a specific cell type but also for the long-term process of commitment to a particular cellular phenotype which was discussed in Section 6.2. Thus, in the case of the imaginal disks

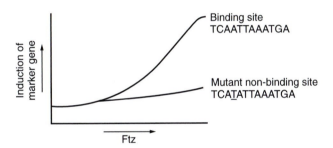

Figure 8.5

Effect of expression of the Ftz protein on the expression of a gene containing its binding site, or a mutated binding site containing a single base pair change which abolishes binding of Ftz.

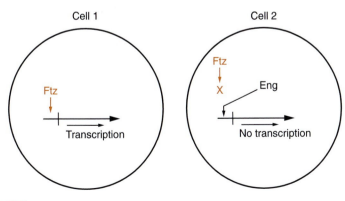

Figure 8.6

Blockage of gene induction by Ftz in cells expressing the engrailed (Eng) protein which binds to the same sequence as Ftz but does not activate transcription.

of *Drosophila* (Section 6.2), commitment to the production of a particular adult structure was maintained through many cell generations in the absence of differentiation. If, during this time, however, a mutation is introduced into one of the homeotic genes in the disk cells, inactivating it, when eventually the cells are allowed to differentiate they will produce the wrong structure. Thus, for example, if the homeotic gene *ultrabithorax* (*ubx*) is inactivated in a disk cell which normally gives rise to the haltere (balancer), these cells will produce wing tissue when allowed to differentiate. Thus the continual expression of homeotic genes within the cells is essential for their commitment to a particular pathway of differentiation.

A possible molecular mechanism for this is provided by the demonstration that the Ubx protein binds to its own promoter and up-regulates its own transcription. Hence once production of this protein has been induced, presumably during the commitment process, it will continue indefinitely and thus maintain this commitment (Fig. 8.7).

In addition to this mechanism, other processes maintain the chromatin structure of the homeotic genes in an active state so that transcription continues once the gene has initially been activated. This is achieved by members of the trithorax group of proteins, which include the GAGA and brahma factors discussed in Section 6.7. These factors bind to active homeotic genes and maintain them in an open chromatin structure, allowing transcription to continue. Conversely, other proteins of the polycomb group bind to inactive homeotic genes and produce an inactive chromatin structure so preventing their inappropriate activation (see Section 7.4) (for reviews see Orlando 2003; Levine *et al.*, 2004; Mohd-Sarip and Verrijzer, 2004). As might be expected, mutations of the genes encoding trithorax or polycomb group proteins result in gross abnormalities in the fly due to, respectively, a failure to activate the appropriate homeotic genes or their inappropriate activation (Fig. 8.8).

Interestingly, as noted in Section 6.2, the changes in commitment which occur in imaginal disks when the process of commitment breaks down after the disks are cultured for long periods are precisely those which occur in homeotic mutations. Hence, a change in the chromatin structure and expression of a specific homeotic gene in the imaginal disk will result in a change in the pattern of commitment similar to that which occurs when this gene is mutated.

The clear evidence that homeotic gene products regulate both their own genes and other genes by binding specifically to DNA has led to extensive

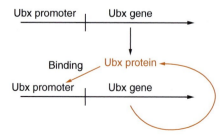

Figure 8.7

The Ubx protein activates its own promoter, producing a positive feedback loop maintaining high-level production of Ubx.

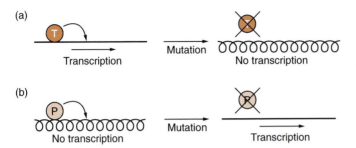

Figure 8.8

(a) Members of the trithorax group of proteins (T) maintain the active chromatin structure (solid line) of transcriptionally active homeotic genes. Inactivation of these proteins by mutation results in an inactive chromatin structure (wavy line) leading to a failure of transcription. (b) Members of the polycomb family (P) maintain the inactive chromatin structure (wavy line) of transcriptionally inactive homeotic genes. Their inactivation by mutation results in an active chromatin structure (solid line) leading to the inappropriate transcription of these genes.

investigation of their structure in order to identify the region that mediates this sequence-specific DNA binding. When the genes encoding these proteins were first cloned, it was found that they each contained a short related DNA sequence of about 180 bp capable of encoding 60 amino acids (Fig. 8.9), which was flanked on either side by sequences that differed dramatically between the different genes. The presence of this sequence, which was named the homeobox or homeodomain (for reviews see Gehring et al., 1994; Kornberg, 1993), in all these genes suggested that it plays a critical role in mediating their regulatory function. This suggestion was confirmed subsequently by the use of the homeobox as a probe to isolate other previously uncharacterized Drosophila regulatory genes.

The role of the homeodomain in DNA binding has been confirmed directly by the synthesis of the homeobox region of the Antennapedia protein without the remainder of the protein, either by expression in bacteria or by chemical synthesis, and showing that it can bind to DNA in the identical sequence-specific manner to that exhibited by the intact protein.

Antp	Arg Lys Arg Gly Arg Gln Thr Tyr Thr Arg Tyr Gln Thr Leu Glu Leu Glu Lys Glu Phe His Phe Asn Arg Tyr Leu Thr Arg Arg Arg
Ubx	Arg Thr His
Ftz	Ser Thr Ile

	Helix	Turn	Recognition helix

Antp	Arg Ile Glu Ile Ala His Ala Leu Cys Leu Thr Glu Arg Gln Ile Lys Ile Trp Phe Gln Asn Arg Arg Met Lys Trp Lys Lys Glu Asn
Ubx	Met Tyr Glu Leu Ile
Ftz	Asp Asn Ser Ser Ser Asp Arg

Figure 8.9

Amino acid sequences of several Drosophila homeodomains, showing the conserved helical motifs. Differences between the sequences of the Ubx and Ftz homeodomains from that of Antp are indicated, a blank denotes identity in the sequence. The helix-turn-helix region is indicated.

The localization of this DNA binding to a short region of the protein, only 60 amino acids in length, allows a detailed structural prediction of the corresponding protein, which reveals that it contains a so-called helix-turn-helix motif which is highly conserved between the different homeobox-containing proteins (for a review see Gehring *et al.*, 1994). In this motif, a short region which can form an α-helical structure is followed by a β-turn and then another α-helical region. The position of these elements in the homeodomain is shown in Fig. 8.9 and a diagram of the helix-turn-helix motif is given in Fig. 8.10.

The prediction that this structure exists in the DNA binding homeo-domain has been directly confirmed by X-ray crystallographic analysis. Moreover, by carrying out this analysis on the homeodomain bound to its DNA binding site, it has been shown that the helix-turn-helix motif does indeed contact DNA, with the second helix lying partly within the major groove where it can make specific contacts with the bases of the DNA (Fig. 8.11). This second helix (labeled the recognition helix in the homeobox sequence in Fig. 8.9) can thus mediate sequence-specific binding.

The presence of this structure therefore indicates how the homeobox proteins can bind specifically to particular DNA binding sequences, which is the first step in transcriptional activation of their target genes. The role of the helix-turn-helix motif in the recognition of specific sequences in the DNA has been demonstrated directly. Thus a mutation which changes a lysine at position nine of the recognition helix in the Bicoid protein to the glutamine found in the equivalent position of the Antennapedia protein results in the protein binding to DNA with the sequence speci-ficity of an Antennapedia rather than a Bicoid protein. Hence, not only does the helix-turn-helix motif mediate DNA binding, but differences in the precise sequence of this motif in different homeoboxes control the precise DNA sequence to which these proteins bind. Clearly, further struc-

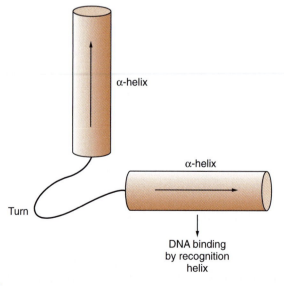

Figure 8.10

The helix-turn-helix motif.

Figure 8.11

Binding of the helix-turn-helix motif to DNA, with the recognition helix in the major groove of the DNA. Redrawn from Schleif, *Science* **241**, 1182–1187 (1988), by permission of Dr R. Schleif and AAAS.

tural and genetic studies of how this is achieved will throw considerable light on the way in which these proteins function.

The obvious importance of the homeobox in *Drosophila* and yeast prompted a search for proteins containing this element in other organisms. Indeed, homeobox-containing proteins which are expressed in specific cell types in the early embryo and play a key regulatory role have now been identified in a number of organisms including mammals (for reviews see Duboule, 2000; Briscoe and Wilkinson, 2004). Hence these proteins play a vital role in the processes regulating development in a variety of different organisms.

The POU domain

As well as the homeodomain proteins, another class of regulatory proteins has been identified which contains the homeobox as one part of a much larger, 150–160 amino acid, conserved region known as the POU domain (for reviews see Verrijzer and Van der Vliet, 1993; Ryan and Rosenfeld, 1997). Unlike the homeobox proteins, these regulatory proteins were not identified by mutational analysis or by homology to other regulatory proteins, but were characterized as transcription factors having a particular pattern of activity.

Thus the mammalian Oct-1 and Oct-2 proteins both bind to the octamer sequence (consensus ATGCAAATNA) in the promoters of genes such as the histone H2B gene and those encoding the immunoglobulins, and mediate the transcriptional activation of these genes. When the genes encoding these proteins were cloned they were found to possess a 150–160 amino acid sequence that was also found in the mammalian Pit-1 protein, which regulates gene expression in the pituitary by binding to a sequence related to, but distinct from, the octamer and in the protein encoded by the nematode gene *unc*-86, which is involved in sensory neuron development.

This POU (Pit-Oct-Unc) domain contains both a homeobox-like sequence and a second conserved domain, the POU-specific domain (Fig. 8.12). Although there are some differences between different POU

```
                                          POU-specific box
Pit        KSKLVEEP IDMDSPE IRELEQFANEFKVRRIKLGYTQTNVGEALAAVHG---SEFSQTTICRFENLQLSFKNACKLKAILSKWLEEAEQV
Oct-1      DTPSLEEPSDLE-----ELEQFAKTFKQRRIKLGFTQGDVGLAMGKLYG---NDFSQTTISRFEALNLSFKNMCKLKPLLEKWLNDAENL
Oct-2      PPSHPEEPSDLE-----ELEQFARTFKQRRIKLGFTQGDVGLAMGKLYG---NDFSQTTISRFEALNLSFKNMCKLKPLLEKWLNDAETM
unc-86     RYPIAPPTSDMDT-DPRQLETFAEHFKQRRIKLGVTQADVGKALAHLKMPGVGSLSQSTICRFESLTLSHNNMVALKPILHSWLEKAEE-
Consensus  ........D........LE FA..FK.RRIKLG.TQ..VG.A............SQ.TI.RFE.L.LS..N...LK..L..WL...AE..

                                          POU Homeo box
Pit        GULYNEK-----------VGAN-ERKRKRRTTISIAAKDALERHFGEHSKPSSQEIMRMAEELNLEKEVVRVWFCNRRQREKRVKTSLNQS
Oct-1      SSDSSLSSPSALNSP--GIEGL-SRRRKKRTSIETNVRFALEKSFLANQKPTSEEITMIADQLNMEKEVIRVWFCNRRQKEKRINPPSSGG
Oct-2      SVDSSLPSPNQLSSPSLGFDGLPGRRRKKRTSIETNVRFALEKSFLANQKPTSEEILLIAEQLHMEKEVIRVWFCNRRQKEKRINPCSAAP
unc-86     -AMKQKDTIGDIN----GILPN TDKKRKRTSIAAPEKRELEQFFKQQPRPSGERIASIADRLDLKKNVVRVWFCNQRQKQKRDFRSQFRA
Consensus  ...........................RT.I........LE..F......P....I...A..L....K.V.RVWFCN.RQ..KR........
```

Figure 8.12

Amino acid sequences of the POU proteins. The homeodomain and the POU-specific domain are indicated. Brown letters indicate regions of identity between the different POU proteins. The final line shows a consensus sequence obtained from the four proteins.

proteins, in general the isolated homeodomain of the POU proteins alone is sufficient for sequence-specific DNA binding, but unlike the classical homeobox, the binding is of relatively low affinity in the absence of the POU-specific domain. Hence both parts of the POU domain are required for high affinity sequence-specific DNA binding, indicating that the POU homeodomain and the POU-specific domain form two parts of a DNA binding element which are held together by a flexible linker sequence.

Like the POU homeodomain, the POU-specific domain can form a helix-turn-helix motif. It has therefore been suggested that the recognition helix from the POU-specific domain and that from the POU homeodomain bind to adjacent regions within the major groove of the DNA.

Interestingly, the POU factors illustrate a novel aspect of gene regulation, namely that the sequence of the DNA binding site to which a factor binds can influence its effect on gene expression. Thus, when the Oct-1 POU factor binds to its target sequence (ATGCAAAT) in cellular genes, it activates transcription only weakly. However, when it binds to its different target sequence (TAATGARAT; R = purine) in the herpes simplex virus (HSV) immediate-early genes, it does so in a distinct configuration. This allows it to bind the HSV protein VP16 which is a strong activator of transcription and hence strong activation occurs (for a review see Wysocka and Herr, 2003) (Fig. 8.13).

This type of effect is not confined to the recruitment of a viral protein by a cellular factor. Thus, when the Pit-1 protein binds to the prolactin gene promoter, it activates transcription. However, the growth hormone promoter has a distinct sequence which binds Pit-1 but produces a different configuration of the Pit-1 protein. This allows it to bind the cellular co-repressor NCo-R (see also Sections 6.6 and 8.3) and hence transcription is repressed (for reviews see Marx, 2000; Latchman, 2001) (Fig. 8.14).

Indeed, this effect is not confined to POU proteins. Thus, the NFκB factor discussed in Section 8.4 has a DNA binding domain unrelated to the POU domain. However, it has recently been shown that NFκB binds to different binding sites in different configurations. In turn, this affects its ability to bind co-activator molecules (see Section 8.3) and, therefore, to activate transcription (Leung et al., 2004; Natoli, 2004).

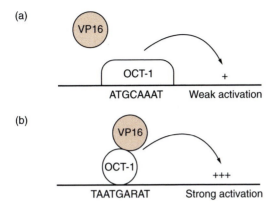

Figure 8.13

The Oct-1 factor binds to its distinct binding sites in cellular (a) or herpes simplex virus (b) genes in different configurations, only one of which allows binding of the viral activator protein VP16 which strongly activates transcription.

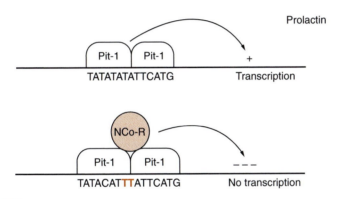

Figure 8.14

The binding sites for two molecules of the Pit-1 factor in the prolactin and growth hormone promoters differ in their sequence. Pit-1 therefore binds in different configurations resulting in activation of the prolactin promoter whilst on the growth hormone promoter, Pit-1 binds the NCo-R co-repressor and hence represses transcription.

Hence, in different types of DNA binding protein, the sequence of the DNA binding site can affect the configuration of the bound factor and, therefore, its ability to recruit other regulatory molecules and to activate or repress transcription.

It is clear, therefore, that the POU proteins represent a family of proteins, related to the homeobox proteins, which are likely to play a critical role in development. Thus inactivation of the Pit-1 gene leads to a failure of pituitary gland development, resulting in dwarfism in mouse and humans, while the *unc-86* mutation results in a failure to form specific neurones in the nematode.

Interestingly, the two highly conserved peptide sequences at either end of the POU domain (Fig. 8.12) were used by He *et al.* (1989) to isolate novel members of the POU family. Thus, they prepared highly degenerate oligonucleotides which contained all the possible DNA sequences able to encode these conserved sequences (Fig. 8.15). They then used these in a polymerase chain reaction (PCR) to amplify cDNA prepared from the mRNA of different tissues. The degenerate oligonucleotides amplified cDNAs derived from the mRNAs of novel POU proteins which, like the original POU proteins, contained the conserved sequences characteristic of such factors. Many of the novel factors isolated in this way have now also been shown to play critical roles in development.

Such sequence homology methods of isolating novel transcription factors are being used increasingly to clone novel factors, considerably supplementing the methods described in Section 8.1. Hence, once several members of a transcription factor family have been identified by conventional means, further members of the family can be identified and characterized in this way. Indeed, as more and more complete genomes are sequenced, they can be scanned electronically to identify novel sequences related to those of known transcription factors, providing an *in silico* approach to the identification of novel transcription factors.

The zinc finger motif

The two cysteine–two histidine zinc finger

As discussed in Section 7.6, one of the earliest gene regulatory systems to be characterized was that of the gene encoding the 5S RNA of the ribosome, in which a transcription factor, TFIIIA, binds to the internal control region of the gene (see Section 3.2). This transcription factor was among the first to be purified (Miller *et al.*, 1985). The pure protein was shown to have a periodic repeated structure and to contain between 7 and 11 atoms of zinc associated with each molecule of the pure protein.

The basis for this repeated structure was revealed when the gene encoding this protein was cloned and used to predict the corresponding amino

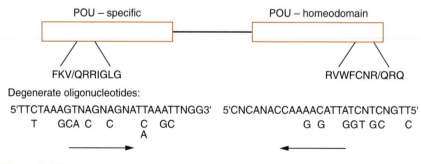

Figure 8.15

Isolation of novel members of the POU family on the basis of two conserved amino acid sequences, one at each end of the POU domain. Degenerate oligonucleotides containing all the sequences able to encode the conserved amino acids are used in a polymerase chain reaction with cDNA prepared from mRNA of an appropriate tissue. Novel POU factor mRNAs expressed in that tissue will be amplified on the basis that they contain the conserved sequences and can then be characterized.

acid sequence. This protein sequence contained nine repeats of a 30 amino acid sequence of the form Tyr/Phe-X-Cys-X-Cys-X$_{2-4}$-Cys-X$_3$-Phe-X$_5$-Leu-X$_2$-His-X$_{3-4}$-His-X$_5$, where X is a variable amino acid. This repeating structure therefore contains two invariant pairs of cysteine and histidine residues which were predicted to bind a single zinc atom, accounting for the multiple zinc atoms bound by the purified protein.

This 30 amino acid repeating unit is referred to as a zinc finger, on the basis that a loop of 12 amino acids, containing the conserved leucine and phenylalanine residues as well as several basic residues, projects from the surface of the protein and is anchored at its base by the conserved cysteine and histidine residues, which directly coordinate an atom of zinc (Fig. 8.16). The binding of zinc by the cysteine and histidine residues has been confirmed directly by X-ray crystallographic analysis of the TFIIIA protein. Such structural studies have also revealed that the finger region consists of two antiparallel β-sheets and an α-helix packed against one of the β-sheets with the α-helix contacting the major groove of the DNA as occurs for the recognition helix of the homeodomain proteins (Fig. 8.17) (for reviews see Rhodes and Klug, 1993; Klug and Schwabe, 1995).

It is clear, therefore, that like the helix-turn-helix motif, the zinc finger represents a protein structure capable of mediating the DNA binding of

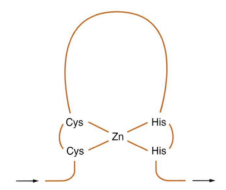

Figure 8.16

Schematic structure of the two cysteine–two histidine zinc finger.

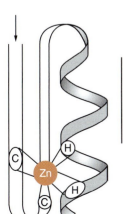

Figure 8.17

Detailed structure of the zinc finger in which two antiparallel β-sheets (solid lines) are packed against an adjacent α-helix (wavy line). The straight line indicates the region which contacts the major groove of the DNA. Redrawn from Evans and Hollenberg, *Cell* **52** 1–3 (1988), by permission of Professor R.M. Evans and Cell Press.

transcription factors. Although originally identified in the RNA polymerase III transcription factor TFIIIA, this motif has now been identified in a number of RNA polymerase II transcription factors and shown to play a critical role in their ability to bind to DNA and thereby influence transcription (Table 8.1; for reviews see Klug and Schwabe, 1995; Turner and Crossley, 1999; Bieker, 2001).

Thus three contiguous copies of the 30 amino acid zinc finger motif are found in the transcription factor Spl whose cloning was discussed earlier (Section 8.1). The sequence-specific binding pattern of the intact Spl protein can be reproduced by expressing in *E. coli* a truncated protein containing only the zinc finger region, confirming the importance of this region in DNA binding (Kadonaga *et al.*, 1987). Similarly, the *Drosophila* Krüppel protein, which is vital for proper thoracic and abdominal development, contains four zinc finger motifs. A single mutation, which results in the replacement of the conserved cysteine in one of these fingers by a serine which could not bind zinc, leads to the complete abolition of the function of the protein, resulting in a mutant fly whose appearance is indistinguishable from that produced by complete deletion of the gene.

The zinc finger therefore represents a DNA binding element which is present in variable numbers in many regulatory proteins. Indeed, the linkage between the presence of this motif and the ability to regulate gene expression is now so strong that, as with the homeobox, it has been used as a probe to isolate the genes encoding new regulatory proteins. The Krüppel zinc finger, for example, has been used in this way to isolate Xfin, a 37 finger protein expressed in the early *Xenopus* embryo (color plate 10) (for a review of Krüppel factors see Turner and Crossley, 1999; Bieker, 2001).

Hence, zinc finger proteins are likely to be involved in controlling development in vertebrates, as well as in *Drosophila*, where numerous proteins involved in regulating development, such as Krüppel, Hunchback and Snail, contain zinc fingers. The interactions of these proteins with the homeobox proteins, which contain the alternative DNA binding helix-turn-helix motif, are of central importance in the development of *Drosophila* and possibly other organisms.

Table 8.1 Transcriptional regulatory proteins containing Cys_2 His_2 zinc fingers

Organism	Gene	Number of fingers
Drosophila	*kruppel*	4
	hunchback	6
	snail	4
	glass	5
Yeast	*ADR1*	2
	Swi5	3
Xenopus	*TFIIIA*	9
	Xfin	37
Mammal	*NGF-1a (Egr1)*	3
	MK1	7
	MK2	9
	Evi 1	10
	Sp 1	3

The multi-cysteine zinc finger

Throughout this work we have noted that the effect of steroid hormones on mammalian gene expression is one of the best characterized examples of gene regulation. Thus the steroid-regulated genes were among the first to be shown to be regulated at the level of gene transcription (Chapter 4) by means of the binding of a specific receptor to a specific DNA sequence (Section 7.2), resulting in a change in chromatin structure and the generation of a DNAase I-hypersensitive site (Section 6.7). When the genes encoding the DNA binding receptors for the various steroid hormones, such as glucocorticoid and estrogen, were cloned, they were found to constitute a family of proteins encoded by distinct but related genes. In turn, these proteins were related to other receptors which mediated the response of the cell to hormones such as thyroid hormone, retinoic acid or vitamin D, leading to the idea of an evolutionarily related family of genes encoding nuclear receptors, known as the steroid–thyroid hormone receptor gene superfamily (for reviews see Weatherman *et al.*, 1999; Khorasanizadeh and Rastinejad, 2001; Olefsky, 2001; McKenna and O'Malley, 2002).

When the detailed structures of the members of this family were compared (Fig. 8.18) it was found that each had a multi-domain structure, which included a central highly conserved domain. On the basis of experiments in which truncated versions of the receptors were introduced into cells and their activities measured, it was shown that this conserved domain mediated the DNA binding ability of the receptor, while the

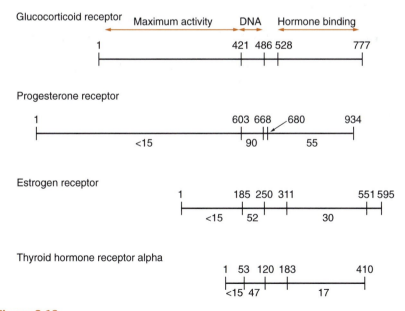

Figure 8.18

Domain structure of individual members of the steroid–thyroid hormone receptor superfamily. The proteins are aligned on the DNA binding domain, which shows the most conservation between different receptors. The percentage homologies in each domain of the receptors to that of the glucocorticoid receptor are indicated.

C-terminal region was involved in the binding of the appropriate hormone and the N-terminal region was involved in producing maximal induction of transcription of target genes.

Sequence analysis of the DNA binding domain in a variety of receptors showed that it conformed to a consensus sequence of the form $Cys\text{-}X_2\text{-}Cys\text{-}X_{13}\text{-}Cys\text{-}X_2\text{-}Cys\text{-}X_{15\text{-}17}\text{-}Cys\text{-}X_5\text{-}Cys\text{-}X_9\text{-}Cys\text{-}X_2\text{-}Cys\text{-}X_4\text{-}Cys$. Like the cysteine–histidine finger described in the previous section, the DNA binding of this element is dependent upon the presence of zinc or a related heavy metal such as cadmium. Moreover, this element can be drawn as two conventional zinc fingers, in which four cysteines replace the two cysteine–two histidine structure of the conventional finger in binding zinc and which are separated by a linker region containing the 15–17 variable amino acids (Fig. 8.19). Such a structure is supported by spectrographic analysis of this region of the receptor, which clearly demonstrates the presence of two zinc atoms each coordinated by four cysteines in a tetrahedral array (color plate 11).

However, such structural analysis also indicates that the two fingers in the steroid receptors constitute a single structural element, unlike the situation in the cysteine–histidine fingers where each finger forms a separate structural unit. Moreover, the multi-cysteine finger cannot be converted to a cysteine–histidine finger by substituting two of its cysteines with histidines. Thus while the multi-cysteine domain is clearly similar to the cysteine–histidine domain in its coordination of zinc, it is distinct in its lack of histidines and conserved phenylalanine and leucine residues, as well as in its structure, and the two elements are unlikely to be evolutionarily related (for reviews see Schwabe and Rhodes, 1991; Rhodes and Klug, 1993).

Whatever its precise relationship to the cysteine–histidine finger, it is clear that, like this type of finger, the multi-cysteine domain in the hormone receptors is involved in mediating DNA binding (color plate 12). Similar single domains containing multiple cysteines separated by non-conserved residues have also been identified in other DNA binding proteins, such as the yeast transcription factors GAL4, PPRI, LAC9, etc., which all contain a cluster of six invariant cysteines, and in the adenovirus transactivator, E1A, which has a cluster of four cysteines within the region that mediates *trans*-activation (Table 8.2).

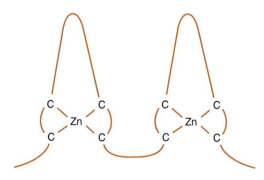

Figure 8.19

Structure of the four-cysteine zinc finger.

Table 8.2 Transcriptional regulatory proteins with multiple cysteine fingers

Finger type	Factor	Species
Cys_4Cys_5	Steroid, thyroid receptors	Mammals
Cys_4	EIA	Adenovirus
Cys_6	GAL4, PPRI, LAC9	Yeast

The existence of a short DNA binding region in a number of different steroid–receptor proteins which bind distinct but related sequences (Section 7.2) has allowed a dissection of the elements in this structure that are important in sequence-specific DNA binding. Thus, as illustrated in Table 7.3, the sequences that confer responsiveness to glucocorticoid or estrogen treatment are distinct but related to one another. If the cysteine-rich region of the estrogen receptor is replaced by that of the glucocorticoid receptor, a chimeric receptor is obtained which has the DNA binding specificity of the glucocorticoid receptor but, because all the other regions of the protein are derived from the estrogen receptor, it continues to bind estrogen. Hence this hybrid receptor induces the expression of glucocorticoid-responsive genes (which carry its DNA binding site) in response to treatment with estrogen (to which it binds) (Fig. 8.20). Further so-called fingerswop experiments using smaller parts of this region have shown that this change in specificity can also be achieved by the exchange of the N-terminal, four-cysteine finger, together with the region immediately following it, which are therefore critical for determining the sequence-specific binding to the DNA.

These findings have been further refined by exchanging individual amino acids in this region of the glucocorticoid receptor for their equivalents in the estrogen receptor. As shown in Fig. 8.21, the alteration of the two amino acids between the third and fourth cysteines

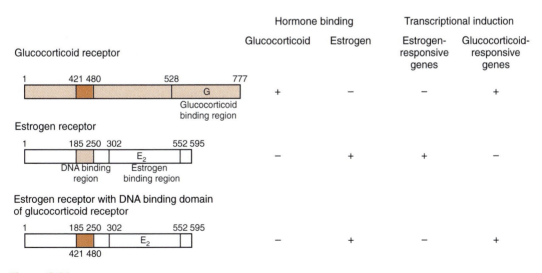

Figure 8.20

Effect of exchanging the DNA binding domain (shaded) of the estrogen receptor with that of the glucocorticoid receptor on the binding of hormone and gene induction by the hybrid receptor.

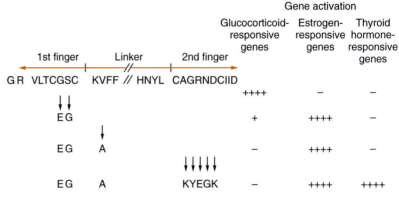

Figure 8.21

Effect of amino acid substitutions in the zinc finger region of the glucocorticoid receptor on its ability to bind to and activate genes which are normally responsive to different steroid hormones.

of the N-terminal finger to their estrogen receptor equivalents results in a glucocorticoid receptor which switches on estrogen-responsive genes. Hence the change of only two critical amino acids within a protein of 777 amino acids can completely change the DNA binding specificity of the receptor.

The specificity of the hybrid receptor for estrogen-responsive genes can be further enhanced by changing another amino acid, which is located in the linker region between the two fingers (Fig. 8.21), indicating that this region also plays a role in controlling the specificity of binding to DNA. Interestingly, this region following the finger can form an α-helical structure similar to the recognition helix seen in the helix-turn-helix motif. Thus, the DNA binding specificity of the steroid receptors appears to involve the cooperation of a zinc finger motif and an adjacent helical motif.

In contrast to the effect of mutations in the first finger and adjacent region, further alteration of five amino acids in the second finger is sufficient to change the binding specificity of the receptor, such that it now recognizes the thyroid hormone-receptor binding sites (Fig. 8.21). Since thyroid hormone-binding sites do not differ from those of the estrogen receptor in sequence but only in the spacing between the two halves of the palindromic DNA recognition sequence (Table 7.3a), this indicates that the second finger is critical for mediating protein–protein interactions between the two copies of the receptor that bind to the two halves of the palindromic sequence (see Section 7.2), and thus for controlling the optimal spacing of these halves for binding of the particular receptor.

Hence by studying the multiple related steroid receptors and their relationship with the related DNA sequences to which they bind, it has been possible to determine the critical role of both the first zinc finger and its adjacent helix in controlling the sequence to which these receptors bind, and of the second zinc finger in determining the spacing of adjacent sequences which is optimal for the binding of each receptor.

As well as determining the nature of the receptor homeodimers which bind to the palindromic repeats illustrated in Table 7.3a, the DNA binding domain also plays a critical role in determining the receptors which bind to the repeat sequences illustrated in Table 7.3b. Thus, when these direct repeats are separated by only one base, they can bind a homeodimer of the retinoid-X-receptor (RXR) and hence confer response to 9-*cis* retinoic acid which binds to this receptor (Fig. 8.22). In contrast, the RXR homeodimer cannot bind to the response elements when the direct repeats are separated by two, three, four or five base pairs. On these response elements the DNA binding domain of RXR interacts with the DNA binding domain of another member of the nuclear receptor family to form heterodimers which can bind to each of these sites. In these heterodimer combinations, the effect of RXR is suppressed and the response of the heterodimer is determined by the other component. Thus a spacing of two or five bases binds a heterodimer of RXR and the retinoic acid receptor (RAR) and results in a response to all *trans*-retinoic acid which binds to RAR. A spacing of four base pairs binds a heterodimer of RXR and the thyroid hormone receptor and therefore responds to thyroid hormone and a spacing of three base pairs binds a heterodimer of RXR and the vitamin D receptor leading to a response to vitamin D (Fig. 8.22) (for reviews see Gronemeyer and Moras, 1995; Mangelsdorf and Evans, 1995).

Hence, the different DNA binding domains of the various nuclear receptors produce different patterns of homeodimer and heterodimer binding to different binding sites, thereby allowing the diverse members of the nuclear receptor family to produce a wide variety of different responses.

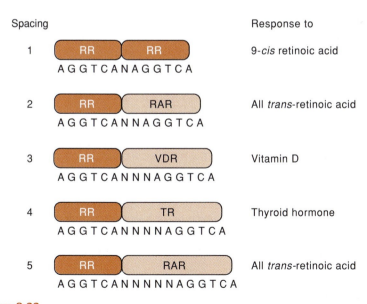

Spacing

		Response to
1	RR RR AGGTCANAGGTCA	9-*cis* retinoic acid
2	RR RAR AGGTCANNAGGTCA	All *trans*-retinoic acid
3	RR VDR AGGTCANNNAGGTCA	Vitamin D
4	RR TR AGGTCANNNNAGGTCA	Thyroid hormone
5	RR RAR AGGTCANNNNNAGGTCA	All *trans*-retinoic acid

Figure 8.22

Patterns of nuclear receptor heterodimers which form on various directly repeated response elements with different spacings (N) between the two halves of the repeat. Note that the response of the element is determined by the nature of the receptor which associates with RXR.

The leucine zipper, the helix-loop-helix motif and the basic DNA binding domain

In earlier sections of this chapter, we have examined how the presence of unusual structural motifs, such as the zinc finger, in several different regulatory proteins led to the identification of the crucial role of these elements in DNA binding. A similar approach has led to the identification of another such motif, the leucine zipper (for reviews see Kerppola and Curan, 1995; Hurst, 1996). Thus, in studies of the gene encoding the transcription factor C/EBPα which is involved in stimulating the expression of several liver-specific genes, it was noted that it contained a region of 35 amino acids, in which every seventh amino acid was a leucine. Similar runs of leucine residues were also noted in the yeast transcriptional regulatory protein GCN4, as well as in the proto-oncogene proteins Myc, Fos and Jun, which were originally identified on the basis of their ability to transform cultured cells to a cancerous phenotype (see Chapter 9) and are believed to act by regulating the transcription of other cellular genes (Fig. 8.23).

It was proposed that the leucine-rich region would form an α-helix in which the leucines would occur every two turns on the same side of the helix. These leucine residues would then facilitate the dimerization of two molecules of the transcription factor by promoting the interdigitation of two such helices, one on each of the individual molecules (Fig. 8.24). In agreement with this idea, replacement of individual leucine residues in C/EBPα with valine or isoleucine residues abolishes the ability of the protein to form a dimer. In turn, these mutations also prevent the binding of the protein to its specific recognition sequence.

Unlike the helix-turn-helix motif, however, the leucine zipper does not bind directly to DNA. Rather, by facilitating the dimerization of the protein it provides the correct protein structure for DNA binding by the

C/EBP	L T S D N D R L R K R V E Q L S R E L D T L R G I F R Q L
Jun B	L E D K V K T L K A E N A G L S S A A G L L R E Q V A Q L
Jun	L E E K V K T L K A Q N S E L A S T A N M L R E Q V A Q L
GCN 4	L E D K V E E L L S K N Y H L E H E V A R L K K L V G E R
Fos	L Q A E T D Q L E D E K S A L Q T E I A N L L K E K E K L
Fra 1	L Q A E T D K L E D E K S G L Q R E I I E L Q K Q K E R L
C-Myc	V Q A E E Q K L I S E E D L L R K R R E Q L K H K L E Q L
N-Myc	L Q A E E H Q L L L L E K E K L Q A R Q Q Q L L K K I E H A
I-Myc	L V G A E K K M A T E K R Q L R C R Q Q Q L Q K R I A Y L

Figure 8.23

Alignment of the leucine-rich region in several cellular transcription factors. Note the conserved leucine residues (L) which occur every seven amino acids.

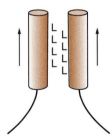

Figure 8.24

Model of the leucine zipper and its role in the dimerization of two molecules of a transcription factor.

adjacent region of the protein, which is rich in basic amino acids that can interact directly with the acidic DNA (Fig. 8.25) (for a review see Ellenberger, 1994). In agreement with this idea, mutations in the basic DNA binding domain abolish the ability of the protein to bind to the DNA without abolishing its ability to dimerize. Similar juxtapositions of a basic DNA binding domain and the leucine zipper are also found in the Fos and Jun oncogene proteins and in the yeast transcription factor GCN4, where a single 60 amino acid region contains all the information needed for both dimerization and sequence-specific DNA binding. Hence the leucine zipper has a role similar to that of the second zinc finger in the steroid receptors (Section 8.2) which modulates the activity of the DNA binding region rather than being involved directly in binding.

Although originally identified in leucine zipper-containing proteins, the basic DNA binding domain has also been identified by homology comparisons in a number of other transcriptional regulatory proteins which lack the leucine zipper. In this case, however, the basic domain is associated with an adjacent region that can form a helix-loop-helix structure (for reviews see Massari and Murre, 2000; Kewley et al., 2004). This motif is distinct from the helix-turn-helix motif described in Section 8.2 and consists of two amphipathic helices (containing all the charged amino acids on one side of the helix) separated by an intervening non-helical loop. Although originally thought to be the DNA binding domain of these proteins, this helix-loop-helix motif is now known to play a similar role to the leucine zipper in mediating protein dimerization and facilitating DNA binding by the adjacent basic DNA binding motif (for reviews see Ellenberger, 1994; Littlewood and Evan, 1995).

The helix-loop-helix motif with its adjacent basic DNA binding region is present in a number of different transcription factors expressed in different tissues. Thus, for example, it is present in a number of factors which are critical for the correct development of the nervous system (for a review see Ross et al., 2003). Similarly, this motif is found in several of the factors which control muscle development, such as the MyoD transcription factor whose

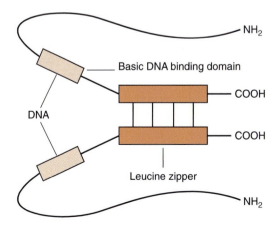

Figure 8.25

Model for the structure of the leucine zipper and the adjacent DNA binding domain following dimerization of the transcription factor C/EBP.

artificial expression in an undifferentiated fibroblast cell line can induce it to differentiate to skeletal muscle cells by activating the expression of muscle-specific genes (for reviews see Buckingham, 1994; Olson and Klein, 1994). Thus, the MyoD gene is likely to be the critical regulatory locus which is activated by treatment of these cells with 5-azacytidine, allowing this agent to induce these cells to differentiate into muscle cells (see Section 6.5).

The leucine zipper and helix-loop-helix structures therefore act to mediate the dimerization of the transcription factors which contain them, so forming a dimeric molecule which is able to bind to DNA via the adjacent basic DNA binding domain. This ability provides an additional aspect to gene regulation by such proteins. Thus, in addition to the formation of a dimer by two identical factors, it is possible to envisage the formation of a heterodimer between two different factors, which might have different properties in terms of sequence-specific binding and gene activation compared with homodimers of one or other of the two factors.

Such homo- and heterodimerization resulting in binding to different response elements occurs in the case of the nuclear receptors as discussed in Section 8.2. In the case of basic domain-containing factors, an example of this type is seen in the case of the related oncoproteins Fos and Jun (see Section 9.4; Kerppola and Curran, 1995). Thus Jun can bind as a homodimer to the AP1 recognition sequence, TGAGTCAG, which mediates transcriptional induction by phorbol esters (Table 7.2). In contrast, Fos cannot bind to DNA alone but can form a heterodimer with the Jun protein. This heterodimer binds to the AP1 recognition site with a 30-fold greater affinity than the Jun homodimer, and is considerably more effective in enhancing transcription of genes containing the binding site. Both hetero- and homodimer formation and DNA binding are dependent on the leucine zipper motif which is found in both proteins. Hence dimerization by the leucine zipper motif allows two different complexes with different binding affinities and different activity to form on the identical DNA binding site (Fig. 8.26).

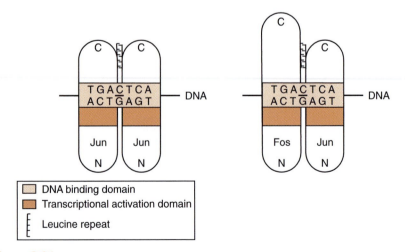

Figure 8.26

Model for DNA binding by the Jun homodimer and the Fos–Jun heterodimer. Redrawn from R. Turner and R. Tjian, *Science* **243**, 1689–1694 (1989), by permission of Professor R. Tjian and AAAS.

The failure of the Fos protein to form homodimers and its inability to bind to DNA in the absence of Jun has been shown to be due to differences in its leucine zipper region from that of Jun. Thus, if the leucine zipper region of Fos is replaced by that of Jun the resulting protein can dimerize. This dimerization allows the chimeric protein to bind to DNA through the basic region of Fos which is therefore a fully functional DNA binding domain.

As well as having a positive role, heterodimerization can also have a negative role. Thus, as discussed in Section 8.3, the ability of the MyoD factor to stimulate gene expression is inhibited by heterodimerization with the Id factor which has a helix-turn-helix motif but no basic domain. Since DNA binding requires the cooperation of two basic domains within the heterodimer, the MyoD-Id heterodimer cannot bind to DNA. Hence MyoD cannot activate the expression of muscle-specific genes and thereby promote the production of skeletal muscle cells in the presence of the Id factor (Fig. 8.27).

It is clear, therefore, that the ability of the leucine zipper and the helix-loop-helix motif to facilitate the formation of different dimeric complexes between different transcription factors is likely to play a crucial role in the regulation of gene expression, by producing complexes with different binding affinities and different activities (for reviews see Jones, 1990; Lamb and McKnight, 1991). Indeed, a recent study of 49 human leucine zipper-containing proteins demonstrated that they show very clear specificities in terms of which proteins pair with one another, even though they have very similar leucine zipper motifs (Newman and Keating, 2003). Hence, such heterodimerization is very specific, further supporting the idea that it plays a critical role in regulating gene expression.

Other DNA binding domains

As the genes encoding more and more transcription factors are isolated and characterized, it has become clear that, while the DNA binding domains of many factors fall into the three classes discussed above, not

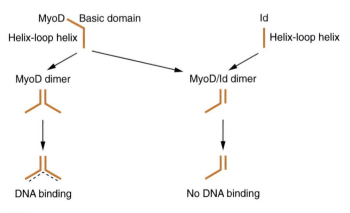

Figure 8.27

DNA binding by the MyoD protein is inhibited by the Id factor which contains the helix-loop-helix dimerization motif but not the basic DNA binding domain.

all do so. Hence additional types of DNA binding motifs exist. Interestingly, however, as the structures of these DNA binding domains are progressively understood, relationships between the different classes of DNA binding domain have emerged. Thus structural analysis of the Ets DNA binding domain, which is present, for example, in the *ets*-1 proto oncogene (see Section 9.4) and the mouse, PU-1 gene has shown it to be identical to the winged helix-turn-helix or fork head DNA binding motif identified in the *Drosophila* fork head factor and the mammalian liver transcription factor HNF-3 (Donaldson *et al.*, 1996). Hence all these factors share a common winged helix-turn-helix motif.

In turn, this motif, as its name suggests, contains a helix-turn-helix structure which is also present in the homeodomain proteins discussed in Section 8.2. These two DNA binding motifs are thus related to one another, although the winged helix-turn-helix motif also contains a β-sheet structure with two loops which appear as wings protruding from the factor, thereby giving it its name (for a review see Brennan, 1993). In the majority of winged helix-containing proteins, it is the helix-turn-helix motif which is responsible for DNA binding. However, in the winged helix-containing protein hRFX1, it is the β-sheet wing structure which binds to the DNA rather than the helix-turn-helix motif, indicating that members of the winged helix family can use one of two distinct structures to bind to DNA (Gajiwala *et al.*, 2000).

In another example of the relationship of different groups of factors, the existence of the POU DNA binding domain, containing a POU-specific domain in addition to a POU homeodomain, is paralleled in the Pax family of vertebrate transcription factors, many of which contain both a homeodomain and a so-called "paired", domain both of which contribute to high affinity DNA binding (for a review see Chi and Epstein, 2002). Unlike the POU-specific domain, however, the paired domain can be found as an isolated DNA binding domain without the homeobox, both in some members of the PAX family and in several *Drosophila* factors, including the paired gene in which it was originally identified.

It is clear therefore that a number of different structures exist which can mediate sequence-specific DNA binding and many of these are related to one another. Each of these DNA binding motifs is common to a number of different transcription factors, with differences in the precise amino acid sequence of the motif in each factor controlling the precise DNA sequence which it binds and hence the target genes for the factor.

8.3 Regulation of transcription

Introduction

Although binding to DNA is generally a necessary prerequisite for the activation of transcription, it is clearly not in itself sufficient for this to occur. Following binding, the bound transcription factor must somehow regulate transcription, for example by directly activating the RNA polymerase itself or by facilitating the binding of other transcription factors and the assembly of a stable transcriptional complex. Although some transcription factors can inhibit transcription, the majority of factors defined so far act to activate transcription. We will therefore discuss in turn the features of activating transcription factors that produce this activation, the manner in which they do so and the manner in which specific factors inhibit transcription.

Activation domains

Identification of activation domains

It is clear from the preceding sections of this chapter that transcription factors have a modular structure in which a particular region of the protein mediates DNA binding while another may mediate binding of a co-factor, such as a hormone, and so on. It seems likely, therefore, that a specific region of each individual transcription factor will be involved in its ability to up-regulate transcription following DNA binding.

In the majority of cases, it is clear that such activation regions are distinct from those which produce DNA binding. This domain-type structure is seen clearly in the yeast transcription factor GCN4, which mediates the induction of the genes encoding the enzymes of amino acid biosynthesis in response to amino acid starvation. Thus, if a 60 amino acid region of this protein, containing the DNA binding region, is introduced into cells, it can bind to the DNA of GCN4-responsive genes but fails to activate transcription. Hence, although DNA binding is necessary for transcriptional activation to occur, it is not sufficient, and gene activation must be dependent upon a region of the protein that is distinct from that mediating DNA binding.

Unlike the DNA binding regions, the region of a transcription factor that mediates gene activation cannot, therefore, be identified on the basis of a simple assay of, for example, the ability to bind to DNA or another protein. Rather, a functional assay of gene activation following binding to DNA is required. Activation regions have therefore been identified on the basis of so-called "domain-swap" experiments, in which the DNA binding region of one transcription factor is combined with various regions of another factor and the ability to activate transcription of a gene containing the binding site of the first factor is assessed (Fig. 8.28). Following binding of the hybrid factor to the target gene binding site, gene activation will occur only if the hybrid factor also contains an activation domain provided by the second factor, and hence the activation domain can be identified.

Thus, in the case of the yeast transcription factor GCN4 discussed above, if a 60 amino acid region, outside the DNA binding domain, is linked to the DNA binding region of the bacterial regulatory protein, Lex A, the hybrid factor will activate transcription in yeast from a gene containing the binding site for Lex A, whereas neither the Lex A DNA binding domain nor the GCN4 region will do so alone. Hence this region of GCN4 contains an activation domain, which can increase transcription following DNA binding and is separate from the region of the protein that normally mediates DNA binding (Fig. 8.29a).

Following its initial use in yeast, similar domain-swapping experiments have also been used to identify the activation domains of mammalian transcription factors. In the glucocorticoid receptor, for example, two independent regions each able to produce gene activation have been identified in this way (Fig. 8.29b).

The success of domain-swap experiments is further proof of the modular nature of transcription factors, allowing the DNA binding domain of

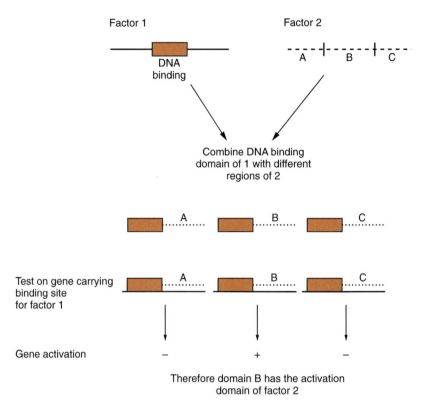

Figure 8.28

Domain swapping experiment, in which the activation domain of factor 2 is mapped by combining different regions of factor 2 with the DNA binding domain of factor 1 and assaying the hybrid proteins for the ability to activate transcription of a gene containing the DNA binding site of factor 1.

one factor and the activation domain of another to cooperate together to produce gene activation. This is analogous to the exchange of the DNA binding domain of the glucocorticoid and estrogen receptors, which allows the creation of a hybrid receptor that binds to estrogen-responsive genes through the DNA binding domain but is responsive to the presence of glucocorticoid through the steroid binding domain of the protein.

An extreme example of this modularity is provided by the herpes simplex virus *trans*-activating protein VP16, discussed in Section 8.2, which transcriptionally activates the viral immediate-early genes during lytic infection of mammalian cells. Thus, although this protein contains a very potent activation region which can strongly induce gene transcription when fused to the DNA binding domain of GAL4, it contains no DNA binding domain and cannot bind to DNA itself. Rather, following infection, it forms a complex with the cellular octamer binding protein, Oct-1 (see Section 8.2). Oct-1 provides the DNA binding domain which allows binding to the sequence TAATGARAT (R = purine) in the viral promoters, and activation is achieved by the activation domain of VP16 (for a review see Wysocka and Herr, 2003). Hence, in this instance, DNA binding and activation motifs actually reside on separate molecules (Fig. 8.30).

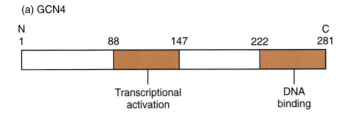

(a) GCN4

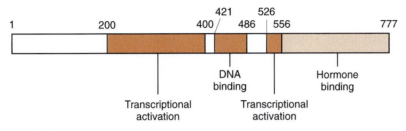

(b) Glucocorticoid receptor

Figure 8.29

Structure of the yeast GCN4 factor (a) and the mammalian glucocorticoid receptor (b), indicating the distinct regions that mediate DNA binding or transcriptional activation.

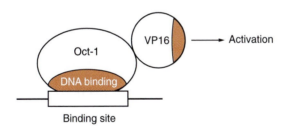

Figure 8.30

Activation of gene transcription by interaction of the cellular factor Oct-1, which contains a DNA binding domain, and the herpes simplex virus VP16 protein, which contains an activation domain but cannot bind to DNA.

Nature of activating regions

Acidic domains

Although the activating regions identified in various transcription factors do not show strong amino acid sequence homology to each other, in many cases they have a very high proportion of acidic amino acids, resulting in a strong net negative charge (for a review see Triezenberg, 1995). Thus, in the N-terminal activating region of the glucocorticoid receptor, 17 acidic amino acids are contained in an 82 amino acid region. Similarly, the activating region of the yeast factor GCN4 contains 17 negative charges in an activating region of only 60 amino acids. This has led to the idea that activation regions consist of so-called "acid blobs" or "negative noodles" which are necessary for the activation of transcription. However, although the net negative charge in acidic activation regions is likely to be of importance in their ability to stimulate transcription, it appears that other features, such as the presence of several conserved hydrophobic residues, are also necessary for transcriptional activation.

Other activating domains

Although acidic domains of the type discussed above have been identified in a wide range of transcriptional activators from yeast to man, it is clear that this type of structure is not the only one which can mediate transcriptional activation. Thus of the two regions of the human Spl transcription factor that could mediate activation of transcription, neither was particularly acidic. Instead, each of these two domains was particularly rich in glutamine residues, and the intactness of the glutamine-rich region was essential for transcriptional activation. Similar sequences have also been identified in the homeotic proteins Antennapedia and Cut, in Zeste, another *Drosophila* transcriptional regulator, and in the POU proteins Oct-1 and Oct-2, indicating that this type of activating region is not confined to a single protein.

A further type of activation domain has been identified in the transcription factor CTF/NF1 which binds to the CCAAT box present in many eukaryotic promoters (Table 7.2). The activation domain of this protein is not rich in acidic or glutamine residues but instead contains numerous proline residues, forming approximately one-quarter of the amino acids in this region. Similar proline-rich regions are found in other transcription factors, such as AP2 and Jun, indicating that, as with glutamine-rich domains, this element is not confined to a single protein.

Hence, as with DNA binding domains, it is clear that several distinct protein motifs are involved in the activation of transcription (for reviews see Mitchell and Tjian, 1989; Triezenberg, 1995).

How is transcription activated?

The widespread interchangeability of activation domains from yeast, *Drosophila* and mammalian transcription factors, discussed in the previous section, suggests that common mechanisms may mediate transcriptional activation in a wide range of organisms. This is supported by the observations that mammalian transcription factors, such as the glucocorticoid receptor, can activate a gene carrying its binding site in cells of *Drosophila*, tobacco plants and mammals. Indeed, yeast and mammalian factors can cooperate together in gene activation, a gene bearing binding sites for GAL4 and the glucocorticoid receptor being synergistically activated by the two factors in mammalian cells so that the activation observed with the two factors together is greater than the sum of that observed with each factor independently. This cooperation between two factors which come from widely different species and would therefore never normally interact, suggests that they both function by interacting with some highly conserved component of the basal transcriptional complex (see Section 3.2) which is involved in the basic process of transcription in all different species.

In fact, there is evidence that activators can interact with a variety of different targets within the basal transcriptional complex and these will be discussed in turn. Clearly such interactions could stimulate transcription by increasing the binding of a particular component of the basal transcriptional complex so enhancing its assembly (Fig. 8.31a). Alternatively, stimulation could occur by the activator altering the conformation of an already bound factor so stimulating the activity and/or stability of the complex (Fig. 8.31b). It is clear that both of these mechanisms are actually used and they could evidently operate whether the complex actually assembles in a fully stepwise fashion in which each component binds sequentially or in a situation where several components bind together in a holoenzyme complex with RNA polymerase itself (see Section 3.2 for further discussion of these alternative models of complex assembly). In discussing the various targets within the complex, however, we shall consider them in the order in which they bind in the stepwise model.

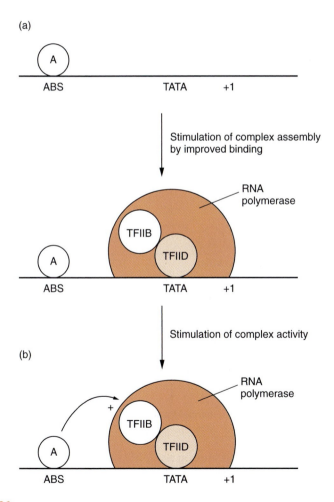

Figure 8.31

An activator (A) bound to its binding site (ABS) can stimulate either complex assembly (a) or the activity of the already assembled complex (b).

TFIID

As described in Section 3.2, the binding of the TFIID complex (containing TBP and associated proteins) to the TATA box is the initial stage of complex assembly in both the stepwise and holoenzyme assembly models. Hence TFIID constitutes an obvious target for activating molecules. Indeed early studies indicated that both the recruitment of TFIID to the promoter and its conformation when bound are affected by activator molecules. Increased binding of TFIID would evidently result in enhanced binding of the other components of the basal transcription complex which bind subsequently while alterations in TFIID configuration might act by improving its ability to recruit these other factors or by directly enhancing its activity within the assembled basal transcription complex (Fig. 8.32). Such findings are complicated, however, by the fact that, as

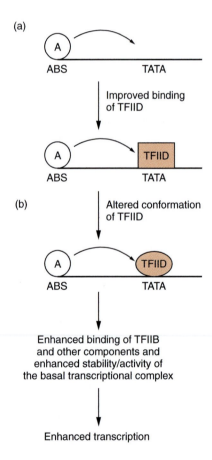

Figure 8.32

An activator (A) bound to its binding site (ABS) can enhance the binding of TFIID to the TATA box, thereby improving the rate of assembly of the basal transcriptional complex by facilitating the subsequent binding of other components which is dependent upon prior binding of TFIID (a). In addition, the activator can also alter the configuration of TFIID so stimulating its activity either by increasing its ability to recruit other components of the complex or by enhancing its ability to stimulate transcription (b).

discussed in Section 3.2, TFIID is a multi-protein complex consisting of the DNA binding component TBP and a variety of TAFs (TBP-associated factors) (for reviews see Hahn, 1998; Green, 2000).

Indeed it is likely that both TBP itself and one or more of the TAFs can be targets for transcriptional activators (Fig. 8.33). Thus, acidic activators have been shown to interact directly with TBP, and single amino acid mutations in such activators which abolish interaction with TBP also abolish the ability to activate transcription, supporting an important role in this effect. It appears that such activators act by enhancing the rate of recruitment of TBP to the promoter, hence increasing the rate of assembly of the basal transcriptional complex. This effect of activators has now been demonstrated in intact cells as well as in cellular extracts (Li *et al.*, 1999; Kuras and Struhl, 1999) using the ChIP assay described in Section 7.2.

Although there is thus evidence that activators can interact directly with TBP, it is clear that in some circumstances activation requires interaction with one or more of the TAFs. Thus, in many cases, stimulation of transcription *in vitro* does not occur with purified TBP alone but only when the full TFIID complex is present, indicating that such activation requires interaction with the TAFs. Most interestingly, different types of activation domain appear to contact different TAFs. Thus, for example, glutamine-rich activators such as Sp1 can bind $TAF_{II}110$ while acid activators bind $TAF_{II}40$ and multiple activators including proline-rich activators bind $TAF_{II}55$. Hence, different components within TFIID can be targeted by different classes of activator (Fig. 8.34). Because TAFs can function as intermediaries between activators and the basal transcription complex they are often referred to as co-activators (see below).

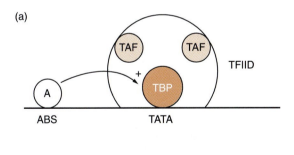

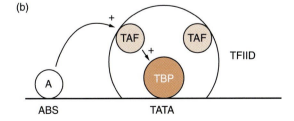

Figure 8.33

An activating molecule (A) can interact with TFIID either by interacting with the TATA box binding protein (TBP) directly (panel a) or by interacting indirectly with TBP via TBP associated factors (TAF) (panel b).

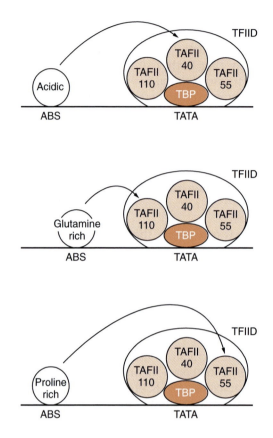

Figure 8.34

Different classes of activator can have distinct targets within TFIID with acidic activators contacting TAF$_{II}$40, glutamine-rich activators contacting TAF$_{II}$110 and proline-rich activators contacting TAF$_{II}$55.

TFIIB

Although there is considerable evidence for the interaction of activators and TFIID, it is clear that activators can also act to stimulate the assembly/activity of the basal transcriptional complex after TFIID has bound. As discussed in Section 3.2 the binding of TFIID to the promoter is followed by the binding of TFIIB. As with TBP and the TAFs, there is clear evidence for the interaction of TFIIB with activators. Thus TFIIB can be purified from a mixture of proteins on a column containing a bound acidic activator. Moreover, this interaction is of importance for activation of transcription since mutations in acidic activators which abolish the interaction with TFIIB prevent the activator from stimulating transcription (for a review see Hahn, 1993).

As with TFIID, such interactions can both stimulate the binding of TFIIB to the promoter and alter its configuration when bound, so improving its ability to recruit the other components of the basal transcriptional complex, such as the RNA polymerase which binds subsequently (Fig. 8.35). Hence, as with the various components of TFIID, the single TFIIB polypeptide is evidently a target for transcriptional activators.

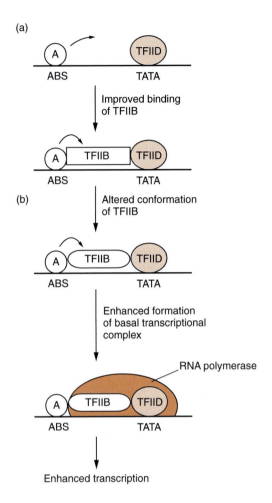

Figure 8.35

The binding of an activator (A) to its appropriate binding site (ABS) can enhance the rate of transcription by enhancing the binding of the TFIIB factor and/or by altering its configuration so that it recruits the other components of the basal transcriptional complex more efficiently. Compare with Fig. 8.32, which shows the same effects for TFIID.

RNA polymerase II and the mediator complex

As discussed in Section 3.2 binding of TFIIB allows the subsequent recruitment of the RNA polymerase itself together with the TFIIF factor. Interestingly, the C-terminal region of the large subunit of RNA polymerase II, which is involved in transcriptional elongation, is also implicated in the response to transcriptional activators. Thus, deletion of this region from RNA polymerase II prevents enhanced transcription in response to transcriptional activators while increasing the number of repeated elements enhances the response to activators.

Despite this, however, it is unlikely that activators contact the RNA polymerase directly. Rather, work in yeast and subsequently in mammalian cells has identified a mediator complex containing over 20 polypeptides which associates with this C-terminal region of the RNA polymerase II and is required for the response to transcriptional activators. It is clear, therefore, that activators can act by interacting with the mediator complex which in turn stimulates RNA polymerase activity (for reviews see Malik and Roeder, 2000; Myers and Kornberg, 2000; Boube *et al.*, 2002) (Fig. 8.36). Interestingly, electron microscope analysis of the mediator complex bound to RNA polymerase II suggests that the mediator partially envelopes the polymerase (Asturias *et al.*, 1999). This would allow it to receive signals from activators and transmit them to the polymerase (Fig. 8.37).

Interestingly, the mediator complex has also been shown to stimulate the ability of TFIIH to phosphorylate the C-terminal region of RNA polymerase. It is possible therefore that activators may enhance the ability of the mediator to stimulate this effect of TFIIH, thereby also stimulating the start of transcriptional elongation which is dependent upon such phosphorylation (Fig. 8.38) (see Section 3.2).

The mediator complex has been shown to be associated with the RNA polymerase holoenzyme prior to binding to the DNA and the interaction between activators and the mediator has been shown to stimulate the

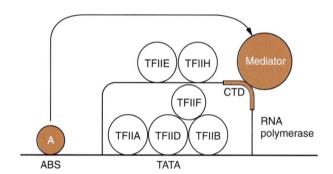

Figure 8.36

Activators appear to interact with the RNA polymerase indirectly via a mediator complex which binds to the C-terminal domain (CTD) of RNA polymerase II.

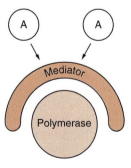

Figure 8.37

Structural analysis of the mediator complex bound to RNA polymerase II indicates that the mediator partially envelops the polymerase. Hence, it could receive signals from activators (A) and transmit them to the polymerase.

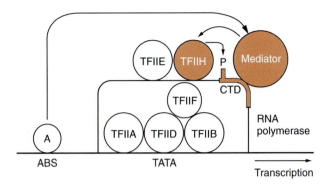

Figure 8.38

As well as affecting transcriptional initiation via the mediator, it is possible that activators act by stimulating the ability of the mediator to interact with TFIIH and stimulate it to phosphorylate the C-terminal domain (CTD) of RNA polymerase II (P), thereby promoting transcriptional elongation.

assembly of the basal transcriptional complex on the DNA (Cantin *et al.*, 2003). Hence, as discussed in Section 3.2 this holoenzyme contains not only RNA polymerase II, and basal transcription factors such as TFIIB, TFIIE, TFIIF and TFIIH, but also contains mediator components which respond to transcriptional activators as well as the SWI/SNF complex which can alter chromatin structure (see Section 6.7).

Co-activators

As discussed above, transcriptional activators often contact their ultimate targets indirectly acting, for example, via TAFs or the mediator complex (see above). This has led to the concept of co-activators which act to transmit the signal from the DNA binding transcriptional activator to the basal transcriptional complex (for review see Spiegelman and Heinrich, 2004).

Indeed, co-activators have been identified in a number of different situations involving specific transcriptional activators, where they are recruited to the DNA by the DNA-bound activator and then activate transcription. Perhaps the best known co-activator is CBP (CREB binding protein: for review see Shikama *et al.*, 1997; Goodman and Smolik, 2000). As its name suggests, CBP was first defined as a protein which binds to the CREB transcription factor (for review of CREB see de Cesare and Sassone-Corsi, 2000; Mayr and Montminy, 2001). Most importantly, CBP only binds to CREB when CREB has been phosphorylated on the serine amino acid at position 133. Since such phosphorylation is necessary for CREB to activate transcription, this indicated that recruitment of CBP was likely to play a key role in mediating transcriptional activation by CREB (see Section 8.4) Hence, CBP does not bind to DNA itself but is recruited to the DNA by DNA-bound phosphorylated CREB and then activates transcription.

CBP has now been shown to play a key role in mediating transcriptional activation via a number of DNA binding transcription factors which are discussed in this book including the steroid–thyroid hormone receptor family (Section 8.2), NFκB (Section 8.4), p53 (Section 9.5) and MyoD

(Section 8.2). Hence, it is clear that CBP and its close relative p300 play a key role as co-activators for a number of transcriptional activators (Fig. 8.39). Indeed, the widespread use of CBP by transcription factors activated by different signaling pathways may result in competition between these pathways for limited amounts of CBP. This in turn would account for the phenomenon in which simultaneous stimulation of two different pathways such as those stimulated by glucocorticoid and phorbol esters results in each pathway inhibiting the other so that no transcriptional activation occurs.

Although CBP can interact with members of the steroid/thyroid hormone receptor family, it is not the only co-activator to do so. In several cases, specific co-activators can bind to the receptors only after hormone treatment. Hence, the binding of specific co-activators to these receptors only after hormone treatment allows the receptors to activate transcription in a hormone-dependent manner (for review see Rosenfeld and Glass, 2001; Nagy and Schwabe, 2004).

It is likely that co-activators can act to stimulate transcription via at least two distinct mechanisms, following recruitment to the DNA by a DNA-bound activator. Both these mechanisms have been shown to operate in the case of CBP. Thus, CBP has been shown to bind to various components of the basal transcription complex such as TBP, TFIIB and the RNA polymerase holoenzyme, indicating that it can bridge the gap between CREB and the basal complex allowing CREB to activate transcription (Fig. 8.40a). Moreover, as discussed in Section 6.6, CBP also has histone acetyltransferase activity, indicating that it can also stimulate transcription by acetylating histones and therefore opening up the chromatin structure (Ogryzko et al., 1996) (Fig. 8.40b). Hence, co-activators can act either via linking the activator to the basal complex or by modifying chromatin structure.

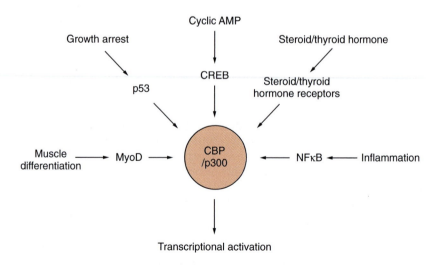

Figure 8.39

A variety of activating transcription factors stimulate transcription via the closely related CBP and p300 co-activators.

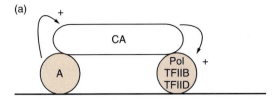

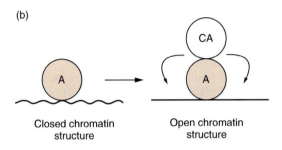

Figure 8.40

A co-activator (CA) may act (a) by linking an activator (A) to the basal transcriptional complex or (b) by promoting the conversion of a closed chromatin structure (wavy line) to a more open structure (solid line). In both cases, the co-activator will be recruited to the DNA by a DNA-bound activator.

A multitude of targets for activators

There exists therefore a bewildering array of potential targets for activator molecules including the RNA polymerase/mediator complex, co-activators, TFIIB and the different components of TFIID. Indeed, other components of the basal transcriptional complex, such as TFIIA, TFIIE and TFIIH, have also been observed to interact directly with transcriptional activators. All these various possibilities are not mutually exclusive however. Thus, it is likely that all these components can act as a target either for the same activating factor or different activating factors.

Several possibilities may account for such a wide range of activator targets. Thus, it is possible that different organisms differ in the preferred target for transcriptional activators. Indeed, while TAFs associated with TBP appear to be of critical importance for transcriptional activation in higher organisms such as *Drosophila*, they may be of much less importance in yeast, Alternatively, it has been suggested that the sole requirement for an activator is to be able to bind to DNA and then bind any component of the basal transcriptional complex and hence recruit it to the DNA (for a review see Ptashne and Gann, 1997). Thus, an activator can function simply by interacting with any component of the complex and thereby enhancing its binding to the promoter.

It is most probable, however, that a multiplicity of targets for activators is required to produce the strong synergistic activation of transcription which is observed when different activating factors are added together, compared with the level observed when each factor is added separately (see above), as well as the strong enhancement of transcription observed when multiple copies of a single factor can bind the target DNA. Thus, different activating factors or different molecules of the same activating factor could contact different targets within the basal transcriptional complex, thus ensuring the great enhancement of transcriptional activity which is the ultimate aim of activating molecules (Fig. 8.41).

Repression of transcription

Although the majority of transcription factors described so far act in a positive manner, a number of cases have now been reported in which a transcription factor exerts an inhibitory effect on transcription and several possible mechanisms by which this can be achieved have been described (for reviews see Hanna-Rose and Hansen, 1996, Latchman, 1996; Smith and Johnson, 2000, Fig. 8.42).

The simplest means of achieving repression is for a repressor to prevent an activating molecule from binding to DNA. This can occur by a negative factor promoting a tightly packed chromatin structure which does not allow activating molecules to bind (Fig. 8.42a) (see Chapter 6 for a discussion of the effect of chromatin structure on gene expression). An example of this effect was discussed in Section 8.2 in which the polycomb repressor binds to *Drosophila* homeotic genes and organizes them into an inactive chromatin structure incapable of binding activators (for reviews see Orlando, 2003; Levine *et al.*, 2004).

Alternatively, the negatively-acting factor can bind specifically to the binding site of the activator so preventing it binding (Fig. 8.42b). This effect is seen in the β-interferon promoter where the binding of several positively acting factors is necessary for gene activation. Another factor acts negatively by binding to this region of DNA and simply preventing the positively acting factors from binding (Fig. 8.42b). In response to viral infection, the negative factor is inactivated, allowing the positively acting factors to bind and transcription occurs. A similar example was described in Section 8.2, whereby the DNA binding of the engrailed gene product inhibits gene activation by preventing the binding of the Ftz activator protein.

In a related phenomenon (Fig. 8.42c), repression is achieved by formation of a complex between the activator and the repressor in solution,

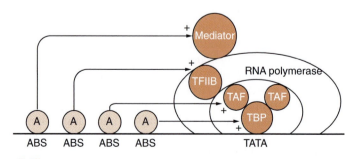

Figure 8.41

Synergistic activation of transcription by activator molecules (A) interacting with different components of the basal transcriptional complex such as the mediator complex, TFIIB or the different components of TFIID (TBP and non-DNA binding TAFs).

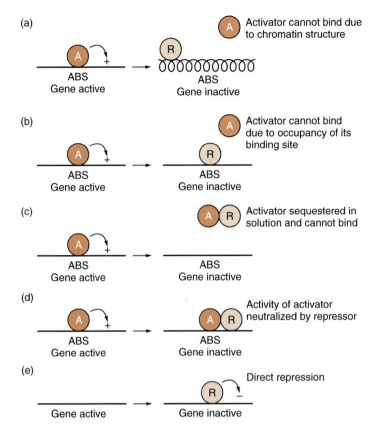

(a)

A Activator cannot bind due to chromatin structure

R

ABS
Gene active

ABS
Gene inactive

(b)

A Activator cannot bind due to occupancy of its binding site

ABS
Gene active

R

ABS
Gene inactive

(c)

A R Activator sequestered in solution and cannot bind

ABS
Gene active

ABS
Gene inactive

(d)

Activity of activator neutralized by repressor

A R

ABS
Gene active

ABS
Gene inactive

(e)

Direct repression

R

Gene active

Gene inactive

Figure 8.42

Mechanisms by which an inhibitory factor (R) can repress transcription. These involve inhibiting an activator (A) binding to its binding site (ABS) in the DNA either by producing an inactive chromatin structure (panel a) or by competing for the DNA binding site (panel b) or by sequestering the activator in solution (panel c); inhibiting the ability of bound activator to stimulate transcription (panel d) or direct repression (panel e).

preventing the activator binding to the DNA. This is seen in the case of the inhibitory factor Id which dimerizes with the muscle determining factor MyoD via its helix-loop-helix motif and prevents it binding to DNA and activating transcription (see Section 8.2).

In addition to inhibiting DNA binding by these different means, a negative factor can also act by interfering with the activation of transcription mediated by a bound factor in a phenomenon known as quenching (Fig. 8.42d). Thus, in the case of the promoter driving expression of the c-*myc* gene, a negatively acting factor *myc*-PRF binds to a site adjacent to that occupied by a positively acting factor *myc*-CF1 and prevents it from activating c-*myc* gene expression.

In all the cases discussed so far the negative factor exerts its inhibitory effect in an essentially passive manner by neutralizing the action of a positively acting factor, by preventing either its DNA binding or its activation of transcription. It is clear, however, that some factors can have an inherently negative action on transcription which does not depend upon the neutralization of a positively acting factor (Fig. 8.42e) (for a review see Hanna-Rose and Hansen, 1996).

Thus, the *Drosophila* even-skipped protein is able to repress transcription from a promoter lacking DNA binding sites for any activating proteins, indicating that it has a directly negative effect on transcription. A similar direct inhibitory effect has been defined in the case of the mammalian c-*erbA* gene which encodes the thyroid hormone receptor, a member of the steroid–thyroid hormone receptor family (see Sections 7.2 and 8.2). Thus, the binding of this receptor to its specific DNA binding site in the absence of thyroid hormone results in the direct inhibition of transcription. This effect is mediated by a co-repressor which binds to the receptor in the absence of hormone and inhibits transcription (for reviews see Rosenfeld and Glass, 2001; Nagy and Schwabe, 2004). In the presence of thyroid hormone, however, the receptor undergoes a conformational change which results in the release of the co-repressor and allows co-activators to bind. This then allows the receptor to activate rather than repress transcription (Fig. 8.43).

This example indicates the critical role of co-activators and co-repressors in regulating gene expression. Thus, thyroid hormone can modulate gene expression by controlling whether its target factor binds a co-activator or a co-repressor. Similarly, as discussed in Section 8.2, the different configurations a factor adopts on different DNA binding sites can alter its effect on transcription by altering its ability to bind co-activators or co-repressors.

The thyroid hormone receptor example also indicates that the distinction between activators and repressors is not a precise one. Thus, in some

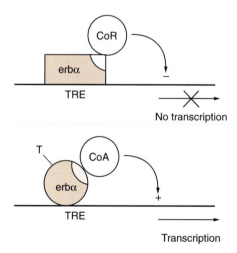

Figure 8.43

The thyroid hormone receptor encoded by the c-*erbA* α gene represses transcription in the absence of thyroid hormone (T) via a specific inhibitory domain (white) which recruits an inhibitory co-repressor (COR). Following binding of hormone, the protein undergoes a conformational change which releases the co-repressor and leads to binding of a co-activator (CoA) which allows the receptor to activate transcription.

cases, the same factor can directly stimulate or directly inhibit transcription depending on the circumstances, with this effect being controlled via the regulated binding of co-repressor and co-activator molecules.

Another example of a co-activator/co-repressor exchange is seen in the LIM homeodomain transcription factor which binds either the RLIM co-repressor or the CLIM co-activator. Interestingly, CLIM promotes the degradation of RLIM, so removing the repressor and allowing activation to occur (Ostendorff *et al.*, 2002) (Fig. 8.44).

In a number of directly acting transcriptional repressors, such as even-skipped, the thyroid hormone repressor and the MeCP2 protein which binds to methylated DNA (see Section 6.5), a specific region of the protein has been shown to be able to confer the ability to inhibit gene expression upon the DNA binding domain of another protein when the two are artificially linked paralleling the similar behavior of activation domains in such domain-swop experiments (see above). As noted in Section 6.6, the nuclear receptor co-repressor (NCo-R), which binds to the inhibitor domain of the thyroid hormone receptor, associates with the Sin3/RPD3 protein complex which can alter chromatin structure by deacetylating histones. A similar ability to recruit a histone deacetylase has also been observed in the case of MeCP2 as discussed in Section 6.5.

Hence, some inhibitory domains may function at least in part by recruiting molecules which can alter chromatin structure. In other cases, however, they are likely to inhibit transcription by reducing the formation and/or stability of the basal transcriptional complex either by interacting directly with the basal complex or by recruiting a co-repressor which then interacts with the basal complex. Such inhibitory domains therefore act in a manner similar to activation domains but reduce rather than enhance the rate of transcription.

Hence, inhibitory factors are likely to play a critical role in the regulation of transcription both by inhibiting the activity of positively acting factors and in other cases by having a direct inhibitory effect on transcription.

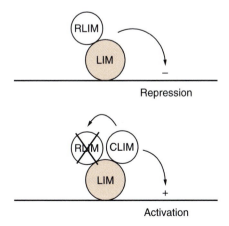

Figure 8.44

Binding of the CLIM co-activator to the LIM transcription factor promotes degradation of the RLIM co-repressor and allows transcriptional activation to occur.

8.4   What regulates the regulators?

Introduction

The crucial role of transcription factors in tissue-specific gene regulation and in the regulation of transcription in response to specific stimuli clearly leads to the question of how such factors are themselves regulated so that they produce gene activation or repression only in a specific tissue or in response to a particular signal.

As discussed in Section 7.1, the Britten and Davidson model envisaged that the products of regulatory or integrator genes which controlled the activity of other genes would be synthesized in response to a particular signal, or in a particular tissue, and would be absent in other tissues. Hence, the activity of the regulated genes would be directly correlated with the presence of the regulatory protein (Fig. 8.45a). An example of this type is provided by the MyoD factor discussed in Section 8.2, which is expressed only in skeletal muscle cells and whose over-expression in other cell types can induce them to differentiate into skeletal muscle cells.

A contrasting example is provided, however, by a factor involved in immunoglobulin gene transcription in B lymphocytes, namely NFκB (for reviews see Serfling *et al.*, 2004; Hayden and Ghosh, 2004). This factor is detected in a form capable of stimulating immunoglobulin gene transcription only in extracts of mature immunoglobulin-producing B cells, no activity being detectable in immature B cells or in non-B cells, such as HeLa cells. The NFκB protein and its corresponding RNA are detectable in all cell types, however, suggesting that this protein exists in an inactive form in most cell types and is activated in mature B cells. In agreement with this, active NFκB capable of stimulating immunoglobulin gene transcription can be induced in other cell types, such as T cells or HeLa cells,

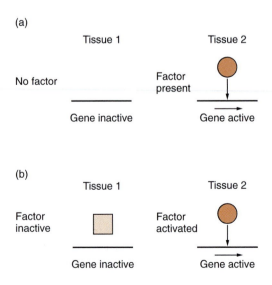

Figure 8.45

Gene activation mediated by the synthesis of a transcription factor only in a specific tissue (a) or its activation in a specific tissue (b).

by treatment with phorbol esters. This effect occurs even under conditions where new protein synthesis cannot occur, indicating that it takes place via the activation of pre-existing NFκB protein.

These two examples illustrate, therefore, how the ability of transcription factors to act only in a particular tissue can be controlled either by tissue-specific synthesis (Fig. 8.45a) or by tissue-specific activation of pre-existing protein (Fig. 8.45b). These two mechanisms will now be considered.

Regulated synthesis of transcription factors

Clearly, all the various levels of gene regulation such as transcription, splicing, translation, etc., which were discussed in Chapter 3, could be used to regulate the expression of the genes encoding transcription factors, and there is evidence that several of these are used. These will be discussed in turn with illustrative examples.

Regulation of transcription

Although the low abundance of many transcription factors makes the demonstration of transcriptional control difficult, it has been demonstrated in the case of the C/EBPα protein which, as previously described (Section 8.2), regulates the transcription of several different liver-specific genes, such as transthyretin and α-1-antitrypsin. This transcription factor is made at high level only in the liver and, in agreement with this, significant transcription of its gene is detectable only in this tissue (Xanthopoulos et al., 1989). Hence, the regulated transcription of the C/EBPα gene controls the production of the corresponding protein which, in turn, directly controls the liver-specific transcription of other genes. Interestingly, such transcriptional control is supplemented by control at the level of mRNA translation which produces activator and repressor forms of the protein (see below).

Regulation of splicing

As discussed in Section 5.2, in mammals alternative splicing is used widely to generate two different forms of a protein, with different functions, from a single gene. A similar theme is seen in the transcription factors, and has been well characterized in the c-erbA α gene, which encodes the receptor mediating gene expression in response to thyroid hormone. As discussed in Section 8.3, gene activation by this protein is dependent on the presence of thyroid hormone, and in the absence of hormone the receptor represses gene expression. Activation by the receptor is dependent upon the presence of a region in the receptor capable of binding thyroid hormone. Interestingly, however, two alternative forms of the protein exist which are encoded by alternatively spliced mRNAs. One of these (α-1) contains the hormone binding domain and can mediate gene activation in response to thyroid hormone, while the other form (α-2) contains another protein sequence instead of part of this domain (Fig. 8.46a). Therefore, the α-2 form cannot respond to the hormone, although, since it contains the DNA binding domain, it can bind to the binding site for the receptor in hormone-responsive genes. By doing so, it prevents binding of the α-1 form and hence the induction of the gene in response

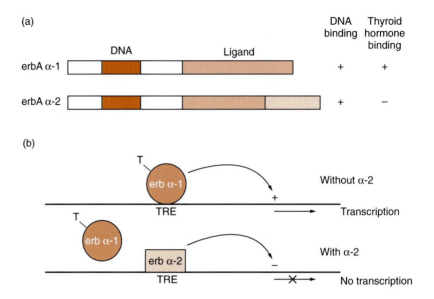

Figure 8.46

(a) Relationship of the erbA α-1 and α-2 proteins. Note that only the α-1 protein has a functional thyroid hormone binding-domain. (b) Inhibition of erbA α-1 binding and of gene activation in the presence of the α-2 protein.

to thyroid hormone (Fig. 8.46b). These two alternatively spliced forms of the transcription factor, which are made in different amounts in different tissues, therefore mediate opposing effects on thyroid hormone-dependent gene expression.

Regulation of translation

As discussed in Section 5.6, the yeast transcription factor GCN4, which induces the expression of genes involved in amino acid biosynthesis in response to amino acid starvation, is itself regulated at the level of translation (see review by Morris and Geballe, 2000). Thus, when amino acids are lacking, translation is initiated preferentially at the start point for GCN4 production, whereas when amino acids are abundant, translation initiates at short open reading frames upstream of the GCN4 start site and does not re-initiate at the GCN4 site (Fig. 5.34). Hence, GCN4 is synthesized in response to amino acid starvation and activates the genes encoding the enzymes required for the biosynthetic pathways necessary to make good this deficiency.

Such regulation of translation by alternative initiation is also observed in mammalian cells in the case of two transcription factors expressed in the liver (C/EBPα and C/EBPβ, see Section 5.6) and supplements the transcriptional control of C/EBPα gene expression described above. In this case, however, following transcription in the liver alternative translational initiation produces two different forms of each protein, known as C/EBP long or liver activator protein (LAP) and C/EBP short or liver inhibitor protein (LIP). Only LAP contains the N-terminal activation domain. In contrast, LIP production involves translational initiation at a downstream AUG residue so that this protein lacks the activation domain and contains only the DNA binding domain and the leucine zipper regions which are also found in LAP (Fig. 8.47). LIP is therefore capable of dimerizing and binding to DNA but obviously cannot activate transcription (Fig. 8.47). It therefore acts as an inhibitory protein which interferes with the activation of transcription by LAP, by competing with it for binding to the DNA. This use of alternative initiation codons to produce different forms of a transcription factor with opposite effects on transcription thus parallels the similar use of alternative splicing to do this in factors such as the thyroid hormone receptor (see above).

Regulated activity of transcription factors

Although, as discussed above, many transcription factors are regulated by controlling their synthesis, a number are controlled by regulating the activity of the protein, which is present in many different cell types. Such a system has obvious advantages in that it allows a direct effect of the agent inducing gene expression on the activity of the factor, either by binding to it or by modifying the protein, for example, by phosphorylation and hence results in a rapid response. Such post-translational modifications therefore allow specific signal transduction pathways to alter the activity of cellular transcription factors and so produce alterations in gene expression (for reviews see Hill and Treisman, 1995; Treisman, 1996). Examples of these kinds of modifications will now be discussed (Fig. 8.48).

Activation by protein–ligand or protein–protein interaction

Perhaps the simplest mechanism for regulating the activity of a transcription factor in response to a specific ligand is for it to bind the ligand directly and undergo a conformational change (Fig. 8.48a). An example of such a protein–ligand interaction activating a transcription factor is provided by the ACE1 factor in yeast, which mediates the induction of the metallothionein gene in response to copper. This protein has been shown to undergo a major conformational change upon binding of copper, which allows it to bind to metallothionein gene regulatory sites

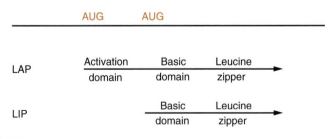

Figure 8.47

Alternative use of different initiation codons results in the generation of the C/EBP long or LAP protein which can activate transcription due to its possession of a functional activation domain and the C/EBP short or LIP protein which lacks this domain and therefore acts as a repressor.

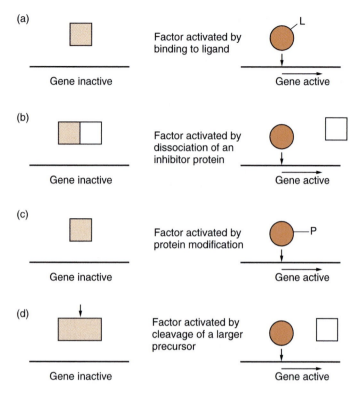

Figure 8.48

Mechanisms by which transcription factors can be activated from an inactive (square) to an active (circle) form by post-translational changes.

and induce transcription. Hence, the activity of this transcription factor is directly modulated by copper, allowing it to mediate gene activation in response to the metal (Fig. 8.49). Similarly, in mammalian cells, the ability of the DREAM transcriptional repressor to bind to DNA is decreased by binding of calcium ions (for a review see Mandel and Goodman, 1999). Hence, in the presence of calcium, the DREAM repressor is removed from the DNA and its target genes are therefore activated in response to calcium.

An interesting variant of this direct regulation of transcription factor activity by a specific stimulus is seen in the Yapl factor which is active under conditions of high oxygen and regulates the expression of anti-oxidant genes. This factor contains disulfide bonds between cysteine amino acids which create a structure for the protein that masks its nuclear export signal (NES: see Section 3.3) and it is therefore retained in the nucleus where it can regulate its target genes. Under conditions of low oxygen, the disulfide bonds are reduced, leading to them breaking and the protein refolds so that the NES is exposed. This results in the export of Yapl to the cytoplasm where it can no longer regulate transcription (Wood *et al.*, 2004) (Fig. 8.50).

Regulation of a transcription factor by its ligand is seen with the thyroid hormone receptor (see Section 8.3). In this case, however, rather than affecting the ability of the receptor to bind to DNA, the ligand induces a conformational change in the receptor which allows it to activate gene

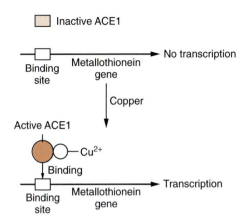

Figure 8.49

Activation of the ACE1 factor in response to copper results in transcription of the metallothionein gene.

expression in response to thyroid hormone treatment by binding a co-activator rather than a co-repressor (Fig. 8.43).

In contrast, however, other members of the steroid–thyroid hormone receptor family do not bind to DNA in the absence of the appropriate hormone. This is seen in the case of the glucocorticoid receptor, for example, where the receptor is normally located in the cytoplasm and only binds to DNA and activates transcription when the hormone is added. Originally it was thought that, as with ACE1, binding of hormone to the receptor activated its ability to bind to DNA and switch on transcription of hormone-responsive genes. However, it has been shown that although the receptor binds to DNA only in the presence of the hormone in the cell, in the test tube it will bind even when no hormone is present. This has led to the idea that, in the cell, the receptor is prevented from binding to DNA by its association with another protein, and that the hormone acts to release it from this association and allow it to fulfil its inherent ability to bind to DNA.

In agreement with this idea, the glucocorticoid receptor, unlike the thyroid hormone receptor, has been shown to be associated in the cytoplasm with a 90 000 molecular weight heat-inducible protein (hsp90). Upon steroid binding the receptor dissociates from the hsp90 and moves to the nucleus, where it activates gene transcription (Fig. 8.51). Hence, the transcription factor is activated by hormone, not by a protein–ligand interaction but by disruption of a protein–protein interaction which inhibits the inherent DNA binding ability of the receptor (Fig. 8.48b).

Activation by protein modification or proteolytic cleavage

Many transcription factors are modified extensively by the addition, for example, of O-linked monosaccharide residues or by phosphorylation. Such modifications represent obvious targets for agents that induce gene activation. Thus, such agents could act by altering the activity of a modifying enzyme, such as a kinase, which in turn would act on the

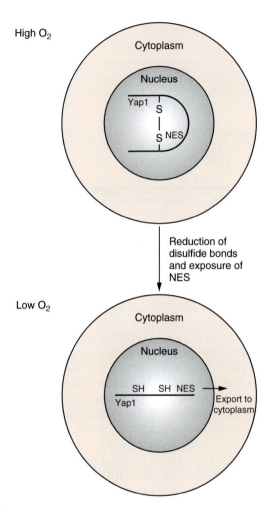

Figure 8.50

Under conditions of high oxygen, the Yap1 protein contains disulfide bonds which mask its nuclear export signal (NES) and allow it to remain active in the nucleus. Under conditions of low oxygen, the disulfide bonds are reduced exposing the NES. This results in export of Yap1 to the cytoplasm preventing it regulating its target genes.

transcription factor, resulting in its activation and the switching on of gene expression (Fig. 8.48c).

Thus, for example, acetylation enhances the DNA binding ability and hence the activity of the p53 transcription factor (Luo *et al.*, 2000) (see Section 9.5). Similarly, the basal transcription factor TFIIB (Section 3.2) has been shown to be able to acetylate itself with a resulting increase in its activity (Choi *et al.*, 2003), indicating that such modifications can also target a component of the basal transcription complex. Hence, processes such as acetylation can affect transcription factors directly as well as targeting the histones within the chromatin structure, as discussed in Section 6.6.

However, the most extensive modification which affects transcription factors is phosphorylation. One of the best-characterized examples of this

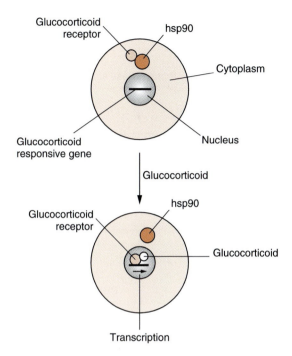

Figure 8.51

Binding of glucocorticoid to the glucocorticoid receptor results in its dissociation from hsp90 and movement of the hormone–receptor complex to the nucleus where it activates the transcription of glucocorticoid-responsive genes.

involves the mammalian CREB factor, which mediates the induction of several genes in response to treatment with cyclic AMP and binds to the cyclic AMP response element in these genes (Table 7.2) (for reviews see de Cesare and Sassone-Corsi, 2000; Mayr and Montminy, 2001). The ability of CREB to stimulate transcription is enhanced greatly via its phosphory-lation by protein kinase A on the serine residue located at amino acid 133 of the protein. Such phosphorylation stimulates the activity of the acti-vation domain located in this region of the protein. Hence, the activation of gene expression by cyclic AMP is mediated by its activation of protein kinase A which, in turn, phosphorylates and activates CREB on serine residue 133. CREB is bound to the cyclic AMP response element (CRE) both before and after cyclic AMP treatment. However, as described in Section 8.3, phosphorylation of CREB allows it to bind the CBP co-acti-vator and thereby recruit it to DNA and activate transcription in a cyclic AMP-dependent manner (Fig. 8.52).

A more complex pattern of activation is seen in the case of the heat-shock factor (HSF-1). As described in Section 7.2, HSF-1 is present in unstressed cells but becomes activated in response to exposure to stresses such as elevated temperature and then activates transcription of genes containing the HSE to which it binds. This activation proceeds in two stages (for a review see Morimoto, 1998) (Fig. 8.53). Firstly, HSF-1 is converted from a monomer to a trimer of three HSF-1 molecules which can bind to the HSE. As with the glucocorticoid receptor, the activation

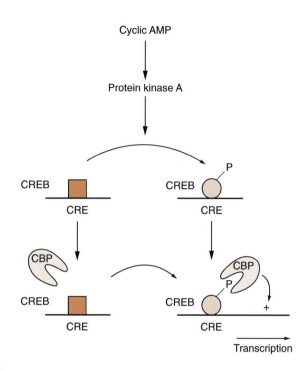

Figure 8.52

Phosphorylation of the CREB transcription factor in response to cyclic AMP stimulates the ability of CREB to bind the CBP co-activator. It therefore results in the
activation of cyclic AMP inducible genes containing the cyclic AMP response
element to which CREB binds.

of HSF-1 involves the dissociation of hsp90 which binds to HSF-1 in
unstressed cells and maintains it in an inactive form which cannot bind
to DNA (Zou *et al.*, 1998).

Interestingly, hsp90, like many of the proteins inducible by heat or
other stresses, functions as a so-called "chaperone" protein, promoting the
folding of unfolded or poorly folded proteins. Clearly, heat or other
stresses will increase the amount of such proteins, with hsp90 being
"called away" to deal with them and thus releasing HSF-1. Hence, this
system neatly couples the effect of the inducing stimulus in producing
unfolded proteins with the activation of HSF-1.

However, although trimerization allows HSF1 to bind to the HSE, this
is not sufficient to promote transcriptional activation. As in the case of
CREB, HSF-1 must be phosphorylated (on serine 230) to allow it to activate transcription (Holmberg *et al.*, 2001). This system therefore combines
the disruption of a protein–protein complex which is observed in the case
of the glucocorticoid receptor with regulation by phosphorylation as in
the case of CREB.

As well as targeting transcription factors themselves, phosphorylation
can also affect the activity of other factors which are associated with transcription factors. Thus, in the absence of calcium, the MEF2 transcription
factor is bound to DNA but it does not activate transcription since it is
bound to histone deacetylase enzymes which produce an inactive chromatin structure (see Section 6.6). Following exposure to calcium, the

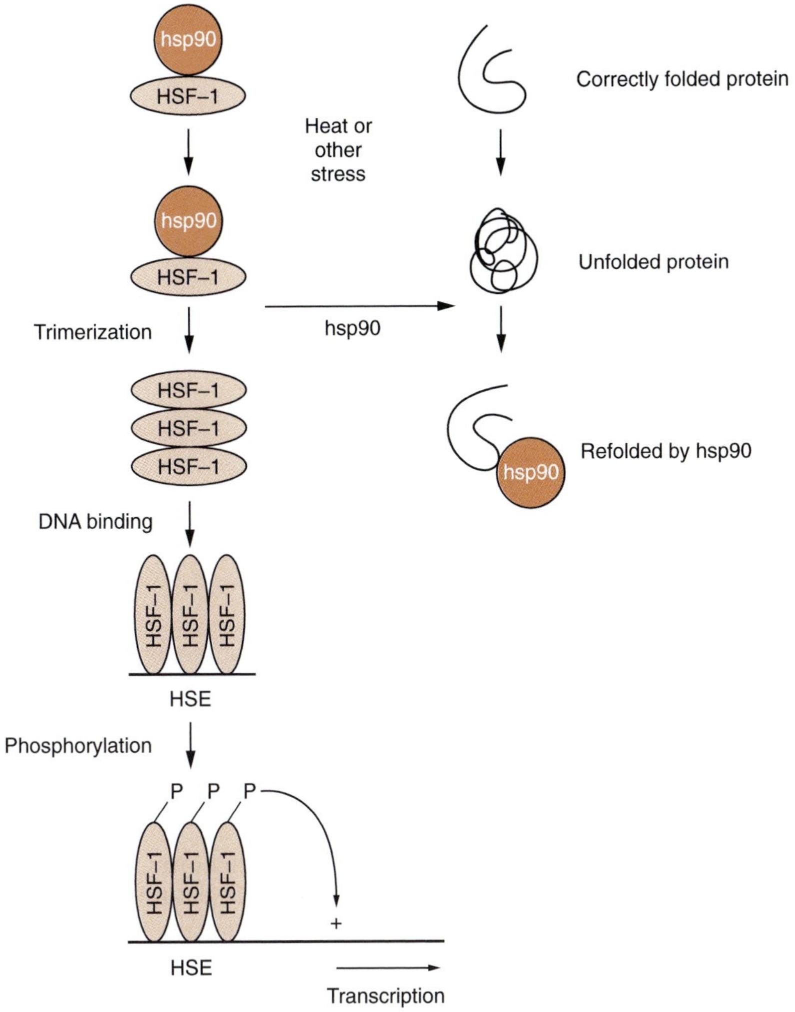

Figure 8.53

In non-stressed cells, HSF-1 exists in a non-DNA binding form associated with hsp90. In response to stresses such as elevated temperatures, hsp90 dissociates from HSF-1 and binds to unfolded proteins produced as a result of the stress. This allows HSF-1 to form a trimer which can bind to DNA but which needs to be further modified by phosphorylation before it can activate transcription.

histone deacetylases are phosphorylated, which disrupts their association with MEF2 and results in export of the histone deacetylase to the cytoplasm, allowing MEF2 to activate transcription (for reviews see Stewart and Crabtree, 2000; McKinsey *et al.*, 2002) (Fig. 8.54). This example obviously links the regulation of chromatin structure and the regulation of transcription factor activity.

A similar example involving transcription factor activation via the phosphorylation of an inhibitory protein is seen in the activation of the NFκB factor in response to phorbol ester treatment, as described above (Section 8.4). In this case, however, the target for phosphorylation is the inhibitory

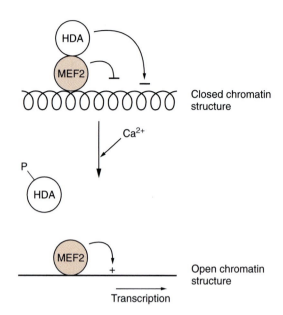

Closed chromatin structure

Ca^{2+}

Open chromatin structure

Transcription

Figure 8.54

In the absence of calcium, the MEF-2 transcription factor cannot activate transcription because it is bound to histone deacetylase enzymes (HDA) which produce an inactive chromatin structure. Calcium treatment results in phosphorylation of the histone deacetylases resulting in their dissociation and allowing MEF-2 to activate transcription.

protein IkB which blocks activation of NFκB by anchoring it in the cytoplasm. Prior to phorbol ester treatment, NFκB is bound to IκB as an inactive complex which is located in the cytoplasm. Following treatment, however, the IκB is phosphorylated by specific kinase enzymes and this renders it a target for proteolytic enzymes which rapidly degrade it. The free NFκB then moves to the nucleus where it activates gene expression (see reviews by Yamamoto and Gaynor, 2004; Hayden and Ghosh, 2004). Such activation is clearly similar to the dissociation of the glucocorticoid receptor from hsp90 following steroid treatment, although in this case phosphorylation rather than steroid binding dissociates the protein–protein complex.

It is clear, therefore, that transcription factors can be regulated by differences in the rate of degradation of the protein such as occurs for phosphorylated IκB or for RLIM following binding of CLIM (Section 8.3). Similarly, as will be discussed in Section 9.5, the p53 protein is rapidly degraded following binding to the MDM2 protein. In addition, however, transcription factor activity can also be regulated by proteolytic cleavage of an inactive precursor (Fig. 8.48d). Thus an NFκB-related protein p105 is synthesized as an inactive precursor in which the active part of the molecule is inhibited by an IκB-like region to which it is linked in a single protein. Following exposure to an activating stimulus, the IκB-like region is phosphorylated, as for IκB itself, and the precursor molecule is cleaved to release the NFκB-like region which is now able to move to the nucleus and activate gene expression (Fig. 8.55).

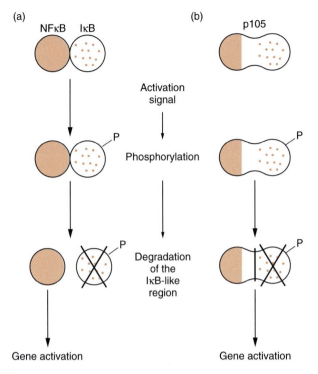

(a) NFκB   IκB

Activation
signal

Phosphorylation

P

P

(b) p105

P

P

Degradation
of the
IκB-like
region

Gene activation

Gene activation

Figure 8.55

Activation mechanisms in the NFκB family. (a) Prior to exposure to an activating signal, the NFκB factor exists as an inactive form complexed to the distinct IκB inhibitory protein. In response to an activating signal the IκB factor is phosphorylated (P), resulting in the release of free NFκB which activates transcription. (b) In the P105 protein the NFκB and IκB-like regions are part of a single large precursor which is inactive. Following exposure to an activating signal, the IκB-like region is phosphorylated and the protein is cleaved releasing the NFκB-like region which is free to activate transcription.

Such regulation by protein precursor cleavage is not confined to the p105 system (for a review see Pahl and Baeurele, 1996). It is also seen, for example, in the SREBP proteins which activate gene expression in response to elevated cholesterol (for a review see Brown and Goldstein, 1997). Thus, in the absence of cholesterol, SREBP is synthesized as an inactive precursor which is anchored within the cytoplasm to the endoplasmic reticulum membrane by a specific region of the protein. Following exposure to cholesterol, this region is cleaved off, freeing the active region and allowing it to move to the nucleus and activate its target genes.

Interestingly, as in the case of SREBP, in many of the cases we have discussed, the regulatory process results in the movement of the transcription factor into or out of the nucleus. This is seen, for example, for the glucocorticoid receptor following its release from hsp90, for the NFκB factor following release from IκB and for the Yap1 factor by masking or exposure of its nuclear export signal.

An interesting example of such regulated entry into the nucleus which does not involve either disruption of a protein–protein interaction or proteolytic cleavage is seen in the case of the Tubby transcription factor (for a review see Cantley, 2001). Thus, the Tubby factor has been shown to bind to phospholipid in the plasma membrane and is therefore anchored in this position via a protein–lipid interaction rather than a protein–protein interaction (Fig. 8.56). Following the activation of G protein receptors by specific stimuli, a phospholipase enzyme is activated. In turn this enzyme cleaves the anchoring phospholipid, releasing Tubby and allowing it to move to the nucleus and activate its target genes.

In this case nuclear localization is therefore achieved by regulation of a protein–lipid interaction rather than by a protein–protein interaction or proteolytic cleavage. Hence, regulation of nuclear localization by a number of different mechanisms is a widely used means of regulating transcription factor activity. Of course, as we have seen, it is not the only mechanism with, for example, direct regulation of DNA binding activity by the ligand being seen in the case of the ACE1 factor, regulation of the ability of an already DNA-bound transcription factor to activate transcription being

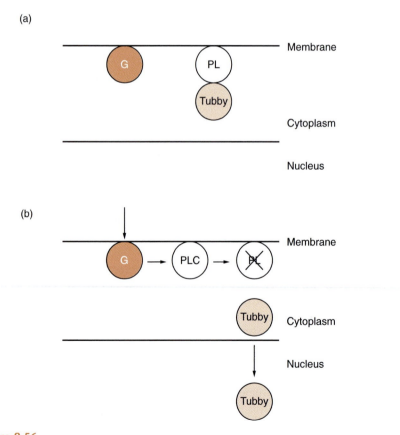

Figure 8.56

(a) The Tubby transcription factor is anchored in the cell membrane by interaction with a phospholipid (PL). (b) Following G protein-coupled receptor activation, the phospholipase C enzyme (PLC) is activated and cleaves the phospholipid. This releases Tubby and allows it to enter the nucleus and activate transcription.

observed in the case of CREB and regulation of protein degradation being seen in the case of the RLIM factor (see Section 8.3) (Fig. 8.57).

Hence, a variety of mechanisms exist for the activation of transcription factors (Fig. 8.48) with many cases involving several of these processes so that, for example, the activation of the glucocorticoid receptor involves both ligand binding and protein–protein interaction, while the NFκB cases involve phosphorylation and protein–protein interaction or proteolytic cleavage. Moreover, these mechanisms can affect the transcription factor in a number of different ways, regulating, for example, its degradation/ stabilization, its location in the cell, its DNA binding ability or the ability of a DNA-bound factor to activate transcription (Fig. 8.57).

8.5 Conclusions

Although the existence of transcription factors had been known or inferred for some time, it is clear that the cloning of the genes encoding these factors has increased our knowledge of their functional properties greatly. A number of different motifs which mediate DNA binding, protein

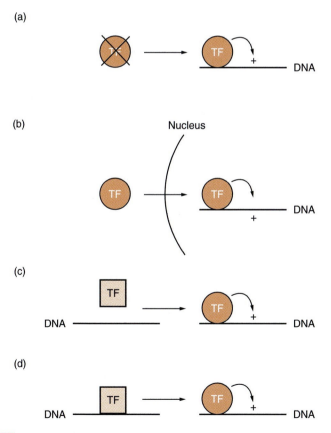

Figure 8.57

Regulatory processes can target the degradation of a transcription factor (a), its location in the cell (b), its ability to bind to DNA (c) or its ability to activate transcription following DNA binding (d)

dimerization and transcriptional activation have been identified and these are listed in Table 8.3.

Interestingly, unlike most cases of gene regulation, the regulation of transcription factors is often modulated by post-transcriptional mechanisms, including regulation of processes such as splicing or translation, as well as post-translational protein modification. Such post-transcriptional regulation is sensible in that transcriptional regulation of the genes encoding these factors would only set the process one step back, leading to the problem of how the genes encoding transcription factors are themselves transcriptionally regulated. In contrast, regulation at the post-transcriptional level allows a direct response to particular stimuli and results, for example, in the rapid activation of the glucocorticoid receptor by dissociation from hsp90 in response to a hormone, and the rapid activation of CREB by phosphorylation in response to cyclic AMP.

Our understanding of the functional domains of these factors and of their activation now opens the way for the detailed molecular modeling of how these factors interact with each other, with co-activators and co-repressors and with the RNA polymerase itself as well as other components of the basal transcriptional complex. Ultimately, this will allow an elucidation of the way in which they stimulate transcription.

Table 8.3 Transcription factor domains

Domain	Role	Factors containing domain	Comments
Homeobox	DNA binding	Numerous *Drosophila* homeotic genes, related genes in other organisms	DNA binding mediated via helix-turn-helix motif
Cysteine–histidine zinc finger	DNA binding	TFIIIA, Kruppel, SP1, etc.	Multiple copies of finger motif
Cysteine–cysteine zinc finger	DNA binding	Steroid–thyroid hormone receptor family	Single pairs of fingers, related motifs in Adenovirus E1A and yeast GAL4, etc.
Basic element	DNA binding	C/EBP, *c-fos, c-jun,* GCN4	Often found in association with leucine zipper
Leucine zipper	Protein dimerization	C/EBP, *c-fos, c-jun,* GCN4, *c-myc*	Mediates dimerization which is essential for DNA binding by adjacent domain
Helix-loop-helix	Protein dimerization	*c-myc, Drosophila* daughterless, MyoD, E12, E47	Mediates dimerization which is essential for DNA binding by adjacent domain
Acidic region	Gene activation	Yeast GCN4, GAL4, steroid—thyroid receptors, etc.	May form α-helical structure
Glutamine-rich region	Gene activation	SP1	Related regions in Oct-1, Oct-2, AP2, etc.
Proline-rich region	Gene activation	CTF/NF1	Related regions in AP2, *c-jun,* Oct-2

References

Asturias, F.J., Jiang, Y.W., Myers, L.C., Gustafsson, C.M. and Kornberg, R.D. (1999). Conserved structures of mediator and RNA polymerase II holoenzyme. *Science* **283**, 985–987.

Bieker, J.J. (2001). Krüppel-like factors: three fingers in many pies. *Journal of Biological Chemistry* **276**, 34355–34358.

Boube, M., Joulia, L., Cribbs, D.L. and Bourbon, H.-M. (2002). Evidence for a mediator of RNA polymerase II transcriptional regulation conserved from yeast to man. *Cell* **110**, 143–151.

Brennan, R.G. (1993). The winged-helix DNA-binding motif: another helix-turn-helix take off. *Cell* **74**, 773–776.

Briscoe, J. and Wilkinson, D.G. (2004). Establishing neuronal circuitry: Hox genes make the connection. *Genes and Development* **18**, 1643–1648.

Brown, M.S. and Goldstein, J.L. (1997). The SREBP pathway: regulation of cholesterol metabolism by proteolysis of a membrane-bound transcription factor. *Cell* **89**, 331–340.

Buckingham, M. (1994). Which myogenic factors make muscle? *Current Biology* **4**, 61–63.

Cantin, G.T., Stevens, J.L. and Berk, A.J. (2003). Activation domain-mediator interactions promote transcription preinitiation complex assembly on promoter DNA. *Proceedings of the National Academy of Sciences of the USA* **100**, 12003–12008.

Cantley, L.C. (2001). Translocating Tubby. *Science* **292**, 2019–2021.

Chi, N. and Epstein, J.A. (2002). Getting your Pax straight: Pax proteins in development and disease. *Trends in Genetics* **18**, 41–47.

Choi, C.H., Hiromura, M. and Usheva, A. (2003). Transcription factor IIB acetylates itself to regulate transcription. *Nature* **424**, 965–969.

De Cesare, D. and Sassone-Corsi, P. (2000). Transcriptional regulation by cyclic AMP-responsive factors. *Progress in Nucleic Acids Research and Molecular Biology* **64**, 343–369.

Donaldson, W., Peterson, J.M., Graves, B.J. and McIntosh, L.P. (1996). Solution structure of the ETS domain from murine Ets-1: a winged helix-turn-helix DNA-binding motif. *EMBO Journal* **15**, 125–134.

Duboule, D. (2000). A Hox by any other name. *Nature* **403**, 607–610.

Ellenberger, T. (1994). Getting a grip on DNA recognition – structures of the basic region leucine zipper and the basic region helix-loop-helix DNA-binding domains. *Current Opinion in Structural Biology* **4**, 12–21.

Gajiwala, K.S., Chen, H., Cornille, F., Roques, B.P., Reith, W., Mach, B. and Burley, S.K. (2000). Structure of the winged-helix protein hRFX1 reveals a new mode of DNA binding. *Nature* **403**, 916–921.

Garvie, C.W. and Wolberger, C. (2001). Recognition of specific DNA sequences. *Molecular Cell* **8**, 937–9461.

Gehring, W.J., Affolter, M., and Burglin, T. (1994). Homeodomain proteins. *Annual Review of Biochemistry* **63**, 487–526.

Goodman, R.H. and Smolik, S. (2000). CBP/p300 in cell growth, transformation and development. *Genes and Development* **14**, 1553–1577.

Green, M.R. (2000). TBP-associated factors (TAF$_{II}$s): multiple, selective transcriptional mediators in common complexes. *Trends in Biochemical Sciences* **25**, 59–63.

Gronemeyer, H. and Moras, D. (1995). How to finger DNA. *Nature* **375**, 190–191.

Hahn, S. (1993). Efficiency in activation. *Nature* **363**, 672–673.

Hahn, S. (1998). The role of TAFs in RNA polymerase II transcription. *Cell* **95**, 579–582.

Hanna-Rose, W. and Hansen, U. (1996). Active repression mechanisms of eukaryotic transcription repressors. *Trends in Genetics* **12**, 229–234.

Hayden, M.S. and Ghosh, S. (2004). Signalling to NF-κB. *Genes and Development* **18**, 2195–2224.

He, X., Treacy, M.N., Simmons, D.M., Ingraham, H.A., Swanson, L.S. and Rosenfeld, M.G. (1989). Expression of a large family of POU-domain regulatory genes in mammalian brain development. *Nature* **340**, 35–42.

Hill, C.S. and Treisman, R. (1995). Transcriptional regulation by extracellular signals: mechanisms and specificity. *Cell* **80**, 199–211.

Holmberg, C.I., Hietakangas, V., Mikhailov, A., Rantanen, J.O., Kallio, M., Meinander, A. *et al.* (2001). Phosphorylation of serine 230 promotes inducible transcriptional activity of heat shock factor 1. *EMBO Journal* **20**, 3800–3810.

Hurst, H. (1996). Leucine zippers. *Protein Profile* **3**, 1–72.

Jones, N. (1990). Transcriptional regulation by dimerization: two sides to an incestuous relationship. *Cell* **61**, 9–11.

Kadonaga, J.T. (2004). Regulation of RNA polymerase II transcription by sequence-specific DNA binding factors. *Cell* **116**, 247–257.

Kadonaga, J.T. and Tjian, R. (1986). Affinity purification of sequence-specific DNA-binding proteins. *Proceedings of the National Academy of Sciences of the USA* **83**, 5889–5893.

Kadonaga, J.T., Carner, K.R., Masiarz, F.R. and Tjian, R. (1987). Isolation of cDNA encoding the transcription factor Sp1 and functional analysis of the DNA-binding domain. *Cell* **51**, 1079–1090.

Kerppola, T. and Curran, T. (1995). Zen and the art of Fos and Jun. *Nature* **373**, 199–200.

Kewley, R.J., Whitelaw, M.L. and Chapman-Smith, A. (2004). The mammalian basic helix-loop-helix/PAS family of transcriptional regulators. *International Journal of Biochemistry and Cell Biology* **36**, 189–204.

Khorasanizadeh, S. and Rastinejad, F. (2001). Nuclear-receptor interactions on DNA-response elements. *Trends in Biochemical Sciences* **26**, 384–390.

Klug, A. and Schwabe, J.W.R. (1995). Zinc fingers. *FASEB Journal* **9**, 597–604.

Kornberg, T.B. (1993). Understanding the homeodomain. *Journal of Biological Chemistry* **268**, 26813–26816.

Kuras, L. and Struhl, K. (1999). Binding of TBP to promoters *in vivo* is stimulated by activators and requires Pol II holoenzyme. *Nature* **399**, 609–613.

Lamb, P. and McKnight, S.L. (1991). Diversity and specificity in transcriptional regulation: the benefits of heterotypic dimerization. *Trends in Biochemical Sciences* **16**, 417–422.

Latchman, D.S. (1996). Inhibitory transcription factors. *International Journal of Biochemistry and Cell Biology* **28**, 965–974.

Latchman, D.S. (ed.). (1999). *Transcription Factors: A Practical Approach*. Second Edition. Oxford University Press, Oxford.

Latchman, D.S. (2001). Transcription factors: bound to activate or repress. *Trends in Biochemical Sciences* **26**, 211–213.

Latchman, D.S. (2004). *Eukaryotic Transcription Factors*. 4th Edition. Elsevier Academic Press, London, San Diego.

Lawrence, P.A. and Morata, G. (1994). Homeobox genes: their function in *Drosophila* segmentation and pattern formation. *Cell* **78**, 181–189.

Leung, T.H., Hoffmann, A. and Baltimore, D. (2004). One nucleotide in a kappaB site can determine cofactor specificity for NF-κB dimers. *Cell* **118**, 453–464.

Levine, S.S., King, I F.G. and Kingston, R.E. (2004). Division of labour in Polycomb group repression. *Trends in Biochemical Sciences* **29**, 478–485.

Li, X.-Y., Virbasius, A., Zhu, X. and Green, M.R. (1999). Enhancement of TBP binding by activators and general transcription factors. *Nature* **399**, 605–609.

Littlewood, T.D. and Evan, G.I. (1995). Helix-loop helix. *Protein Profile* **2**, 621–702.

Luo, J., Chen, D., Shiloh, A. and Gu, W. (2000). Deacetylation of p53 modulates its effect on cell growth and apoptosis. *Nature* **408**, 377–381.

Malik, S. and Roeder, R.G. (2000). Transcriptional regulation through mediator-like coactivators in yeast and metazoan cells. *Trends in Biochemical Sciences* **25**, 277–283.

Mandel, G. and Goodman, R.H. (1999). DREAM on without calcium. *Nature* **398**, 29–30.

Mangelsdorf, D.J. and Evans, R.M. (1995). The RXR heterodimers and orphan receptors. *Cell* **83**, 841–850.

Massari, M.E. and Murre, C. (2000). Helix-loop-helix proteins: regulators of transcription in eucaryotic organisms. *Molecular and Cellular Biology* **20**, 429–440.

Marx, J. (2000). New clues to how genes are controlled. *Science* **290**, 1066–1067.

Mayr, B. and Montminy, M. (2001).Transcriptional regulation by the phosphorylation-dependent factor CREB. *Nature Reviews Molecular Cell Biology* **2**, 599–609.

McKenna, N.J. and O'Malley, B.W. (2002). Combinational control of gene expression by nuclear receptors and coregulators. *Cell* **108**, 465–474.

McKinsey, T.A., Zhang, C.L. and Olson, E.N. (2002). MEF2: a calcium-dependent regulator of cell division, differentiation and death. *Trends in Biochemical Sciences* **27**, 40–47.

Miller, J., McLachlan, A.D. and Klug, A. (1985). Repetitive zinc-binding domains in the protein transcription factor III A from *Xenopus* oocytes. *EMBO Journal* **4**, 1609–1614.

Mitchell, P.J. and Tjian, R. (1989). Transcriptional regulation in mammalian cells by sequence specific DNA binding proteins. *Science* **245**, 371–378.

Mohd-Sarip, A. and Verrijzer, C.P. (2004). A higher order of silence. *Science* **306**, 1484–1485.

Morimoto, R.I. (1998). Regulation of the heat shock transcriptional response: cross talk between a family of heat shock factors, molecular chaperones and negative regulators. *Genes and Development* **12**, 3788–3796.

Morris, D.R. and Geballe, A.P. (2000). Upstream open reading frames as regulators in mRNA translation. *Molecular and Cellular Biology* **20**, 8635–8642.

Myers, L.C. and Kornberg, R.D. (2000). Mediator of transcriptional regulation. *Annual Reviews of Biochemistry* **69**, 729–749.

Nagy, L. and Schwabe, J.W. (2004). Mechanism of the nuclear receptor molecular switch. *Trends in Biochemical Sciences* **29**, 317–324.

Natoli, G. (2004). Little things that count in transcriptional regulation. *Cell* **118**, 406–408.

Newman, J.R.S. and Keating, A.E. (2003). Comprehensive identification of human bZIP interactions with coiled-coil arrays. *Science* **300**, 2097–2101.

Ogryzkio, V.V., Schiltz, R.L., Russanova, V., Howard, B.H. and Nakatani, Y. (1996). The transcriptional coactivators p300 and CBP are histone acetyltransferases. *Cell* **87**, 953–959.

Olefsky, J.M. (2001). Nuclear receptor minireview series. *Journal of Biological Chemistry* **276**, 36863–36864.

Olson, E.N. and Klein, W.H. (1994). bHLH factors in muscle development: dead lines and commitments what to leave in and what to leave out. *Genes and Development* **8**, 1–8.

Orlando, V. (2003). Polycomb, Epigenomes and control of cell identity. *Cell* **112**, 599–606.

Ostendorff, H.P., Peirano, R.I., Peters, M.A., Schlüter, A., Bossenz, M., Scheffner, M. and Bach, I. (2002). Ubiquitination-dependent cofactor exchange on LIM homeodomain transcription factors. *Nature* **416**, 99–103.

Pahl, H.L. and Baeurele, P.A. (1996). Control of gene expression by proteolysis. *Current Opinion in Cell Biology* **8**, 340–347.

Ptashne, M. and Gann, A. (1997). Transcriptional activation by recruitment. *Nature* **386**, 569–577.

Rhodes, D. and Klug, A. (1993). Zinc finger structure. *Scientific American* **268**, 32–39.

Rosenfeld, M.G. and Glass, C.K. (2001). Coregulator codes of transcriptional regulation by nuclear receptors. *Journal of Biological Chemistry* **276**, 36865–36868.

Ross, S.E., Greenberg, M.E. and Stiles, C.D. (2003). Basic helix-loop-helix factors in cortical development. *Neuron* **39**, 13–25.

Ryan, A.K. and Rosenfeld, M.G. (1997). POU domain family values: flexibility, partnerships and developmental codes. *Genes and Development* **11**, 1207–1225.

Schwabe, J.W.R. and Rhodes, D. (1991). Beyond zinc fingers: Steroid hormone receptors have a novel structural motif for DNA recognition. *Trends in Biochemical Sciences* **16**, 291–296.

Serfling, E., Berberich-Siebelt, F., Avots, A., Chuvpilo, S., Klein-Hessling, S., Jha, M.K. *et al.* (2004). NFAT and NF-[kappa]B factors – the distant relatives. *International Journal of Biochemistry and Cell Biology* **36**, 1166–1170.

Shikama, N., Lyon, J. and La Thangue, N.B. (1997). The p300/CBP family: integrating signals with transcription factors and chromatin. *Trends in Cell Biology* **7**, 230–236.

Singh, H., LeBowitz, J.H, Baldwin, A.S. and Sharp, P.A. (1988). Molecular cloning of an enhancer binding protein isolation by screening of an expression library with a recognition site DNA. *Cell* **52**, 415–429.

Smith, R.L. and Johnson, A.D. (2000). Turning genes off by Ssn6-Tup1: a conserved system of transcriptional repression in eukaryotes. *Trends in Biochemical Sciences* **25**, 325–330.

Spiegelman, B.M. and Heinrich, R. (2004). Biological control through regulated transcriptional coactivators. *Cell* **119**, 157–167.

Stewart, S. and Crabtree, G.R. (2000). Regulation of the regulators. *Nature* **408**, 46–47.

Treisman, R. (1996). Regulation of transcription by MAP kinase cascades. *Current Opinion in Cell Biology* **8**, 205–215.

Triezenberg, S.J. (1995). Structure and function of transcriptional activation domains. *Current Opinion in Genetics and Development* **5**, 190–196.

Turner, J. and Crossley, M. (1999). Mammalian Krüppel-like transcription factors: more than just a pretty finger. *Trends in Biochemical Sciences* **24**, 236–240.

Verrijzer, C.P. and Van der Vliet, P.C. (1993). POU domain transcription factors. *Biochimica et Biophysica Acta* **1173**, 1–21.

Weatherman, R.V., Fletterick, R.J. and Scanlan, T.S. (1999). Nuclear-receptor ligands and ligand-binding domains. *Annual Reviews of Biochemistry* **68**, 559–581.

Wood, M.J., Storz, G. and Tjandra, N. (2004). Structural basis for redox regulation of Yap1 transcription factor localization. *Nature* **430**, 917–921.

Wysocka, J. and Herr, W. (2003). The herpes simplex virus VP16-induced complex: the makings of a regulatory switch. *Trends in Biochemical Sciences* **28**, 294–304.

Xanthopoulos, K.G., Mirkovitch, J., Decker, T., Kuo, C.F. and Darnell, J.E., Jr. (1989). Cell-specific transcriptional control of the mouse DNA-binding protein mc/EBP. *Proceedings of the National Academy of Sciences of the USA* **86**, 4117–4121.

Yamamoto, Y. and Gaynor, R.B. (2004). IkappaB kinases: key regulators of the NF-kappaB pathway. *Trends in Biochemical Sciences* **29**, 72–79.

Zou, J., Guo, Y., Guettouche, T., Smith, D.F. and Voellmy, R. (1998). Repression of heat shock transcription factor HSF1 activation by HSP90 (HSP90 complex) that forms a stress-sensitive complex with HSF1. *Cell* **94**, 471–480.

Gene regulation and human disease

9

SUMMARY

- Many human diseases involve alterations in gene expression caused by changes in transcription factors, co-activators and chromatin-modifying factors or in post-transcriptional processes.
- Widespread changes in gene regulation are seen in cancer resulting from the enhanced expression/activity of oncogenes or the reduced expression/activity of anti-oncogenes.
- A number of oncogenes and anti-oncogenes encode transcription factors which play a key role in regulating normal cellular processes and in cancer.
- Study of these factors can enhance our understanding of normal cellular gene regulation as well as of the abnormalities observed in cancer.
- Several human diseases are already treated by drugs which target transcription factors, but these were isolated on the basis of their efficacy against the disease and their ability to target transcription factors was only demonstrated subsequently.
- Recent advances indicate the potential of new therapeutic drugs designed specifically to target particular aspects of gene regulation.

9.1 Gene regulation and human disease

As we have discussed in the preceding chapters, the regulation of gene expression in higher eukaryotes is a highly complex process. It is not surprising, therefore, that this process can go wrong, and the identification of the molecular basis of many human diseases has shown some to be due to defects in gene regulation. A number of different processes have been shown to be affected in different human diseases and these will be discussed in turn in this section (for reviews see Engelkamp and Van Heyningen, 1996; Latchman, 1996) (Fig. 9.1). In discussing these diseases, it should be noted that only gene mutations which are compatible with life will manifest as human diseases, however severe these may be. Mutations in factors which are incompatible with survival at least to birth will evidently not be detected and it is likely that many mutations affecting gene regulation fall into this category.

Transcriptional regulators

Transcription factors (Fig. 9.1a)

A number of congenital diseases which are detectable at birth or shortly after involve mutations in the genes encoding specific transcription factors. Thus, for example, mutations in individual members of the PAX transcription factor family, discussed in Section 8.2, have been shown to be involved in a number of congenital eye disorders, while mutations in the gene encoding the Pit-1 member of the POU family of transcription factors (see Section 8.2) result in a failure of pituitary gland development and consequent dwarfism in both mice and humans. Similarly, mutations in the MEF2A transcription factor (see Section 8.4) have been shown to produce coronary artery disease in middle-aged patients, indicating that mutations in transcription factors can produce disease later in life as well as in early development (Wang et al., 2003).

As well as such defects in the development or functioning of specific organs, mutations in the genes encoding the members of the steroid/thyroid hormone receptor family, discussed in Section 8.2, can result in a failure to respond to the specific hormone which normally binds to the receptor and thereby regulates gene expression. Thus, for example, mutations in the gene encoding the glucocorticoid receptor result in a syndrome of steroid resistance in which the patients do not respond to glucocorticoid.

Interestingly, mutations in the peroxisome proliferator-activated receptor gamma (PPARγ), which is also a member of the steroid/thyroid hormone receptor family, have been found in a few human individuals with insulin resistance leading to type 2 diabetes. Although such cases are rare, they suggest that PPARγ plays a key role in insulin responses and that it may be a valuable therapeutic target to enhance such responses in other cases of diabetes and in obesity (for reviews see Kersten et al., 2000; Rosen and Spiegelman, 2001; Evans et al., 2004) (see also Section 9.7).

All the examples discussed above involve transcription factors which regulate transcription by RNA polymerase II. However, specific diseases have also been identified where the abnormal protein is involved in transcription by the other RNA polymerases. Thus, in the case of RNA polymerase I, mutations in the CSB protein, which forms a complex with RNA polymerase I and other proteins, cause the disorder known as Cockayne's syndrome with abnormalities in the nervous system and skeleton (Bradsher et al., 2002). Similarly, the abnormal craniofacial development characteristic of Treacher Collins syndrome has been shown to result from mutations in the gene encoding a protein that interacts with the RNA polymerase I basal transcription factor UBF (Valdez et al., 2004) (see Section 3.2).

Co-activators (Fig. 9.1b)

As described in Section 8.3, co-activators play a key role in the action of transcriptional activators. In particular, the CBP co-activator is of vital importance in a number of different signaling pathways, linking the transcription factor which is activated by the pathway to either the basal transcriptional complex or altering chromatin structure. It is not surprising, therefore, that mutations which result in individuals lacking

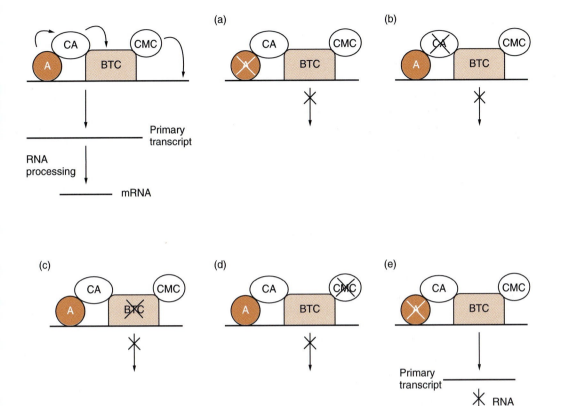

Figure 9.1

Mutations can affect a number of components of the gene expression processes including (a) transcriptional activators (A); (b) co-activators (CA); (c) the basal transcriptional complex (BTC); (d) chromatin modeling complexes (CMC); or (e) factors involved in RNA processing.

functional CBP are incompatible with life. Indeed, even having a single inactive CBP gene whilst retaining a single functional gene results in the severe human disease Rubinstein–Taybi syndrome, which is characterized by mental retardation and physical abnormalities (for a review see D'Arcangelo and Curran, 1995).

Hence, even if a single CBP gene remains functional, the inactivation of the second copy produces disease. This is likely to be due to the amount of CBP in normal cells being low with various transcription factors competing for it (see Section 8.3). Hence, the reduction in the level of CBP due to loss of one gene copy results in disease (Fig. 9.2).

CBP can also be involved in the disease processes even when it is intact and unmutated. Thus, it has been shown that individuals with the neurodegenerative disease Huntington's chorea produce an abnormal form of a protein known as Huntingtin, in which a CAG sequence in the protein coding sequence has expanded, producing a run of glutamine amino acids (encoded by the CAG triplet) in the Huntingtin protein. This abnormal form of Huntingtin binds to the normal CBP protein and

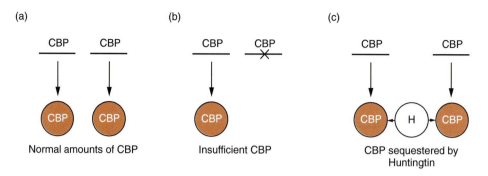

Figure 9.2

To function effectively each cell needs to have two intact genes encoding the CBP co-activator protein (panel a). Specific diseases result when one gene is inactive resulting in reduced levels of CBP (panel b) or when CBP is intact but is sequestered by an abnormal Huntingtin protein (H) so that it cannot fulfil its co-activator role (panel c).

prevents it from regulating transcription by depositing it in insoluble aggregates (Nucifora *et al.*, 2001) (Fig. 9.2). Interestingly, mutant Huntingtin has also been shown to interfere with the functioning of the Sp1 transcription factor and the TBP component of the basal transcriptional complex, indicating that it can disrupt the function of different classes of transcriptional regulators (Schaffar *et al.*, 2004; for a review see Sugars and Rubinsztein, 2003) (Fig. 9.3).

The production of disease by the removal of transcription factors from their normal sites within the cell is also seen in the muscle disease, type 1 myotonic dystrophy. As with Huntington's disease, this disease involves the presence of abnormal numbers of a triplet sequence, in this case CUG, in a specific gene. However, this run of multiple CUG triplets is located in the 3′ untranslated region of the DM protein kinase gene rather than in its protein coding sequence. The abnormal RNA containing the multiple CUG sequence has been shown to bind transcription factors such as Sp1, STAT1, STAT3 and the retinoic acid receptor, preventing them from functioning

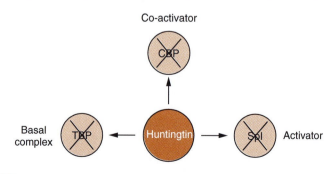

Figure 9.3

The abnormal Huntingtin protein found in Huntington's disease can interfere with the functioning of a transcriptional activator, a co-activator and a component of the basal transcriptional complex.

normally (Ebralidze *et al.*, 2004). Hence, the triplet repeat abnormalities in Huntington's disease and myotonic dystrophy act in different ways at the RNA or protein level to produce the same effect of sequestering transcription factors (for a review see Ranum and Day, 2004) (Fig. 9.4).

Basal transcriptional complex (Fig. 9.1c)

The effect of Huntingtin on TBP indicates the ability of disease processes to disrupt the functioning of the basal transcriptional complex which is one of the major targets for activators and co-activators (see Section 8.3). Given the critical role for this complex (see Section 3.2), it is likely, however, that most mutations in components of the complex would be incompatible with survival to birth and would not therefore be observed as post-natal human diseases. However, mutations in components of TFIIH have been observed in the human disease Xeroderma pigmentosum, which results in skin defects and a higher risk of cancer. Importantly, these mutations have been shown to result in defective responses to transcriptional activators and repressors rather than affect-

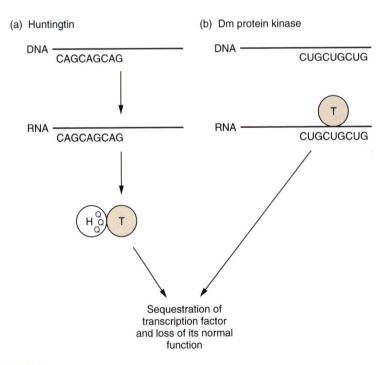

Figure 9.4

The expansion of a specific triplet sequence in the Huntington gene in Huntington's disease or in the DM protein kinase gene in myotonic dystrophy both produce disease by binding transcriptional regulators (T) and preventing them performing their normal function. In the case of Huntingtin, the repeated sequence is found in the protein coding sequence and produces a mutant Huntington protein (H) with a run of glutamine residues (Q) which binds transcriptional regulators. However, in the DM protein kinase case, the repeat is in the 3′ untranslated region and is not translated into protein. Rather, it is the mRNA containing the repeat which binds the transcriptional regulators.

ing the basal activity of TFIIH (Liu *et al.*, 2001) and this is presumably why they are compatible with survival, albeit with abnormal functioning which results in disease.

Chromatin remodeling factors (Fig. 9.1d)

As discussed in Chapter 6, changes in chromatin structure are an essential prerequisite for activation of transcription. It is not surprising, therefore, that human diseases result when these processes are disrupted. Thus, mutations in the MeCP2 factor which prevent it from specifically recognizing methylated DNA result in the neurological disease Rett syndrome (see Section 6.5). Similarly, mutations in a DNA methyltransferase enzyme, which normally adds methyl groups to C residues, have been implicated in the human ICF syndrome (immunodeficiency, centromeric instability, facial anomalies; Okano *et al.*, 1999).

In addition, as well as resulting from alteration in methylation recognition proteins or methylases, disease can also result from changes in methylation patterns on the DNA. Thus, in the human fragile X syndrome a triplet sequence CGG, which is present in 10–50 tandem copies in the first exon of the FMR-1 gene, is amplified so that over 230 tandem copies are present in patients with this syndrome. This amplified sequence is then heavily methylated on the C residues (see Section 6.5), resulting in transcriptional silencing of the FMR-1 gene. The protein normally produced by the FMR-1 gene acts as a regulator of the translation of other mRNAs and its absence in turn produces the mental retardation and other symptoms characteristic of the fragile X syndrome (for reviews see Trottier *et al.*, 1993; Jin *et al.*, 2004). This represents another mechanism by which an expansion in a three-base sequence can cause disease, in addition to the different mechanisms involving sequestration of transcription factors which were discussed above). Altered methylation patterns have also been observed in other human diseases such as cancer (for reviews see Baylin and Herman, 2000; Feinberg and Tycko, 2004; Lund and van Lohuizen, 2004).

Other aspects of chromatin remodeling can also be affected in human diseases. Thus, the Williams syndrome transcription factor, which is mutated in the human disease Williams syndrome, is a component of the WINAC chromatin remodeling complex. This complex interacts with the vitamin D receptor, which is a member of the steroid/thyroid hormone receptor family of transcription factors (see Sections 7.2 and 8.2). Chromatin remodeling by the WINAC complex is essential for recruitment of the vitamin D receptor to its target genes and this process is defective in Williams syndrome producing mental retardation and growth deficiency (Kitagawa *et al.*, 2003; for a review see Belandia and Parker, 2003).

Similarly, a mutation in the SNF2 factor which is part of the chromatin remodeling SWI/SNF complex discussed in Section 6.7, results in a severe form of α-thalassemia that not only results in lack of globin gene expression as in other thalassemias (see, for example, Section 7.5) but also leads to other symptoms such as mental retardation, indicating that the activity of this factor is necessary for opening the chromatin structure of both the α-globin genes and a number of other genes, so preventing their transcription if it is absent (Gibbons *et al.*, 1995).

Interestingly, in another type of α-thalassemia, changes in chromatin structure are also involved but in a completely different way. In this case, a deletion removes one of the two copies of the α-globin gene, leaving the other copy on the same chromosome intact. However, the deletion also removes the 3′ end of the LUC7L gene, which is adjacent to the α-globin genes but is transcribed from the opposite strand of the DNA (Fig. 9.5). This removes the polyadenylation signal terminating the RNA of the LUC7L gene and results in its transcript continuing through the remaining α-globin gene, where it is transcribed off the opposite, anti-sense strand to that from which the α-globin gene is transcribed. By a mechanism which is not understood but may be related to that of small interfering RNAs (see Section 6.6), this antisense RNA somehow induces C methylation of the α-globin gene. In turn this results in transcriptional silencing of the remaining α-globin gene, leading to thalassemia, even though all the regulatory elements of the gene and its coding sequence are intact (Tufarelli *et al.*, 2003; for a review see Kleinjan and van Heyningen, 2003) (Fig. 9.5).

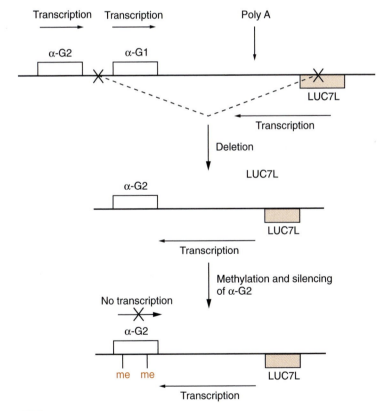

Figure 9.5

In a particular form of α-thalassemia, one of the two α-globin genes (α-G1) and part of the LUC7L gene, including its polyadenylation site (Poly A) are deleted. This results in the LUC7L RNA transcript extending into the remaining α-globin gene (α-G2) in an anti-sense orientation. In turn, this results in the methylation (me) and transcriptional silencing of the α-G2 gene.

Post-transcriptional events (Fig. 9.1e)

As well as affecting transcriptional processes, mutations can also affect gene regulation at post-transcriptional stages (see Chapters 3 and 5 for an account of these processes and their role in gene regulation).

A number of such mutations which affect the processes of RNA splicing or alternative splicing have been described and fall into two classes (for a review see Faustino and Cooper, 2003). The first class involves a mutation in the splice site of a particular gene which results in a failure properly to splice its RNA or to produce one of the alternatively spliced mRNAs (Fig. 9.6a). Thus, for example, the failure to produce one of the two alternatively spliced mRNAs derived from the porphobillinogen deaminase gene is the cause of one form of the disease acute intermittent porphyria.

In the second class, the mutation affects a splicing factor and therefore affects a number of genes whose splicing requires this factor (Fig. 9.6b). Hence, the effect of this class of mutation is not confined to a single gene. An example of this second class is seen in the SMN gene which was originally defined on the basis of its mutation in the disease spinal muscular atrophy and has now been shown to encode a protein which is essential for RNA splicing. Hence diseases can result both from mutations in RNA splicing proteins and from mutations in specific target RNAs which affect their splicing pattern.

Although splicing represents the post-transcriptional stage of gene expression which has been most characterized in terms of disease-causing mutations, other post-transcriptional stages also appear to be affected in specific diseases. Thus, for example, a defect in RNA editing of the glutamate receptor mRNA has been suggested to be involved in motor neurone disease (Kawahara *et al.*, 2004). Similarly, mutations in the translational initiation factor eIF2B has been shown to be the cause of the neurologi-

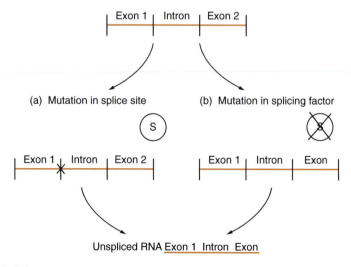

Figure 9.6

Splicing can be disrupted by a mutation (X) in a splice site of a particular RNA (panel a) or in a splicing factor (S) required for its correct splicing.

cal disease leukoencephalopathy with vanishing white matter (for a review see Abbott and Proud, 2004).

Although a wide variety of human diseases arise due to defects in gene regulation, the human disease that exhibits the most extensive malregulation of gene expression is cancer. Thus, not only does cancer often result from the over-expression of certain cellular genes (known as proto-oncogenes) due to errors in their regulation, but several of these genes themselves actually encode transcription factors and cause the disease by affecting the expression of other genes. Subsequent sections of this chapter will therefore focus on the connection between cancer and the malregulation of gene expression, and will illustrate how our increasing knowledge of this connection has aided our understanding both of the disease and of the processes that regulate gene expression in normal and transformed cancer cells.

9.2 Proto-oncogenes

In order to provide a background to our discussion of cancer and gene regulation, it is necessary to discuss briefly the nature of cancer-causing oncogenes and the process that led to their discovery (see reviews by Broach and Levine, 1997; Hunter, 1997; Vogelstein and Kinzler, 1992).

As long ago as 1911, Peyton Rous showed that a connective tissue cancer in the chicken was caused by an infectious agent. This agent was subsequently shown to be a virus and was named Rous Sarcoma virus (RSV) after its discoverer. This cancer-causing, tumorigenic virus is a member of a class of viruses called retroviruses, whose genome consists of RNA rather than DNA as in most other organisms.

The majority of viruses of this type do not cause cancers but simply infect a cell and produce a persistent infection with continual production of virus by the infected cell. In the case of RSV, however, such infection also results in the conversion of the cell into a cancer cell, capable of indefinite growth and eventually killing the organism containing it.

In the case of non-tumorigenic retroviruses, the genome contains only three genes, which are known as *gag*, *pol* and *env*, and which function in the normal life cycle of the virus (Fig. 9.7). Following cellular entry, the viral RNA is converted into DNA by the action of the Pol protein, and this DNA molecule then integrates into the host chromosome. Subsequent transcription and translation of this DNA produces the viral structural proteins Gag and Env which coat the viral RNA genome, yielding viral particles that leave the cell to infect other susceptible cells.

Inspection of the RSV genome (Fig. 9.8) reveals an additional gene, known as the *src* gene, which is absent in the other viruses, suggesting that this gene is responsible for the ability of the virus to cause cancer. This idea was confirmed subsequently by showing that, if the *src* gene alone was introduced into normal cells, it was able to transform them to a cancerous phenotype. This gene was therefore called an oncogene (from the Greek *onkas* for mass or tumor) or cancer-causing gene.

Following the identification of the *src* oncogene in RSV, a number of other oncogenes were identified in other oncogenic retroviruses infecting both chickens and mammals, such as the mouse or rat. Over 20 cellular proto-oncogenes which were originally identified in this way are now known (Table 9.1).

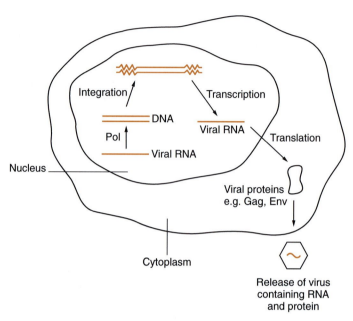

Figure 9.7

Life cycle of a typical non-tumorigenic retrovirus.

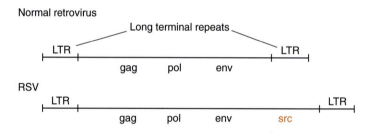

Figure 9.8

Comparison of the genome of a non-tumorigenic retrovirus with that of the tumorigenic retrovirus RSV.

The identification of individual genes that are able to cause cancer obviously opened up many avenues of investigation for the study of this disease. From the point of view of gene regulation, however, the most exciting aspect of oncogenes was provided by the discovery that these cancer-causing genes are derived from genes present in normal cellular DNA. Thus, using Southern blotting techniques (Methods Box 2.1) Takeya and Hanafusa (1985) detected a cellular equivalent of the viral *src* gene in the DNA of both normal and cancer cells (Fig. 9.9), and also showed that an mRNA capable of encoding the Src protein was produced in normal cells. Subsequent studies identified cellular equivalents of all the retroviral oncogenes. Many of these cellular equivalents of the viral oncogenes have now been cloned and shown to encode proteins identical or closely related to those present in the retroviruses. To avoid confusion, the viral

Table 9.1 Proto-oncogenes and their functions

Oncogene	Species infected by virus	Normal function of protein
abl	Mouse	Tyrosine kinase
erbA	Chicken	Transcription factor and hormone receptor
erbB	Chicken	Receptor for epidermal growth factor, tyrosine kinase
ets	Chicken	Transcription factor
fes	Cat	Tyrosine kinase
fgr	Cat	Tyrosine kinase
fms	Cat	Tyrosine kinase, receptor for colony-stimulating factor
fos	Mouse	Transcription factor
jun	Chicken	AP1-related transcription factor
kit	Cat	Tyrosine kinase
lck	Chicken	Tyrosine kinase
mos	Mouse	Serine/threonine kinase
myb	Chicken	Transcription factor
myc	Chicken	Transcription factor
raf	Chicken	Serine/threonine kinase
ras	Rat	GTP-binding protein
rel	Turkey	Transcription factor
ros	Chicken	Tyrosine kinase
sea	Chicken	Tyrosine kinase
sis	Monkey	Platelet-derived growth factor B chain
ski	Chicken	Nuclear protein
src	Chicken	Tyrosine kinase
yes	Chicken	Tyrosine kinase

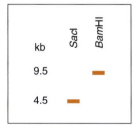

Figure 9.9

Southern blot showing hybridization of the RSV *src* gene with normal chicken DNA cut with the restriction enzymes *Sac*I or *Bam*HI. Redrawn from Takeya and Hanafusa, *Cell* **32**, 881–890 (1983), by permission of Professor H. Hanafusa and Cell Press.

oncogenes are given the prefix v, as in v-*src*, while their cellular equivalents are designated proto-oncogenes and given the prefix c, as in c-*src*.

The presence of cellular equivalents of the retroviral oncogenes suggests that the viral genes have been picked up from normal cellular DNA following integration of a non-tumorigenic virus next to the proto-oncogene. The subsequent incorrect excision of the viral genome resulted in its picking up the gene and converting it into a tumorigenic virus (Fig. 9.10). Hence the oncogene in the virus is derived from a cellular gene which is present in the DNA of normal cells.

Paradoxically, the same gene can be present in normal cells without any apparent adverse effect and yet can cause cancer when incorporated into a virus. It is now clear that this conversion of the proto-oncogene into a cancer-causing oncogene is caused by one of two processes. Either the gene is mutated in some way within the virus, so that an abnormal product is formed, or alternatively the gene is expressed within the virus at much higher levels than are achieved in normal cells, and this high level of the normal product results in transformation (Fig. 9.11).

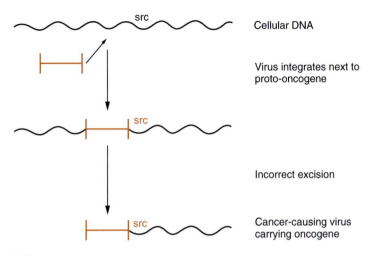

Figure 9.10

Model for the way in which a retrovirus could have picked up the cellular *src* gene by integration adjacent to the gene and subsequent incorrect excision.

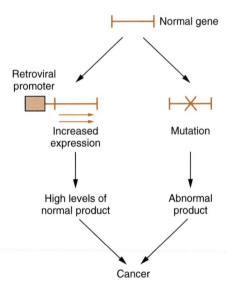

Figure 9.11

A cellular proto-oncogene can be converted into a cancer-causing oncogene by increased expression or by mutation.

Such conversion of a proto-oncogene into a cancer-causing oncogene is not confined to viruses, however. Cases of cancer which have no viral involvement in humans have now been shown to be due either to the over-expression of individual cellular proto-oncogenes or to their alteration by mutation within the cellular genome. Hence these genes play an important role in the generation of human cancer.

The potential risk of such proto-oncogenes causing cancer raises the question of why these genes have not been deleted during evolution. In fact, proto-oncogenes have been highly conserved in evolution, equivalents to mammalian and chicken oncogenes having been found not only in other vertebrates but also in invertebrates, such as *Drosophila*, and even in single-celled organisms, such as yeast.

The extraordinary evolutionary conservation of many of these genes, despite their potential danger, led to the suggestion that their products were essential for the processes regulating the growth of normal cells, and that their malregulation or mutation therefore results in abnormal growth and cancer. This idea has been confirmed abundantly as more and more proto-oncogenes have been characterized and shown to encode growth factors that stimulate the growth of normal cells, cellular receptors for growth factors and other cellular proteins involved in transmitting the growth signal within the cell, either by acting as a protein kinase enzyme or by binding GTP.

Ultimately the growth regulatory pathways controlled by oncogene products end in the nucleus, with the activation of genes whose corresponding proteins are required by the growing cell. It is not surprising, therefore, that several proto-oncogenes have been shown to encode transcription factors that regulate the expression of genes activated in growing cells (for reviews see Lewin, 1991; Forrest and Curran, 1992) (Table 9.2).

Hence, oncogenes present two aspects of importance from the point of view of the regulation of gene expression. Firstly, since cancer is often caused by elevated expression of cellular oncogenes, the processes whereby this occurs are of interest, both from the point of view of the etiology of cancer and for the light they throw on the mechanisms which regulate gene expression. This topic is discussed in Section 9.3. Second, the study of the transcription factors encoded by a few proto-oncogenes has led to a better understanding of the processes regulating gene expression in cells growing normally and in cancer cells, and this is discussed in Section 9.4.

9.3   Elevated expression of oncogenes

The products of cellular proto-oncogenes play a critical role in cellular growth control and, in many cases, are synthesized only at specific times and in small amounts. It is not surprising, therefore, that transformation into a cancer cell can result when these genes are expressed at high levels in particular situations.

The simplest example of such over-expression occurs in the case of retroviruses where, as we have already discussed, the oncogene comes under the

Table 9.2   Oncogenes and transcription factors

Oncogene	Comments
erbA	Mutant form of the thyroid hormone receptor
ets	Binding site often found in association with AP1 site
fos	Binds to AP1 site as Fos-Jun dimer
jun	Can bind to AP1 site alone as Jun-Jun dimer
mdm2	Inhibits gene activation by p53
myb	–
myc	Requires Max protein to bind to DNA
rel	Member of the NFκB family
spi-1	Identical to PU.1 transcription factor

influence of the strong promoter contained in the retroviral long terminal
repeat (LTR) region and is hence expressed at a high level (Fig. 9.11).

A similar up-regulation due to the activity of a retroviral promoter is
also seen in the case of avian leukosis virus (ALV) of chickens. Unlike the
retroviruses described so far, however, this virus does not carry its own
oncogene. Rather, it transforms by integrating into cellular DNA next to
a cellular oncogene, the c-*myc* gene (Fig. 9.12a) (for reviews of c-*myc* see
Eisenman, 2001; Levens, 2003; Nilsson and Cleveland, 2003). The expression of the c-*myc* gene is brought under the control of the strong promoter
in the retroviral LTR, and it is hence expressed at levels 20–50 times higher
than normal, producing transformation. This process is known as
promoter insertion.

In other cases of this type, the ALV virus has been shown to have integrated downstream rather than upstream of the c-*myc* gene. Hence, the
elevated expression of the c-*myc* gene in these cases cannot be due to
promoter insertion. Rather, it involves the action of an enhancer element
in the viral LTR which activates the c-*myc* gene's own promoter. As
discussed in Section 7.3, enhancers, unlike promoters, can act in either
orientation and at a distance, and hence the ALV enhancer can activate
the *myc* promoter from a position downstream of the gene (Fig. 9.12b).

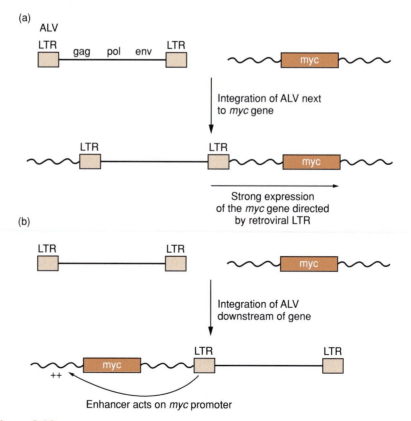

Figure 9.12

Avian leukosis virus (ALV) can increase expression of the *myc* proto-oncogene
either via promoter insertion (a) or by the action of its enhancer (b).

These cases are of interest from the point of view of gene regulation, as indicating how viral regulatory systems can subvert cellular control processes. Of potentially greater interest, however, are the cases where up-regulation of a cellular oncogene occurs through the alteration of internal cellular regulatory processes, rather than through viral intervention. The best-studied example of this type concerns the increased expression of the c-*myc* oncogene which occurs in the transformation of B cells in cases of Burkitt's lymphoma in humans or in the similar plasmacytomas that occur in mice.

When these tumors were studied, it was noted that they commonly contained very specific chromosomal translocations which involved the exchange of genetic material between chromosomes 8 and 14 (Fig. 9.13). Most interestingly, the region of chromosome 8 involved includes the c-*myc* gene, and the translocation results in the *myc* gene being moved to chromosome 14, where it becomes located adjacent to the gene encoding the immunoglobulin heavy chain. This translocation results in the increased expression of the *myc* gene which is observed in the tumor cells.

Such translocation of specific oncogenes to the immunoglobulin locus resulting in increased expression is not unique to the *myc* gene but has been observed for a variety of other oncogenes in different B-cell leukemias, while similar translocations involving oncogene translocation to the T-cell receptor gene locus (which is highly expressed in T lymphocytes) have been observed in T-cell leukemias (for reviews see Rabbits, 1994; Latchman, 1996) (see also Section 9.4). The mechanisms responsible for the observed up-regulation of oncogene expression have been best described, however, in the case of the *myc* oncogene and these will now be discussed.

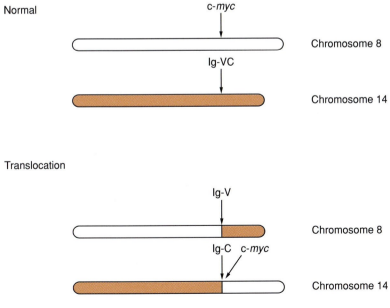

Figure 9.13

Translocation of the c-*myc* gene from chromosome 8 to the immunoglobulin heavy-chain gene locus on chromosome 14 that occurs in cases of Burkitt's lymphoma.

Detailed study of the processes mediating increased *myc* gene expression has indicated that it is produced by different mechanisms in different lymphomas, depending on the precise break points of the translocation within the c-*myc* and immunoglobulin genes. In all cases studied, however, the break point of the translocation occurs within the immunoglobulin gene, resulting in a truncated gene lacking its promoter being linked to the c-*myc* gene. This fact, together with the fact that the genes are always linked in a head-to-head orientation (Fig. 9.14), indicates that the up-regulation of the c-*myc* gene does not occur via a simple promoter insertion mechanism (see Fig. 9.12a) in which it comes under the control of the immunoglobulin promoter. In some cases, however, the B-cell-specific enhancer element, which is located between the joining and constant regions of the immunoglobulin genes (see Sections 2.4 and 7.3), is brought close to the *myc* promoter by the translocation (Fig. 9.15). This enhancer element is highly active in B cells and can activate the *myc* promoter in a manner analogous to the enhancer of ALV (see Fig. 9.12b).

Hence, in this case the c-*myc* gene is up-regulated by the action of the B-cell-specific regulatory mechanisms of the immunoglobulin gene, the immunoglobulin enhancer activating the c-*myc* promoter rather than its own promoter, which has been removed by the translocation. In other cases, however, this does not appear to be the case, the break point of the translocation having removed both the immunoglobulin promoter and enhancer. This leaves the c-*myc* gene adjacent to the constant region of

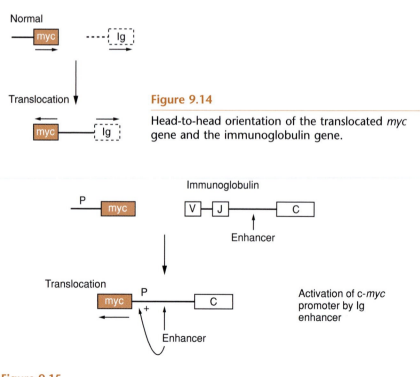

Normal

myc   Ig

Translocation

myc   Ig

Figure 9.14

Head-to-head orientation of the translocated *myc* gene and the immunoglobulin gene.

Immunoglobulin

P myc   V J   C

Enhancer

Translocation

P myc   +   C

Enhancer

Activation of c-*myc* promoter by Ig enhancer

Figure 9.15

In some cases the enhancer of the immunoglobulin heavy-chain gene activates the *myc* gene promoter.

the immunoglobulin gene without any obvious B-cell-specific regulatory elements. In these cases it is likely that the up-regulation of the c-*myc* gene arises from its own truncation in the translocation. This results in the removal of negative regulatory elements that normally repress its expression, such as the upstream silencer element which normally represses the c-*myc* promoter (see Section 7.4 and Fig. 9.16).

More extensive truncation of the c-*myc* gene, involving the removal of transcribed sequences rather than upstream elements, has also been observed in some tumors. Frequently, this involves the removal of the first exon of the c-*myc* gene which does not contain any protein-coding information. This exon may thus fulfil a regulatory role by modulating the stability of the c-*myc* RNA or by affecting its translatability. Hence, its removal could enhance the level of c-*myc* protein by increasing the stability or the efficiency of translation of the c-*myc* RNA produced at a constant level of transcription. A similar increase in gene expression could be achieved by the removal of sequences within the first intervening sequence which inhibit the transcriptional elongation of the nascent c-*myc* transcript (see Section 4.3).

The increased expression of an oncogene produced by the removal of sequences that negatively regulate it is also seen in the case of the *lck* proto-oncogene, which encodes a tyrosine kinase related to the c-*src* gene product. In this case, activation of the oncogene in tumors is accompanied by the removal of sequences within its 5′ untranslated region, upstream of the start site of translation. The removal of these sequences results in a 50-fold increase in the initiation of translation of the *lck* mRNA into protein. Most interestingly, the region removed contains three AUG translation initiation codons which are located upstream of the correct initiation codon for production of the Lck protein (Fig. 9.17). The elimination of these codons results in increased translation initiation from the correct AUG, suggesting that initiation at the upstream codons inhibits correct initiation. This is exactly analogous to the regulation of translation of the GCN4 protein, which was discussed in Section 5.6.

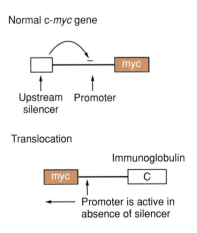

Figure 9.16

Activation of *myc* gene expression can be achieved by removal of the upstream silencer element.

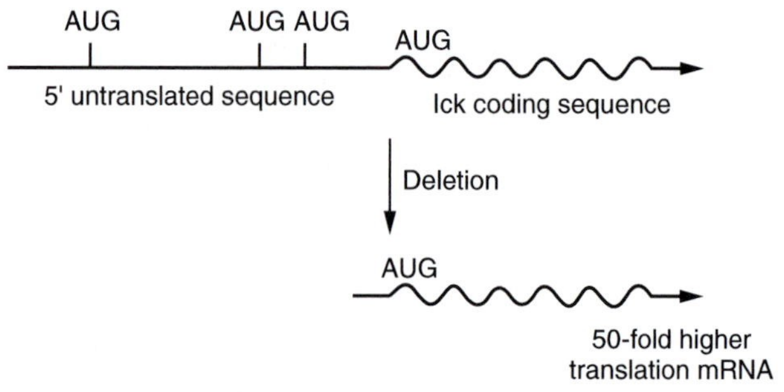

Figure 9.17

Increased translation of the *lck* proto-oncogene mRNA can be achieved by deletion of AUG translational start sites upstream of the AUG that initiates the coding sequence of the Lck protein.

It is likely that the processes regulating the expression of oncogenes such as c-*myc* and *lck*, which are revealed by studying their over-expression in tumors, also play a role in their normal pattern of regulation during cellular growth. Hence, their further study will throw light not only on the mechanisms of tumorigenesis but also on the processes regulating gene expression in normal cells.

Other examples of up-regulation of oncogene expression in tumorigenesis may occur, however, by mechanisms that are unique to the transformed cell. Thus, as discussed in Section 2.3 DNA amplification is relatively rare in normal cells. In tumors, however, it is observed frequently for specific oncogenes and results in the presence of regions of amplified DNA which are visible in the microscope as homogeneously staining regions or as double minute chromosomes. Such amplification is especially common in human lung tumors and brain tumors, and frequently involves the c-*myc*-related genes, N-*myc* and L-*myc*. The expression of these genes in the tumor is increased dramatically due to the presence of up to 1000 copies of the gene in the tumor cell.

A variety of mechanisms, involving both the subversion of normal control processes or abnormal events occurring in the tumor cell, therefore result in the observed over-expression of oncogenes in tumor cells. In turn, such over-expression is critical for tumor formation. Thus, for example, if the high levels of c-*myc* in a mouse model of liver cancer are reduced, the tumor cells lose their cancerous phenotype and differentiate into hepatocytes. If over-expression of c-*myc* is then restored, the cells return to a cancerous phenotype (Shachaf *et al.*, 2004). When taken together with the production of abnormal oncogene products due to mutations, it is likely that the overexpression of specific oncogenes is involved in a wide variety of human cancers.

9.4 Transcription factors as oncogenes

As described in Section 9.2, the isolation of the cellular genes encoding particular oncogene products led to the realization that they were involved in many of the processes regulating cellular growth. Ultimately the onset and continuation of cellular growth is likely to involve the activation of cellular genes that are not expressed in quiescent cells. It is not surprising, therefore, that several proto-oncogenes have been shown to encode transcription factors which regulate the transcription of genes activated in growing cells (Table 9.2) and several of these cases will be discussed (for reviews see Lewin, 1991; Forrest and Curran, 1992).

Fos, Jun and AP1

The chicken retrovirus, avian sarcoma virus AS17, contains an oncogene, v-*jun*, whose equivalent cellular proto-oncogene encodes a nuclearly located DNA binding protein. Sequence analysis of this protein revealed that it showed significant homology to the DNA binding domain of the yeast transcription factor GCN4, suggesting that it might bind to similar DNA sequences (Fig. 9.18). Interestingly, GCN4 itself had been shown previously to bind to similar sequences to those bound by a factor, AP1 (activator protein 1), which had been detected in mammalian cell extracts by its DNA binding activity (Fig. 9.19).

This relationship of Jun and AP1 to the sequence and binding activity, respectively, of GCN4 led to the suggestion that Jun might be related to AP1. This was confirmed by the findings that antibody to Jun reacted with purified AP1 preparations, and that Jun expressed in bacteria was capable of binding to AP1 binding sites in DNA. Moreover, Jun was capable of stimulating transcription from promoters containing AP1 binding sites but not from those which lacked these sites. Hence, the *jun* oncogene encodes a sequence-specific DNA binding protein capable of stimulating transcription of genes containing its binding site, which is identical to the AP1 binding site.

Although Jun undoubtedly binds to AP1 binding sites, preparations of AP1 purified on the basis of this ability contain several other proteins in addition to c-Jun. Several of these are encoded by genes related to *jun*, but

```
jun    206 P L F P I DM E S Q E R I K A E R K R M R N R I A A S K S R K
GCN4   216 P L S P I V P E S S D P    A A L K R A R N T E A A R R S R A

           R K L E R I A R L E E K V K T L K A Q N S E L A S T A N M L R
           R K L Q R M K Q L E D K V              E E L L S K N Y H L E

           E Q V A Q L K Q K V M N H V N S G C Q L M L T Q Q L G T F 296
           N E V A R L K K L V G E R 281
```

Figure 9.18

Comparison of the C-terminal amino acid sequences of the chicken Jun protein and the yeast transcription factor GCN4. Boxes indicate identical residues.

```
                        DNA binding site

            GCN 4  5' T G A C/G T C A T 3'
            AP 1   3' T G A  G  T C A G 3'
```

Figure 9.19

Relationship of the DNA binding sites for the yeast transcription factor GCN4 and the mammalian transcription factor AP1.

another is the product of a different proto-oncogene, namely c-*fos*. Interestingly, however, although Fos is present in AP1 preparations, it does not bind to DNA when present alone but requires the product of the c-*jun* gene for DNA binding. Hence, in addition to its ability to bind to AP1 sites alone, Jun can also form a complex with Fos that binds to this site. As discussed in Section 8.2, this association takes place through the leucine zipper domains of the two proteins and results in a heterodimer complex which binds to the AP1 binding site with much greater affinity than the Jun homodimer.

Hence, both Fos and Jun, which were identified originally through their association with oncogenic retroviruses, are also cellular transcription factors (for reviews see Karin *et al.*, 1997; Jochum *et al.*, 2001; Shaulian and Karin, 2002). Such a finding raises the question of the normal role of these factors and how they can cause cancer. In this regard it is of obvious interest that the AP1 binding site is involved in mediating the induction of genes that contain it in response to treatment with phorbol esters, which are also capable of promoting cancer. Thus, not only do many phorbol ester-inducible genes contain AP1 binding sites (see Table 7.2), but, in addition, transfer of AP1 binding sites to a normally non-inducible gene renders that gene inducible by phorbol esters. Increased levels of Jun and Fos are also observed in cells after treatment with phorbol esters. Hence, these substances act by increasing the levels of Fos and Jun, which in turn cause increased transcription of other genes containing AP1 binding sites that mediate induction by the Fos-Jun complex.

Most interestingly, increased levels of Jun and Fos are also produced by treatment with serum or growth factors which stimulates the growth of quiescent cells. Hence, the transduction of the signal to grow, which begins with the growth factors and their cellular receptors and continues with intra-cytoplasmic signal transducers such as protein kinases and GTP binding proteins, ends in the nucleus with the increased level of the transcription factors Jun and Fos (Fig. 9.20). These proteins will then activate the genes whose products are necessary for the process of growth itself.

Clearly, it is relatively easy to fit the oncogenic properties of Jun and Fos into this framework. Thus, if these proteins are normally produced in response to growth-inducing signals and activate growth, their continual abnormal synthesis will result in a cell which will be stimulated to grow continually and will not respond to growth-regulating signals. Such continuous uncontrolled growth is characteristic of the cancer cell.

In agreement with this idea, mutations in the leucine zipper region of Fos, which abolish its ability to dimerize with Jun and induce genes containing AP1 sites, also abolish its ability to transform cells to a cancerous phenotype. Hence, the ability of Fos to cause cancer is directly linked to its ability to act as a transcription factor for genes containing the appropriate binding site.

Interestingly, AP1 sites in growth-regulated genes are often located close to binding sites for the Ets protein, another transcription factor which is encoded by a cellular proto-oncogene (see Table 9.2). Hence several different oncogenic transcription factors may cooperate to produce high-level transcription of specific genes in actively growing cells.

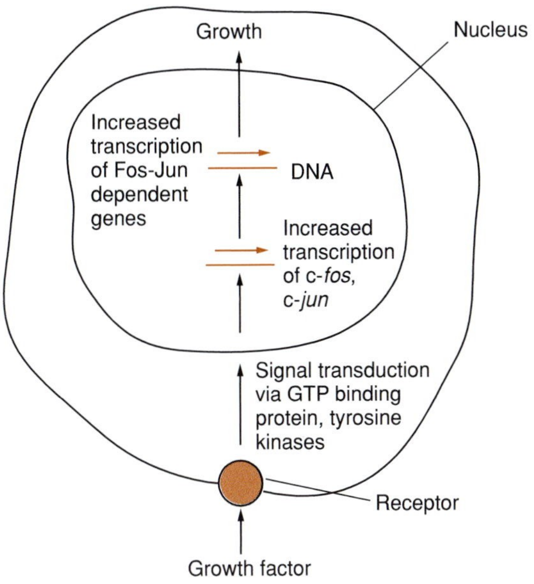

Figure 9.20

Growth factor stimulation of cells results in increased transcription of the c-*fos* and c-*jun* genes, which in turn stimulates transcription of genes which are activated by the Fos-Jun complex.

v-erbA and the thyroid hormone receptor

Unlike most other retroviruses, avian erythroblastosis virus (AEV) carries two cellular oncogenes, v-*erb*A and v-*erb*B. When c-*erb*A, the cellular equivalent of v-*erb*A, was cloned, it was shown to encode the cellular receptor that mediates the response to thyroid hormone.

As discussed in Sections 8.3 and 8.4, this receptor is a member of the superfamily of steroid–thyroid hormone receptors which, following binding of a particular hormone, induce the transcription of genes containing a binding site for the hormone–receptor complex. In the case of ErbA, the protein contains a region that can bind thyroid hormone. Following such binding, the hormone–receptor complex induces transcription of thyroid hormone-responsive genes, such as those encoding growth hormone or the heavy chain of the myosin molecule (Fig. 9.21).

The finding that the cellular homolog of the v-*erb*A oncogene is a hormone-responsive transcription factor provides a further connection between oncogenes and cellular transcription factors. It raises the question, however, of the manner in which the transfer of the thyroid hormone receptor to a virus can result in transformation. To answer this question it is necessary to compare the protein encoded by the virus with its cellular counterpart. As shown in Fig. 9.22 the ErbA protein has the typical structure of a member of the steroid–thyroid hormone family (see also Fig. 8.18), containing both DNA binding and hormone binding regions. The viral ErbA protein is generally similar except that it is fused to a portion of the retroviral gag protein at its N-terminus. It also contains

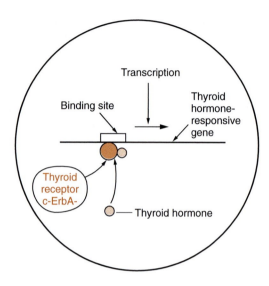

Figure 9.21

The c-*erb*A gene encodes the thyroid hormone receptor and activates transcription in response to thyroid hormone.

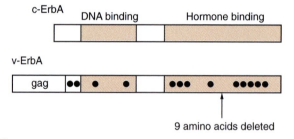

Figure 9.22

Relationship of the cellular ErbA protein and the viral protein. The black dots indicate single amino acid differences between the two proteins while the arrow indicates the region where nine amino acids are deleted in the viral protein.

a number of mutations in both the DNA binding and hormone binding regions, as well as a small deletion in the hormone binding domain.

Of these changes, it is the alterations in the hormone binding domain which have the most significant effects on the function of the protein and which are thought to be critical for transformation. Thus, these changes abolish the ability of the protein to bind thyroid hormone and activate transcription. However, the protein retains the inhibitory domain which allows the Erb-A protein to repress gene transcription in the absence of thyroid hormone (see Section 8.3). Thus, the viral protein is functionally analogous to the alternatively spliced form of the c-*erb*A gene product discussed in Section 8.4, which lacks the hormone binding domain and dominantly represses the ability of the hormone binding receptor form to activate thyroid hormone-responsive genes.

The idea that the non-hormone binding viral ErbA protein might also be able to do this has been confirmed by studying the effect of this oncogene on thyroid hormone-responsive genes. As expected, the v-*erb*A gene product was able to abolish the responsiveness of such genes to thyroid hormone by binding to the thyroid hormone-response elements in their promoters and preventing activation by the cellular ErbA protein–thyroid hormone complex (Fig. 9.23).

Interestingly, however, this repression by v-*erb*A does not simply involve the passive blockage of binding by c-*erb*A following thyroid hormone treatment. Thus, mutations in the inhibitory domain described in Section 8.3 can abolish the oncogenic activity of v-*erb*A. These mutations do not affect the ability of v-*erb*A to bind to DNA but prevent it from recruiting the inhibitory co-repressor which is essential for its ability to actively repress transcription (see Section 8.3) (for a review see Perlmann and Vennstrom, 1995). Hence the ability actively to repress transcription is essential for transformation by v-*erb*A (Fig. 9.24).

The explanation of how such gene repression by viral ErbA can result in transformation is provided by the observation that the introduction of the viral gene into cells can repress transcription of the avian erythrocyte anion transporter gene. This gene is one of those which is switched on when chicken erythroblasts differentiate into erythrocytes. It has been known for some time that the viral ErbA protein can block this process, and it is now clear that this is achieved by blocking the induction of the genes needed for this to occur. In turn, such blockage of differentiation allows the cells to continue to proliferate. When this is combined with the introduction of the v-*erb*B gene, which encodes a truncated form of the epidermal growth factor receptor and renders cell growth independent of external growth factors, transformation results (Fig. 9.25).

Transformation caused by v-*erb*A thus represents an example of the activation of an oncogene by mutation resulting, in this case, in its losing the

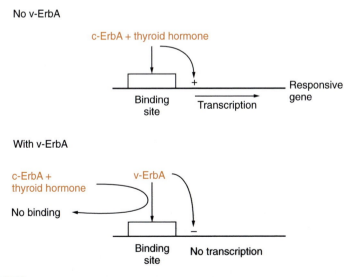

Figure 9.23

Inhibitory effect of the viral ErbA protein on gene activation by the cellular protein, in response to thyroid hormone. Note the similarity to the action of the α-2 form of the c-ErbA protein, illustrated in Fig. 8.46.

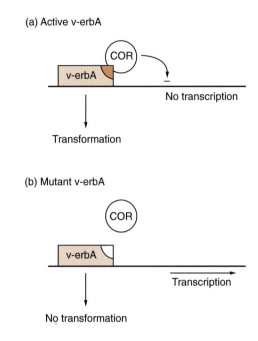

(a) Active v-erbA

COR

v-erbA

No transcription

Transformation

(b) Mutant v-erbA

COR

v-erbA

Transcription

No transformation

Figure 9.24

The ability of v-*erb*A to transform cells to a cancerous phenotype requires it to be able actively to inhibit transcription via its inhibitory domain (brown) which acts by recruiting an inhibitory co-repressor (COR) (panel a). Thus mutations in the v-*erb*A inhibitory domain (white) which abolish its ability to bind the co-repressor also abolish its ability to transform cells even though it can still bind to DNA (panel b).

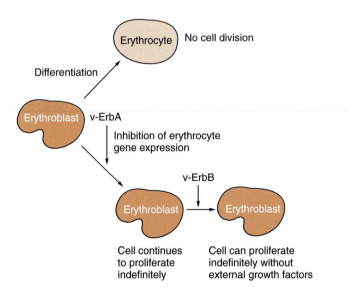

Erythrocyte   No cell division

Differentiation

Erythroblast   v-ErbA

Inhibition of erythrocyte
gene expression

v-ErbB

Erythroblast ──→ Erythroblast

Cell continues   Cell can proliferate
to proliferate   indefinitely without
indefinitely   external growth factors

Figure 9.25

Inhibition of erythrocyte-specific gene expression by the v-ErbA protein prevents erythrocyte differentiation and allows transformation by the v-ErbB protein.

ability to activate transcription but retaining its ability to act as a dominant repressor of transcription. As discussed in Section 8.4, however, one alternatively spliced transcript of the c-erbA gene is also able to do this and, like the v-erbA gene product, cannot bind thyroid hormone. Hence, this repression of transcription by a non-hormone binding form of the receptor is likely to be of importance in normal cells also.

Other transcription factor-related oncogenes

Although the fos-jun and erbA cases represent the best characterized examples of the connection between oncogenes and transcription factors, several other cellular oncogenes encode transcription factors (see Table 9.2; for a review see Lewin, 1991).

One of these, the Myc protein, has been studied intensively in view of its over-expression in many human tumors (see Section 9.3) (for reviews see Eisenman, 2001; Levens, 2003; Nilsson and Cleveland, 2003). The Myc protein clearly encodes a transcription factor which contains the leucine zipper motif characteristic of many transcription factors, including Fos and Jun (see Section 8.2), as well as a helix-loop-helix motif (see Section 8.2). Moreover, mutations in the leucine zipper region of the protein abolish its oncogenic ability to transform normal cells, suggesting that the ability to act as a transcription factor is essential for Myc-induced transformation.

Despite all this evidence, for many years the actual role of Myc in transcriptional control remained unclear. This was because it was not possible to demonstrate the binding of Myc to a specific DNA sequence in the manner that had been shown to occur for Jun and ErbA. This problem was resolved, however, by the finding that Myc has to heterodimerize with a second factor, Max, in order to bind to DNA (for reviews see Baudino and Cleveland, 2001; Levens, 2003). Hence, Myc resembles Fos in requiring another factor for sequence-specific binding. This finding indicates once again the importance of heterodimerization in regulating the activity of transcription factors and also illustrates how the function of a particular factor can remain obscure simply because its partner has not yet been isolated.

Although our discussion of oncogenes as transcription factors has focused on the cellular genes which are initially identified in RNA tumor viruses, a number of oncogenes encoding transcription factors have also been identified on the basis of their involvement in the chromosomal translocations which occur in human leukemias (see Section 9.3). Thus, as well as involving previously characterized oncogenes, such as the myc gene, these translocations can involve genes not previously shown to produce cancer when over-expressed. For example, a member of the homeobox family of transcription factors (see Section 8.2), Hox-11, has been shown to be involved in cases of acute childhood leukemia where its expression is activated by translocation to the T-cell receptor gene locus.

As well as resulting in increased expression of a proto-oncogene by translocation to a highly active immunoglobulin or T-cell receptor gene locus, such chromosomal translocations can also cause leukemias by producing fusion genes between two genes which were previously located on separate chromosomes (for reviews see Rabbits, 1994; Latchman, 1996;

Look, 1997) (Fig. 9.26). Such fusion proteins presumably produce cancer because the fusion protein has cancer-causing properties distinct from those of either intact protein alone. Thus, the fusion protein which is present in 15% of human acute myeloid leukemias involves the AML and ETO transcription factors. This AML-ETO fusion protein is able to bind to basic helix-loop-helix factors (see Section 8.2) and prevent them binding the CBP co-activator, thereby acting as a transcriptional repressor (Zhang et al., 2004).

As in the case of translocations causing over-expression, such translocations can involve genes encoding transcription factors previously identified within RNA tumor viruses or those which have not previously been shown to have an oncogenic effect. Thus, the c-ets-1 proto-oncogene originally identified in a chicken retrovirus (see Table 9.1), which was

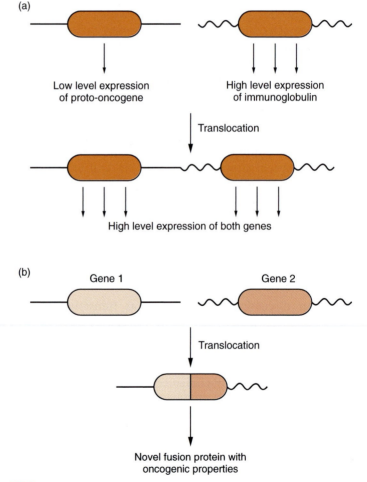

(a)

Low level expression
of proto-oncogene

High level expression
of immunoglobulin

Translocation

High level expression of both genes

(b)

Gene 1

Gene 2

Translocation

Novel fusion protein with
oncogenic properties

Figure 9.26

Activation of an oncogene through chromosomal translocation can occur if its expression is increased by translocation to the actively transcribed immunoglobulin or T-cell receptor gene loci (panel a) or if a fusion gene encoding a novel protein with oncogenic properties is created (panel b).

discussed in Section 8.2, becomes fused to the platelet-derived growth factor receptor gene in chronic myelomonocytic leukemia. In contrast, in acute promyelocytic leukemia (PML) the translocation involves the fusion of two transcription factor genes not characterized in oncogenic retroviruses, namely the retinoic acid receptor α gene (a member of the steroid/thyroid hormone receptor family discussed in Section 8.2 and the PML transcription factor, a zinc finger transcription factor which was originally characterized on the basis of its involvement in this disease.

As in the case of the AML-ETO fusion protein, the RAR-PML fusion protein acts as a transcriptional repressor, even though the retinoic acid receptor alone can act as an activator. This inhibitory activity of RAR-PML involves its ability to recruit histone deacetylases and produce an inactive chromatin structure (Grignani *et al.*, 1998; for a review see Lin *et al.*, 2001). Indeed, the AML-ETO fusion protein can also bind histone deacetylases (in addition to its effect on CBP binding by basic helix-loop-helix factors). This suggests that enhanced deacetylation induced by oncogenic fusion proteins may play a critical role in human leukemia (for reviews see Minucci *et al.*, 2001; Hake *et al.*, 2004) (Fig. 9.27) and may be a target for therapy (see Section 9.7).

Interestingly, the ELL gene which is involved in chromosomal translocations leading to fusion protein products in acute myeloid leukemia is a transcription elongation factor (for a review see Conaway and Conaway, 1999). This indicates that oncogenes can encode factors involved in transcriptional elongation as well as transcriptional initiation and alteration of chromatin structure (see Section 4.3 for discussion of transcriptional elongation and its control).

In addition to such cellular oncogenes identified in RNA tumor viruses or at sites of chromosomal translocations, it is worth noting that DNA viruses that can cause cancer also encode oncogenes capable of regulating

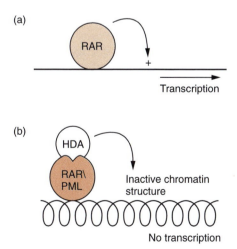

Figure 9.27

The RAR/PML fusion protein found in promyelocytic leukemia can recruit a histone deacetylase which produces an inactive chromatin structure. It therefore acts as a transcriptional repressor (panel b) unlike the normal retinoic acid receptor (panel a).

cellular gene expression. In particular, both the large T oncogenes of the small DNA viruses, SV40 and polyoma, and the E1A protein of adenovirus are capable of affecting the transcription of specific cellular genes, and this ability is critical for the ability of these proteins to transform cells. Unlike the oncogenes of RNA viruses, however, the genes encoding these viral proteins do not appear to have specific cellular equivalents, and are likely to have evolved within the virus rather than having been picked up from the cellular genome.

The similar ability of oncogenes from DNA and RNA tumor viruses to affect cellular gene expression, despite their very different origins, suggests that such modulation of cellular gene expression is critical for the transforming ability of these viruses.

9.5 Anti-oncogenes

Nature of anti-oncogenes

Following the discovery of cellular oncogenes it was rapidly shown that they encoded proteins which promoted cellular growth so that their activation by over-expression or by mutation resulted in abnormal growth leading to cancer. Subsequently, however, it became clear that cancer could also result from the deletion or mutational inactivation of another group of genes. This indicated that these genes encoded products which normally restrained cellular growth so that their inactivation would result in abnormal, unregulated growth. These genes were therefore named anti-oncogenes or tumor suppressor genes (for reviews see Knudson, 1993; Weinberg, 1993; Vogelstein and Kinzler, 1992).

Clearly the inhibitory role of anti-oncogenes indicates that cancers will involve the deletion of these genes or the occurrence of mutations within them which inactivate their protein product, rather than their over-expression or mutational activation as occurs in the oncogenes (Fig. 9.28). Thus far a number of genes of this type have been defined on the basis of their mutation or deletion in specific tumor types (Table 9.3). Most interestingly, the three best-defined anti-oncogene products all encode transcription factors. Two of these, p53 and the Wilms' tumor gene product, appear to act by binding to target sites in the DNA of specific genes

Table 9.3 Anti-oncogenes and their functions

Anti-oncogene	Tumors in which gene is mutated	Nature of protein product
APC	Colon carcinoma	Cytoplasmic protein
BRCA-1	Breast cancer	DNA repair, transcription factor?
BRCA-2	Breast cancer	DNA repair, transcription factor?
DCC	Colon carcinoma	Cell adhesion molecule
NF1	Neurofibromatosis	Activator of Ras GTPase activity
NF2	Schwannomas, meningiomas	Cytoplasmic protein
p53	Sarcomas, breast carcinomas, leukemia, etc.	Transcription factor
RB1	Retinoblastoma, osteosarcoma, small lung cell carcinoma	Transcription factor
VHL	Pheochromocytoma, kidney carcinoma	Transcriptional elongation regulator
WT1	Wilms' tumor	Transcription factor

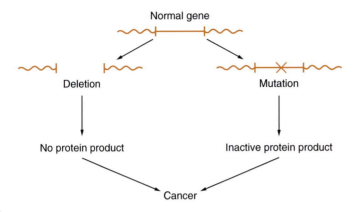

Figure 9.28

Deletion of an anti-oncogene or its inactivation by mutation can result in cancer.

and regulating their expression while the third, the retinoblastoma gene product (Rb-1), acts predominantly via protein–protein interactions with other regulatory transcription factors. The p53 and Rb-1 proteins will therefore be discussed as examples of these two types (for a review of the Wilms' tumor gene product see Hastie, 2001).

p53

The p53 protein was originally identified as a 53 kDa protein which bound to the large T-oncogene protein of the small DNA virus SV40. Subsequent studies have shown that the gene encoding this protein is mutated in a very wide variety of human tumors, especially carcinomas (for reviews of p53, see Vogelstein *et al.*, 2000; Haupt *et al.*, 2002; Sharpless and DePinho, 2002). In normal cells, p53 is induced in response to DNA damage and its activation results in growth arrest and/or cell death. Most interestingly, inactivation of the p53 gene in mice does not prevent the normal development of the animal. Rather, it results in an abnormally high rate of tumor formation which results in the early death of the animal. This has led to the idea that the p53 gene product normally acts to prevent cells with damaged DNA from proliferating by inducing their growth arrest or death. In the absence of p53 such cells proliferate and form tumors, which occur at high frequency when p53 is inactivated by mutation.

The detailed characterization of the p53 gene product has shown that it is a transcription factor capable of binding to a specific DNA sequence and activating the expression of specific genes. The mutations which occur in human tumors result in a loss of the ability to bind to DNA, indicating that the ability of p53 to do this is crucial for its ability to control cellular growth and suppress cancer. The p53 protein thus functions, at least in part, by activating the expression of specific genes whose protein products act to inhibit cellular growth (Fig. 9.29i). Its inactivation by gene deletion (Fig. 9.29ii) or by mutation (Fig. 9.29iii) thus leads to a failure to express these genes, leading to uncontrolled growth.

Interestingly, a failure of p53-mediated gene activation can also occur even in the presence of functional p53 (Fig. 9.29iv). Thus, many human

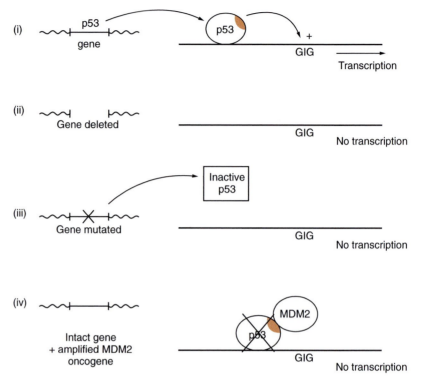

Figure 9.29

The functional p53 protein acts to stimulate the transcription of genes (GIG) whose protein products inhibit growth (i). This effect can be prevented, however, by the deletion of the p53 gene (ii) or by its inactivation by mutation (iii), as well as by the MDM2 oncoprotein which binds to p53 and inactivates it (iv). The activation domain of p53 is indicated by the colored area.

soft tissue sarcomas contain intact p53 protein but have amplified the cellular *mdm2* oncogene. The MDM2 oncoprotein binds to p53 and promotes its modification by the addition of ubiquitin. Interestingly, the addition of a single ubiquitin residue to p53 promotes its export from the nucleus whilst the addition of two such residues promotes its degradation (Li *et al.*, 2003) (Fig. 9.30). Either effect evidently prevents p53 activating its target genes in the nucleus (for a review see Yang *et al.*, 2004).

The interaction of p53 with oncogenic proteins is not unique to the MDM2 oncoprotein. Thus, as noted above, it was the interaction of p53 with the SV40 large T oncoprotein which led to the original identification of the p53 protein. As in the case of the MDM2 protein, the interaction of p53 with either the large T protein or the transforming proteins of several other DNA viruses prevents p53 from activating its target genes. Such functional inactivation of p53 appears to play a critical role in the ability of these viruses to transform cells to a cancerous phenotype. When taken together with the action of the MDM2 protein, this indicates that functional interactions between oncogene and anti-oncogene products are likely to be critical in the control of cellular growth. Hence changes in this balance due to the over-expression of specific oncogenes or the loss of anti-oncogenes will result in cancer.

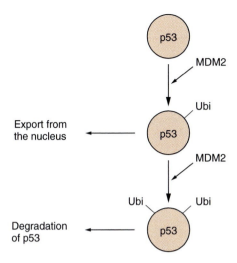

Figure 9.30

MDM2 catalyzes the addition of one or two ubiquitin residues to p53 resulting in its inactivation by respectively promoting its export from the nucleus or its degradation.

As well as being modified by ubiquitination, the activity of p53 is also regulated by other post-translational modifications (for reviews see Mayo and Donner, 2002; Prives and Manley, 2001). Thus, the phosphorylation of p53 at its N-terminus prevents its interaction with MDM2, thereby stabilizing the protein, whilst the addition of acetyl residues at its C-terminus enhances its DNA binding ability (Luo et al., 2004). Similarly, methylation of p53 on a specific lysine residue stimulates its nuclear localization and enhances its stability (Chuikov et al., 2004).

Hence, as in the case of the histones discussed in Section 6.6, the activity of p53 is modulated by phosphorylation, methylation and acetylation (Fig. 9.31). Interestingly, the RAR-PML fusion protein found in promyelocytic leukemia (see Section 9.4) can induce deacetylation of p53, thereby inhibiting its activity, providing an example of the activity of an anti-oncogene being targeted by an oncogenic fusion protein (Insinga et al., 2004).

The critical role of p53 in regulating the expression of genes encoding growth inhibitory proteins focuses attention on the nature of these growth inhibitory genes. A number of potential target genes for p53 have now been identified. One of these encodes a protein (p21) that acts as an inhibitor of cyclin-dependent kinases. Cyclin-dependent kinases are enzymes that stimulate cells to enter cell division. Hence, the identification of a p53-regulated gene as an inhibitor of these enzymes immediately suggested that p53 acts by stimulating the expression of this inhibitory factor, which in turn inhibits the cyclin-dependent kinases and therefore prevents cells replicating their DNA and undergoing cell division (Fig. 9.32).

Similarly, the observation that p53 also stimulates the expression of the bax gene, whose protein product stimulates programmed cell death or apoptosis, supports a role for p53 in promoting the death of cells which have become abnormal and can no longer divide normally.

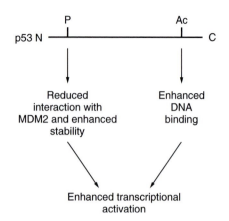

Figure 9.31

The addition of phosphate residues at the N-terminus of the p53 protein reduces the interaction with MDM2 and improves its stability whilst the addition of acetyl residues at the C-terminus enhances its DNA binding ability.

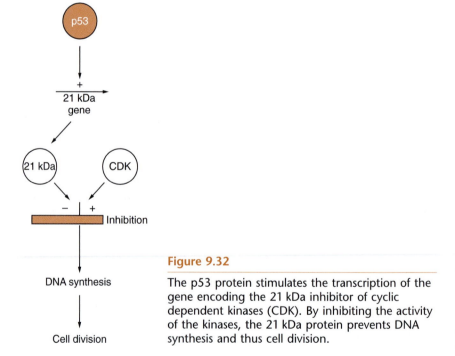

Figure 9.32

The p53 protein stimulates the transcription of the gene encoding the 21 kDa inhibitor of cyclic dependent kinases (CDK). By inhibiting the activity of the kinases, the 21 kDa protein prevents DNA synthesis and thus cell division.

It is likely, therefore, that p53 acts as a sensor for damage to cellular DNA, for example, by irradiation, which could result in mutations occurring. When it is activated, growth arrest genes are stimulated and the cell ceases to divide so that it can repair the damage. If the damage to the cell is irreparable however, genes inducing apoptosis are activated by p53 and the cell dies. Evidently, in the absence of p53, cells with DNA damage or mutations will continue to replicate and, if the mutations activate specific oncogenes, then a cancer will result (Fig. 9.33).

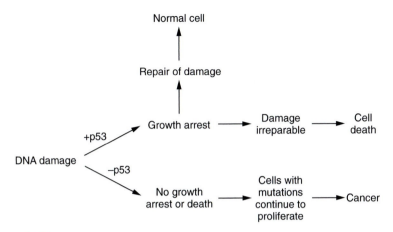

Figure 9.33

In normal cells, p53 is activated by damage to DNA. The cell is then induced to growth arrest, allowing the damage to be repaired. If the damage cannot be repaired, p53 activates cell death and the damaged cell dies. In the absence of p53, the damaged cell continues to proliferate and, if it has mutations in onco-genes, will form a cancer.

Hence, the p53 gene product plays a key role in regulating cell division, acting by regulating the expression of specific target genes. Its inactiva-tion by mutations or by specific oncogene products is likely to play a critical role in the majority of human cancers. Although p53 was initially believed to be a unique protein, unrelated to any others, two proteins which are related to p53 have now been identified and have been named p63 and p73 on the basis of their molecular weights (for reviews see Lohrum and Vousden, 2000; Davis and Dowdy, 2001; Yang *et al.*, 2002). Neither p63 or p73 appears to be commonly mutated in human cancer but they do appear to be required for normal development. Thus, loss or mutation of p63 leads to the human disease known as EEC syndrome (ectrodactyly, ectodermal dysplasia and cleft lip) in which patients have limb defects and facial clefts (Celli *et al.*, 1999).

The retinoblastoma protein

The retinoblastoma gene (Rb-1) was the first anti-oncogene to be identi-fied. This was on the basis that its inactivation results in the formation of eye tumors known as retinoblastomas (for a review of Rb see Harbour and Dean, 2000). As with p53, the Rb-1 protein acts as an anti-oncogene by regulating the expression of specific target genes. Unlike p53, however, it appears to act primarily via protein–protein interactions with other tran-scription factors.

In particular, Rb-1 has been shown to interact with the cellular tran-scription factor E2F. E2F normally stimulates the transcription of several growth-promoting genes such as the cellular oncogenes c-*myc* and c-*myb* (see Section 9.4), and the genes encoding DNA polymerase α and thymi-dine kinase (for a review see Bracken *et al.*, 2004). The interaction of Rb-1 with E2F does not affect the ability of E2F to bind to its target sites in the

DNA but prevents it stimulating transcription. This is due to Rb-1 binding to the transcriptional activation domain of E2F and blocking its activity via a quenching type mechanism (see Section 8.3). In addition, however, it has been shown that Rb-1 can bind to both histone deacetylases (Luo *et al.*, 1998) and histone methylases (Ringrose and Paro, 2001) and thereby promote histone deacetylation and methylation. As discussed in Section 6.6, both these modifications promote a more closed chromatin structure incompatible with transcription.

Hence, Rb-1 can block the action of E2F both by inhibiting its ability to activate transcription and via promoting a closed chromatin structure (Fig. 9.34). Hence Rb-1 acts as an anti-oncogene by preventing the transcription of several growth-promoting genes, including oncogenes such as c-*myc*, which themselves encode transcription factors.

During the normal cell cycle of dividing cells, the Rb-1 protein becomes phosphorylated and this prevents its interacting with E2F, allowing the E2F factor to activate the growth-promoting genes whose protein products are necessary for cell cycle progression (Fig. 9.35a). Such an effect can also be achieved by the deletion of the Rb-1 gene or its inactivation by mutation which prevents the production of a functional Rb-1 protein (Fig. 9.35b). Similarly, it is also possible for such an absence of functional Rb-1 protein to arise from transcriptional inactivation of the Rb-1 gene which is otherwise intact. Thus six out of 77 retinoblastomas were found to have a heavily methylated Rb-1 gene which would result in a failure of its transcription (see Section 6.5), even though the gene itself was theoretically capable of encoding a functional protein. In addition, as with the p53 protein, the Rb-1 protein can be inactivated by protein–protein interaction with the oncogene products of DNA viruses. In this case, however, the association of the viral protein with Rb-1 dissociates the Rb-1–E2F complex, releasing free E2F which can then activate gene expression (Fig. 9.35c)

Thus, the Rb-1 protein plays a critical role in cellular growth regulation, modulating the expression of specific oncogenes and acting as a target for the transforming oncoproteins of specific viruses. Interestingly, Rb-1 has also been shown to interact with the RNA polymerase I transcription factor UBF (see Section 3.2) and thereby repress the transcription of the genes encoding ribosomal RNA. In addition, it can also repress the transcription of the 5S rRNA and tRNA genes by RNA polymerase III (see Section 3.2)

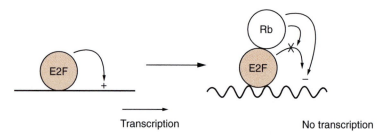

Transcription   No transcription

Figure 9.34

Following binding of Rb to the DNA-bound E2F transcription factor, transcription is inhibited since Rb both blocks the ability of E2F to activate transcription and promotes the organization of a tightly packed chromatin structure (wavy line) which does not allow transcription to occur.

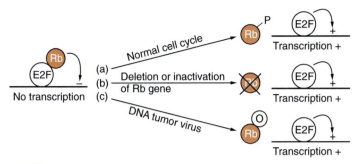

Figure 9.35

The retinoblastoma protein (Rb) binds to the E2F transcription factor and represses transcription. This inhibition can be relieved allowing E2F to activate transcription when: (a) the Rb protein is phosphorylated which occurs during the normal cell cycle and prevents its interacting with E2F; (b) the gene encoding Rb is deleted or inactivated by mutation; (c) the Rb protein binds to the product of a DNA tumor virus oncogene (o) which releases it from E2F.

by interacting with TFIIIB (for a review see White, 2004). Since transcription of the RNAs involved in protein synthesis is evidently necessary for cellular growth, Rb-1 can act as a global inhibitor of cellular growth by preventing the transcription of all the genes transcribed by RNA polymerases I and III as well as those of specific genes essential for growth by RNA polymerase II (Fig. 9.36). Interestingly, p53 can also inhibit RNA polymerase III transcription by targeting TFIIIB (Crighton *et al.*, 2003), suggesting that this may represent a common effect of different anti-oncogenes (for a review see White, 2004).

Rb-1 itself is inhibited by association with the Id2 transcription factor, a member of the inhibiting class of helix-loop-helix proteins which was discussed in Chapter 8 (Section 8.2). In tumor cells over-expressing the N-*myc* oncogene transcription factor (see Section 9.3), the Id2 gene is transcriptionally activated by N-*myc*. The excess Id2 binds to Rb-1 and inactivates it, allowing the tumor to grow (Lasorella *et al.*, 2000). This therefore represents another example of oncogenes and anti-oncogenes interacting antagonistically to regulate cellular growth, with the over-expression of an oncogene protein resulting in the inactivation of an anti-oncogene protein.

Together with p53, therefore, Rb-1 plays a key role in restraining cellular growth. Indeed many parallels exist between these two proteins. Thus, as well as being regulated by tumor virus proteins, Rb-1, like p53 (see above), is inhibited by interaction with the MDM2 protein. Moreover, acetylation of Rb-1 enhances its interaction with MDM2 (Chan *et al.*, 2001) indicating that both p53 and Rb-1 can be regulated via phosphorylation and acetylation.

In addition, there is evidence that the p53 and Rb-1 pathways interact with one another. Thus, the phosphorylation of Rb-1 which regulates its activity is carried out by the cyclin-dependent kinases. Hence, the action of p53 which stimulates the p21 inhibitor of these kinases will reduce their activity and hence maintain Rb-1 in its non-phosphorylated growth inhibitory form (Fig. 9.37).

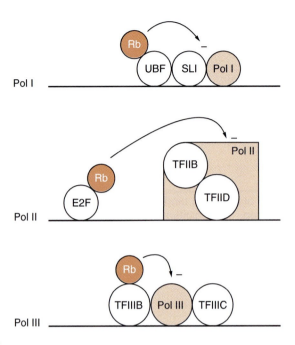

Figure 9.36

By binding to the polymerase transcription factor UBF, the polymerase II transcription factor E2F and the polymerase III transcription factor TFIIIB, Rb-1 can
repress transcription of the ribosomal RNA genes by RNA polymerase I, the transcription of E2F-dependent genes by RNA polymerase II and the transcription of
the tRNA and 5S rRNA genes by RNA polymerase III.

Other anti-oncogene transcription factors

In addition to p53, Rb-1 and the Wilms' tumor gene product, which are
well-defined transcription factors, it is probable that other anti-oncogenes
are also transcription factors. Thus, the products of the BRCA-1 and BRCA-
2 genes which are mutated in many cases of breast cancer appear to act
primarily by regulating the rate at which damaged DNA is repaired. They
also contain regions which can act as activation domains (see Section 8.3)
and it has therefore been suggested that they also act as transcriptional
regulators (for a review see Marx, 1997). In agreement with this idea,
BRCA-1 has been shown to interact with the C-terminal domain of RNA
polymerase II and modulate its phosphorylation which is critical for transcriptional elongation (Moisan *et al.*, 2004) (for discussion of the
C-terminal domain of RNA polymerase II and its role in transcriptional
elongation see Sections 3.2 and 4.3).

The vital role of p53, Rb-1 and the other anti-oncogene products, indicates that, although discovered later than the oncogenes, the
anti-oncogenes are likely to play as critical a role in regulating cellular
growth in general and the pattern of gene expression in particular. Hence,
the precise rate of cellular growth is likely to be controlled by the balance
between interacting oncogene and anti-oncogene products with cancer
resulting from a change in this balance by activation or over-expression
of oncogenes or inactivation of anti-oncogenes.

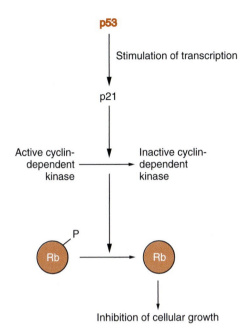

p53

Stimulation of transcription

p21

Active cyclin-dependent kinase → Inactive cyclin-dependent kinase

P

Rb → Rb

Inhibition of cellular growth

Figure 9.37

Activation of the gene encoding the p21 inhibitor of cyclin-dependent kinases by p53 results in the inhibition of these kinases which in turn maintains Rb-1 in its non-phosphorylated growth inhibitory form.

9.6   Oncogenes/anti-oncogenes: the relationship of cancer and normal cellular function

The key role of gene regulation in the control of normal development and functioning of the adult organism is indicated by the very large number of diseases which involve the mutation of genes encoding transcription factors or other proteins involved in gene regulation. However, the disease which most frequently results from alterations in gene regulation is cancer. Indeed, the study of the processes whereby increased expression or mutation of certain cellular genes or the deletion or inactivation of other genes can cause cancer has greatly increased our knowledge of the process of transformation, whereby normal cells become cancerous. Similarly, the recognition that the products of these cellular genes play a critical role in the growth regulation of normal cells has allowed insights obtained from studies of their activities in tumor cells to be applied to the study of normal cellular growth control.

Such a reciprocal exchange of information is well illustrated in the case of the oncogenes and anti-oncogenes that encode cellular transcription factors. Thus, the study of the Rb-1 and p53 proteins has greatly enhanced our knowledge of the processes regulating the transcription of genes whose protein products enhance or inhibit cellular growth.

Similarly, the isolation of the *fos* and *jun* genes in tumorigenic retroviruses has aided the study of the effects of growth factors on normal cells, while the recognition that the v-*erbA* gene product is a truncated form of the thyroid hormone receptor has allowed the elucidation of its role in transformation via the inhibition of erythroid differentiation. These two cases also illustrate the two mechanisms by which cellular genes can become oncogenic – namely mutation or over-expression. Thus, in the case of the v-erbA gene, mutations have rendered the protein different from the corresponding c-erbA gene from which it was derived. These alter the properties of the protein so it cannot bind thyroid hormone, and it behaves as a dominant repressor of transcription.

Interestingly, in a situation where an oncogene product and an anti-oncogene product interact, cancer can result from mutations which enhance the activity of the oncogenic protein or decrease the activity of the anti-oncogene protein. This is seen in the case of the adenomatous polyposis coli (APC) anti-oncogene protein (see Table 9.3), which is not a transcription factor and the β-catenin oncogene protein which plays a key role in cell–cell adhesion but also acts as a transcription factor (for reviews see Taipale and Beachy, 2001; Nelson and Nusse, 2004). These two factors normally interact and this interaction results in the export of β-catenin to the cytoplasm and its rapid degradation (Rosin-Arbesfeld et al., 2000; Fig. 9.38a). This prevents it from moving to the nucleus and interacting with the LEF-1 transcription factor which was discussed in Section 7.3, and stimulating its ability to activate transcription (for reviews see Hunter, 1997; Nusse, 1997). In normal cells, in response to specific signals, β-catenin is stabilized and can stimulate the ability of LEF-1 to activate transcription resulting in cellular proliferation (Fig. 9.38b). However, such stabilization can also be achieved either by the inactivation of APC by mutation or by mutation of β-catenin itself resulting in its enhanced stability (Fig. 9.38c).

Hence, normal cellular growth is controlled by interaction of the anti-oncogene product APC and the oncogene product β-catenin and can be stimulated by inactivating mutations in the anti-oncogenic partner or acti-vating mutations in the oncogenic partner. In addition to illustrating another example of oncogene activation by mutation, β-catenin also illus-trates a unique dual role, acting both as a factor critical for cell-to-cell adhesion and as an oncogenic transcription factor capable of regulating gene activity. Indeed, it is likely that this dual role allows it to transmit signals from cell adhesion components to the nucleus, resulting in changes in gene expression in response to extracellular events.

In contrast to the situation with oncogenes encoding proteins which pre-exist in an inactive form, and can be activated by mutation, onco-genes whose products are made only in response to a particular growth signal and whose activity then mediates cellular growth can cause cancer simply by the normal product being made at an inappropriate time. Thus, in the case of Fos or Jun, which are synthesized in normal cells in response to treatment with growth-promoting phorbol esters or growth factors, their continuous synthesis is sufficient to transform the cell. Interestingly, it has recently been shown that the c-Jun protein antagonizes the pro-apoptotic activity of p53 (see Section 9.5), providing a further example of proto-oncogene/anti-oncogene antagonism (Eferl et al., 2003).

Such cases of high-level expression of a normal oncogene product caus-ing cancer also occur in a number of cases without any evidence of retroviral involvement. These examples of alterations in cellular regula-tory processes producing increased expression of particular genes obviously provide another aspect to the connection between cancer and

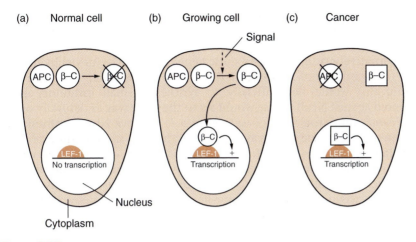

(a) Normal cell (b) Growing cell (c) Cancer

Figure 9.38

Interaction of the oncogene product β-catenin (β–C) and the anti-oncogene product APC (APC). In normal cells APC and β-catenin are associated in the cytoplasm and this results in the rapid degradation of β-catenin (panel a). In response to specific stimuli, however, β-catenin is stabilized and it can move to the nucleus where it interacts with the LEF-1 transcription factor (dark brown) and stimulates its ability to activate its target genes (panel b). This effect can also be achieved in cancer by the inactivation of APC by mutation or the activating mutation of β-catenin which converts it to a form (square) which is not susceptible to degradation. In either case the β-catenin moves to the nucleus and interacts with LEF-1 stimulating its activity (panel c).

gene regulation. Thus, information on the origin of the translocations of the c-*myc* gene in Burkitt's lymphoma is obviously important for the study of cancer etiology. Similarly, the fact that such translocations increase expression in some cases by removing elements that normally inhibit c-*myc* expression, allows the characterization of such negative elements and their role in regulating c-*myc* expression in normal cells.

Hence, the study of cellular oncogenes and anti-oncogenes has contributed greatly to our knowledge both of cancer and of cellular growth-regulatory processes, and is likely to continue to do so in the future.

9.7   Gene regulation and therapy of human disease

As the role of aberrant gene regulation in cancer and many other diseases is better understood, it is becoming clear that such increased understanding could lead to improved therapy for such diseases based on the manipulation of gene expression.

In some cases, such therapy might involve artificially increasing the expression of a regulatory protein. Thus, for example, as described in Section 8.4, activation of the NFκB transcription factor plays a key role in the immune response and blocking its activity would be of value in treating human diseases involving damaging inflammation. This could potentially be achieved by enhancing the levels of the inhibitory IκB protein either by identifying drugs which can switch on the patient's own

IκB gene or by using a gene therapy procedure to deliver exogenous copies of the IκB gene (for review of gene therapy procedures see French Anderson, 1998; 2000) (Fig. 9.39a).

Alternatively, a similar therapeutic benefit could be achieved by inhibiting the synthesis of NFκB itself (Fig. 9.39b). One potential method of doing this would be artificially to synthesize small inhibitory RNAs directed against NFκB. These would act in the same way as naturally occurring small inhibitory RNAs (see Section 5.7) to block NFκB synthesis (for reviews of the potential therapeutic applications of small inhibitory RNAs see Hannon and Rossi, 2004; Rossi, 2004).

Although these potential therapeutic approaches are of interest, they are not specific to gene regulatory proteins but could be used for any protein where increased or decreased expression was desired. Moreover, currently they suffer from the limitation that methods to deliver an exogenous gene or small inhibitory RNAs safely and efficiently to human patients are still

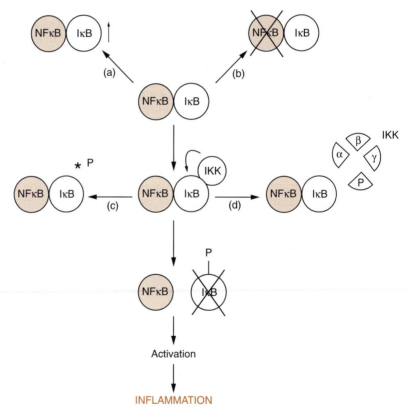

Figure 9.39

Potential therapeutic strategies for disrupting the activation of NFκB.
(a) Enhanced expression of IκB; (b) Reduced expression of NFκB; (c) and (d) Inhibiting the phosphorylation of IκB which is essential for its activation either by directly inhibiting phosphorylation using a drug such as salicylate (c) or by using a small peptide (P) to disrupt the interaction between the different proteins in the IκB kinase complex (IKK) which is responsible for phosphorylating IκB (see also Fig. 8.55).

being developed. Similarly, considerable effort will be required to identify drugs which can alter the expression of the endogenous gene encoding, for example, a transcription factor in a safe, specific and effective manner.

Interestingly however, a number of therapeutic drugs which are currently taken by many patients act by targeting transcription factors (for a review see Latchman, 2000). In many cases, these drugs were introduced on the basis of their efficacy in a particular situation and were only shown to act via a transcription factor many years later. Thus, for example, salicylate (aspirin), one of the most commonly used drugs, was introduced many years ago. Much more recently, it was shown that it can inhibit IκB phosphorylation, thereby promoting the association of IκB with NFκB and having an anti-inflammatory effect (Pierce et al., 1996) (Fig. 9.39c).

This example suggests that targeting the post-translational modification of transcription factors which is often essential for their activation (see Section 8.4) may have therapeutic potential. Indeed, when the mechanism of action of the commonly used anti-inflammatory drugs cyclosporin and FK506 (tarcolimus) was characterized, they were found to block the dephosphorylation of the NF-AT transcription factor which is required for an effective immune response.

As well as targeting post-translational modification, it is also possible to target protein–protein interactions which can also play a role in transcription factor activation (see Section 8.4). Thus, it is now relatively easy to map the region of a specific protein which interacts with another protein and to confirm this by structural studies. Once this has been achieved, small peptides can be prepared which mimic the binding region. These therefore compete for binding and can be used to disrupt the protein–protein interaction.

Thus, for example, a small peptide has been used to disrupt the protein–protein interactions in the IκB kinase (IKK) which is a multi-protein complex required for the phosphorylation of IκB. This peptide therefore blocked IkB phosphorylation (which is required for NFκB activation) and produced an anti-inflammatory effect in vivo (Jimi et al., 2004) (Fig. 9.39d).

Hence, a variety of potential therapeutic methods for inhibiting NFκB exist with salicylate treatment being used clinically (Fig. 9.39). This potential modulation of NFκB is of particular importance in view of its key role in inflammation and more recent indications that it is involved in certain types of cancer (for reviews see Balkwill and Coussens, 2004; Marx, 2004b; Perkins, 2004).

Although the identification of peptides which can disrupt protein–protein interactions is of value, their potential therapeutic application is limited by the ability to deliver them efficiently to the target cells in the patient's body. Ideally, it would be preferable to develop small diffusible molecules which would be simple to deliver and would disrupt specific protein–protein interactions. This has been achieved in the case of the interaction between the anti-oncogene p53 and the MDM2 which normally promotes degradation of p53 (see Section 9.5). Thus, structural studies indicated that part of the p53 protein inserts itself into a deep pocket in the MDM2 molecule. A chemical compound designed to fill this pocket was able to block p53/MDM2 binding and thereby activated p53 leading to reduced tumor growth both in culture and in the intact animal (Vassilev et al., 2004; for reviews see Lane and Fischer, 2004; Marx, 2004a)

(Fig. 9.40) (color plate 13). This type of drug which targets protein–protein interaction is therefore likely to be of value clinically, paralleling the use of drugs such as salicylate or cyclosporin/FK506 which target phosphorylation.

As well as transcription factor activity being regulated by phosphorylation or protein–protein interaction, it can also be regulated by ligand binding (see Section 8.4). This is seen particularly in the members of the steroid/thyroid hormone family of transcription factors which are activated by their appropriate ligand. This effect has been exploited therapeutically in different situations either to stimulate a receptor or inhibit it. Thus, for example, in breast cancers which are dependent on estrogen for their growth, therapy can involve the drug tamoxifen. This competes with estrogen for binding to the estrogen receptor but does not induce gene activation by the receptor following binding.

Conversely, as discussed above (Section 9.1), the PPARγ member of the steroid/thyroid hormone receptor family is mutated in a few patients with diabetes, indicating its normal role in preventing the disease. In the vast majority of patients, who have a normal PPARγ receptor and whose disease is due to other causes, the disease can be treated by stimulating the activity of the receptor. This is achieved by using synthetic drugs, known as thiazolidinediones, which bind to the receptor and stimulate its activity (for reviews see Kersten *et al.*, 2000; Rosen and Spiegelman, 2001; Evans *et al.*, 2004).

Hence, in different therapeutic applications, treatment can be carried out by either inhibiting the activity of a member of the steroid/thyroid hormone receptor family of transcription factors (Fig. 9.41a) or artificially stimulating it (Fig. 9.41b).

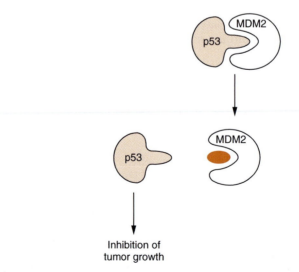

Figure 9.40

Binding of p53 to the MDM2 inhibitor involves a deep pocket in MDM2 into which p53 inserts itself. Filling this pocket with a specific synthetic chemical known as nutlin (solid), prevents binding of p53 to MDM2 and allows p53 to inhibit growth.

(a)   (b)

Inactive   Inactive   Active

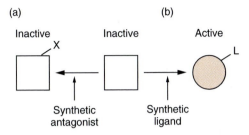

Synthetic   Synthetic
antagonist   ligand

Figure 9.41

Manipulating the activity of members of the steroid/thyroid hormone receptor family of transcription factors. Panel (a), the activity of a receptor can be inhibited by using a synthetic antagonist (X) which competes with its normal ligand for binding but which does not activate the receptor. Panel (b), the activity of a receptor can be stimulated by using a synthetic ligand (L) which binds to the receptor and activates it.

A similar therapeutic approach has also been used in cases of promyelocytic leukemia involving the oncogenic fusion protein linking the PML protein with the retinoic acid receptor γ (RARγ) protein, which is a member of the steroid/thyroid hormone receptor family. As discussed in Section 9.4, this fusion protein acts as a repressor of transcription, unlike the parental RARγ protein, which stimulates transcription following activation by retinoic acid.

One means of treating this form of leukemia, therefore, involves the administration of retinoic acid, to stimulate the gene activation properties of the retinoic acid receptor portion of the fusion protein and overcome the transcriptional inhibition normally produced by the fusion protein (Fig. 9.42a).

Although this treatment is effective in some cases, especially when combined with chemotherapy, it does not produce a long-term cure in many cases. This has led to the idea of an alternative therapy based on the finding that the RAR/PML fusion protein represses gene expression by recruiting histone deacetylases which produce an inactive chromatin structure (see Section 9.4). Thus, histone deacetylase inhibitors have been developed and may ultimately have a clinical role in leukemia caused by the RAR/PML fusion protein, acting by blocking the gene repression caused by the fusion protein (Fig. 9.42b).

As noted in Section 9.4, histone deacetylation is involved in gene repression induced by other oncogenic fusion proteins such as AML-ETO. Hence, this approach may be applicable in a number of different human leukemias (for reviews see Lin *et al.*, 2001; Minucci *et al.*, 2001). Indeed, the involvement of changes in chromatin structure, histone modification and DNA methylation in a number of human diseases (Section 9.1) indicates that therapies involving the manipulation of chromatin structure by chemicals which alter histone acetylation or DNA methylation may have widespread applicability in the treatment of cancer and other human diseases (for a review see Egger *et al.*, 2004).

Hence, a wide range of actual and potential therapies exist involving the manipulation of chromatin structure or of transcription factor activation, whether achieved by ligand binding, post-translational modification

(a)   (b)

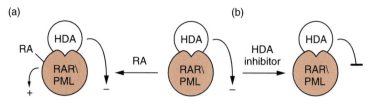

Figure 9.42

Alternative approaches to overcoming the transcription inhibitory effect of the RAR/PML fusion protein. Panel (a), treatment with retinoic acid stimulates gene activation by the RAR component. Panel (b), treatment with histone deacetylase inhibitors blocks the inhibiting effect of histone deacetylases which bind to the RAR/PML fusion protein. Compare with Fig. 9.27.

or protein–protein interaction. These methods are evidently targeted against a specific gene or genes which are regulated by a particular transcription factor or which show alteration in their chromatin structure in specific diseases.

However, another potential therapeutic approach exists which takes advantage of a specific property of transcription factors in order to manipulate the expression of any gene in the genome. Thus, as described in Section 8.2, the two cysteine–two histidine zinc finger has an α-helical region which contacts the DNA. It has been shown that the nature of the amino acids at the N-terminus of this α-helix determines the exact DNA sequence to which the zinc finger binds. Moreover, it is now possible to predict the exact DNA sequence which will be bound by a finger with particular amino acids at the N-terminus of the α-helix.

On the basis of this, zinc fingers can be designed which will bind to a specific target sequence that is present in a particular gene of interest. If such a zinc finger is introduced into cells, it will bind specifically to the target gene and not to all the other genes in the cell which do not contain its target sequence.

If this designer zinc finger is linked to the inhibitory domain of a transcription factor (see Section 8.3) it will deliver this domain to the target gene and specifically inhibit its transcription (Fig. 9.43a). Hence, this method offers an exciting means of specifically switching off a target gene. It has been used, for example, to inhibit the expression of specific viral genes and thereby block infection of cultured cells with viruses causing human disease such as herpes simplex virus or human immunodeficiency virus (Papworth et al., 2003; Reynolds et al., 2003).

As well as targeting viral gene expression, this method can also be used to target an individual cellular gene. Thus, Tan et al. (2003) used this approach to inhibit the expression of the CHK2 gene (which is involved in cell proliferation and cancer) and used gene chip technology (see Section 1.3) to show that whilst CHK2 mRNA levels were reduced, the levels of all other RNAs were unchanged.

Evidently, as well as using this approach to inhibit the expression of a single gene, it is also possible to activate specifically an individual gene by linking the designer finger to the activation domain of a transcription factor (see Section 8.3) (Fig. 9.43b). This approach has been used to activate the gene encoding the VEGF growth factor in the intact animal *in*

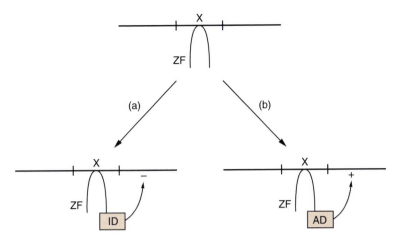

Figure 9.43

It is possible to design zinc fingers (ZF) which will bind specifically to a DNA sequence (X) in a target gene. Linking this finger to a transcriptional inhibitory domain can be used to specifically switch off the target gene (panel a), whilst linkage to a transcriptional activation domain can specifically switch on the target gene (panel b).

vivo. The enhanced levels of VEGF produced were functional and were able to produce increased blood vessel growth, indicating that this effect could be of therapeutic value in diseases where patients suffer from poor blood supply (for a review see Pasqualini *et al.*, 2002).

The effective use of this approach, like gene therapy or small inhibitory RNAs, requires the development of effective means of delivering the designer finger or the DNA encoding it to the patient. Nonetheless, it has considerable therapeutic potential, since unlike the other methods we have discussed, it can target any gene regardless of its normal method of regulation. The use of this method, together with the other more specific methods discussed in this section, thus offers real hope for improved therapy of many human diseases.

9.8 Conclusions

As discussed in this chapter, many human diseases including cancer are caused by alterations in the processes which regulate gene expression due to either the loss of a particular factor, its enhanced expression or to the presence of a mutant form. Often the study of such diseases can throw light not only on the mechanisms causing the disease itself but also on the processes regulating gene expression in normal cells which have been disrupted in the disease.

Ultimately, such an enhanced understanding of the processes normally regulating gene expression and the way they are altered in disease will lead to a new generation of therapies for human diseases. Thus, currently, many thousands of patients take therapeutic drugs such as salicylate, FK506 or tamoxifen which target transcription factors but all these drugs were isolated on the basis of their efficacy and their mechanism of action only determined subsequently. The recent identification of small molecules which can block the p53/MDM2 interaction and the production of designer zinc fingers targeted at particular genes offers hope of a new generation of therapeutic drugs designed specifically to target particular aspects of gene regulation.

References

Abbott, C.M. and Proud, C.G. (2004). Translation factors: in sickness and in health. *Trends in Biochemical Sciences* **29**, 25–31.

Balkwill, F. and Coussens, L.M. (2004). An inflammatory link. *Nature* **431**, 405–406.

Baudino, T.A. and Cleveland, J.L. (2001). The Max network gone mad. *Molecular and Cellular Biology* **21**, 691–702.

Baylin, S.B. and Herman, J.G. (2000). DNA hypermethylation in tumorigenesis: epigenetics joins genetics. *Trends in Genetics* **16**, 168–174.

Belandia, B. and Parker, M.G. (2003). Nuclear receptors: a rendezvous for chromatin remodeling factors. *Cell* **114**, 277–280.

Bracken, A.P., Ciro, M., Cocito, A. and Helin, K. (2004). E2F target genes: unravelling the biology. *Trends in Biochemical Sciences* **29**, 409–417.

Bradsher, J., Auriol, J., Proietti, de Santis, L., Iben, S., Vonesch, J.L. et al. (2002). CSB is a component of RNA pol I transcription. *Molecular Cell* **10**, 819–829.

Broach, J.R. and Levine, A.J. (1997). Oncogenes and cell proliferation. *Current Opinion in Genetics and Development* **7**, 1–6.

Celli, J., Duijf, P., Hamel, B.C.J., Bamshad, M., Kramer, B., Smits, A.P.T. et al. (1999). Heterozygous germline mutations in the p53 homologue p63 are the cause of EEC syndrome. *Cell* **99**, 143–153.

Chan, H.M., Krstic-Demonacos, M., Smith, L., Demonacos, C. and La Thangue, N. B. (2001). Acetylation control of the retinoblastoma tumour-suppressor protein. *Nature Cell Biology* **3**, 667–674.

Chuikov, S., Kurash, J.K., Wilson, J.R., Xiao, B., Justin, N., Ivanov, G.S. et al. (2004). Regulation of p53 activity through lysine methylation. *Nature* **432**, 353–360.

Conaway, J.W. and Conaway, R.C. (1999). Transcription elongation and human disease. *Annual Reviews of Biochemistry* **68**, 301–319.

Crighton, D., Woiwode, A., Zhang, C., Mandavia, N., Morton, J.P., Warnock, L.J. et al. (2003). p53 represses RNA polymerase III transcription by targeting TBP and inhibiting promoter occupancy by TFIIIB. *EMBO Journal* **22**, 2810–2820.

D'Arcangelo, G. and Curran, T. (1995). Smart transcription factors. *Nature* **376**, 292–293.

Davis, P.K. and Dowdy, S.F. (2001). p73. *International Journal of Biochemistry and Cell Biology* **33**, 935–939.

Ebralidze, A., Wang, Y., Petkova, V., Ebralidse, K. and Junghans, R.P. (2004). RNA leaching of transcription factors disrupts transcription in myotonic dystrophy. *Science* **303**, 383–387.

Eferl, R., Ricci, R., Kenner, L., Zenz, R., David, J.-P., Rath, M. and Wagner, E.F. (2003). Liver tumour development: c-Jun antagonises the proapoptotic activity of p53. *Cell* **112**, 181–192.

Egger, G., Liang, G., Aparicio, A. and Jones, P.A. (2004). Epigenetics in human disease and prospects for epigenetic therapy. *Nature* **429**, 457–463.

Eisenman, R.N. (2001). Deconstructing Myc. *Genes and Development* **15**, 2023–2030.

Engelkamp, D. and van Heyningen, V. (1996). Transcription factors in disease. *Current Opinion in Genetics and Development* **6**, 334–342.

Evans, R.M., Barish, G.D. and Wang, Y.X. (2004). PPARs and the complex journey to obesity. *Nature Medicine* **10**, 355–361.

Faustino, N.A. and Cooper, T.A. (2003). Pre-mRNA splicing and human disease. *Genes and Development* **17**, 419–437.

Feinberg, A.P. and Tycko, B. (2004). The history of cancer epigenetics. *Nature Reviews Cancer* **4**, 143–153.

Forrest, D. and Curran, T. (1992). Crossed signals : oncogenic transcription factors. *Current Opinion in Genetics and Development* **2**, 19–27.

French Anderson, W. (1998). Human gene therapy. *Nature* **392** (Suppl), 25–30.

French Anderson, W. (2000). Gene therapy scores against cancer. *Nature Medicine* **6**, 862–863.

Gibbons, R.J., Pickets, D.S., Villard, L. and Higgs, D.R. (1995). Mutations in a putative global transcriptional regulator cause X-linked mental retardation with a-Thalassaemia (ATR-X syndrome). *Cell* **80**, 837–845.

Grignani, F., De Matteis, S., Nervi, C., Tomassoni, L., Gelmetti, V., Cioce, M. et al. (1998). Fusion proteins of the retinoic acid receptor-alpha recruit histone deacetylase in promyelocytic leukaemia. *Nature* **391**, 815–818.

Hake, S.B., Xiao, A. and Allis, C.D. (2004). Linking the epigenetic 'language' of covalent histone modifications to cancer. *British Journal of Cancer* **90**, 761–769.

Hannon, G.J. and Rossi, J.J. (2004). Unlocking the potential of the human genome with RNA interference. *Nature* **431**, 371–378.

Harbour, J.W. and Dean, D.C. (2000). The Rb/E2F pathway: expanding roles and emerging paradigms. *Genes and Development* **14**, 2393–2409.

Hastie, N.D. (2001). Life, sex, and WT1 isoforms – three amino acids can make all the difference. *Cell* **106**, 391–394.

Haupt, Y., Robles, A.I., Prives, C. and Rotter, V. (2002). Deconstruction of p53 functions and regulation. *Oncogene* **21**, 8223–8231.

Hunter, T. (1997). Oncoprotein networks. *Cell* **88**, 333–346.

Insinga, A., Monestiroli, S., Ronzoni, S., Carbone, R., Pearson, M., Pruneri, G. et al. (2004). Impairment of p53 acetylation, stability and function by an oncogenic transcription factor. *EMBO Journal* **23**, 1144–1154.

Jimi, E., Aoki, K., Saito, H., D'Acquisto, F., May, M.J., Nakamura, I. et al. (2004). Selective inhibition of NF-kappa B blocks osteoclastogenesis and prevents inflammatory bone destruction in vivo. *Nature Medicine* **10**, 617–624.

Jin, P., Alisch, R.S. and Warren, S.T. (2004). RNA and microRNAs in fragile X mental retardation. *Nature Cell Biology* **6**, 1048–1053.

Jochum, W., Passegué, E. and Wagner, E.F. (2001). AP-1 in mouse development and tumorigenesis. *Oncogene* **20**, 2401–2412.

Karin, M., Liu, Z.-G. and Zandi, E. (1997). AP-1 function and regulation. *Current Opinion in Cell Biology* **9**, 240–246.

Kawahara, Y., Ito, K., Sun, H., Aizawa, H., Kanazawa, I. and Kwak, S. (2004). RNA editing and death of motor neurons. *Nature* **427**, 801.

Kersten, S., Desvergne, B. and Wahli, W. (2000). Roles of PPARs in health and disease. *Nature* **405**, 421–424.

Kitagawa, H., Fujiki, R., Yoshimura, K., Mezaki, Y., Uematsu, Y., Matsui, D. et al. (2003). The chromatin-remodeling complex WINAC targets a nuclear receptor to promoters and is impaired in Williams syndrome. *Cell* **113**, 905–917.

Kleinjan, D.A. and Van Heyningen, V. (2003). Turned off by RNA. *Nature Genetics* **34**, 125–126.

Knudson, A.G. (1993). Anti-oncogenes and human cancer. *Proceedings of the National Academy of Sciences of the USA* **90**, 10914–10921.

Lane, D.P. and Fischer, P.M. (2004). Turning the key on p53. *Nature* **427**, 789–790.

Lasorella, A., Noseda, M., Beyna, M. and Lavarone, A. (2000). Id2 is a retinoblastoma protein target and mediates signalling by Myc oncoproteins. *Nature* **407**, 592–598.

Latchman, D.S. (1996). Transcription factor mutations and disease. *New England Journal of Medicine* **334**, 28–33.

Latchman, D.S. (2000). Transcription factors as potential targets for therapeutic drugs. *Current Pharmaceutical Biotechnology* **1**, 57–61.

Levens, D.L. (2003). Reconstructing MYC. *Genes and Development* **17**, 1071–1077.

Lewin, B. (1991). Oncogenic conversion by regulatory changes in transcription factors. *Cell* **64**, 303–312.

Li, M., Brooks, C.L., Wu-Baer, F., Chen, D., Baer, R. and Gu, W. (2003). Mono- versus polyubiquitination: differential control of p53 fate by Mdm2. *Science* 302, 1972–1975.

Lin, R.J., Sternsdorf, T., Tini, M. and Evans, R.M. (2001). Transcriptional regulation in acute promyelocytic leukemia. *Oncogene* **20**, 7204–7215.

Liu, J., Akoulitchev, S., Weber, A., Ge, H., Chuikov, S., Libutti, D. *et al.* (2001). Defective interplay of activators and repressors with TFIIH in Xeroderma Pigmentosum. *Cell* **104**, 353–363.

Lohrum, M.A.E. and Vousden, K.H. (2000). Regulation and function of the p53-related proteins: same family, different rules. *Trends in Cell Biology* **10**, 197–202.

Look, A.T. (1997). Oncogenic transcription factors in the human acute leukaemias. *Science* **278**, 1059–1064.

Lund, A.H. and van Lohuizen, M. (2004). Epigenetics and cancer. *Genes and Development* **18**, 2315–2335.

Luo, J., Li, M., Tang, Y., Laszkowska, M., Roeder, R.G., and Gu, W. (2004). Acetylation of p53 augments its site-specific DNA binding both *in vitro* and *in vivo*. *Proceedings of the National Academy of Sciences of the USA* **101**, 2259–2264.

Luo, R.X., Postigo, A.A. and Dean, D.C. (1998). Rb interacts with histone deacetylase to repress transcription. *Cell* **92**, 463–473.

Marx, J. (1997). Possible function found for breast cancer genes. *Science* **276**, 531–532.

Marx, J. (2004a). Drug candidate bolsters cell's tumour defences. *Science* **303**, 23–25.

Marx, J. (2004b). Inflammation and cancer: the link grows stronger. *Science* **306**, 966–968.

Mayo, L.D. and Donner, D.B. (2002). The PTEN, Mdm2, p53 tumour suppressor-oncoprotein network. *Trends in Biochemical Sciences* **27**, 462–467.

Minucci, S., Nervi, C., Lo, Coco, F. and Pelicci, P.G. (2001). Histone deacetylases: a common molecular target for differentiation treatment of acute myeloid leukemias? *Oncogene* **20**, 3110–3115..

Moisan, A., Larochelle, C., Guillemette, B. and Gaudreau, L. (2004). BRCA1 can modulate RNA Polymerase II carboxy-terminal domain phosphorylation levels. *Molecular and Cellular Biology* **24**, 6947–6956.

Nelson, W.J. and Nusse, R. (2004). Convergence of Wnt, beta-catenin, and cadherin pathways. *Science* **303**, 1483–1487.

Nilsson, J.A. and Cleveland, J.L. (2003). Myc pathways provoking cell suicide and cancer. *Oncogene* **22**, 9007–9021.

Nucifora, F.C., Jr., Sasaki, M., Peters, M.F., Huang, H., Cooper, J.K., Yamada, M. *et al.* (2001). Interference by Huntingtin and Atrophin-1 with CBP-mediated transcription leading to cellular toxicity. *Science* **291**, 2423–2428.

Nusse, R. (1997). A versatile transcriptional effector of wingless signalling. *Cell* **89**, 321–323.

Okano, M., Bell, D.W., Haber, D.A. and Li, E. (1999). DNA methyltransferases Dnmt3a and Dnmt3b are essential for de novo methylation and mammalian development. *Cell* **99**, 247–257.

Papworth, M., Moore, M., Isalan, M., Minczuk, M., Choo, Y. and Klug, A. (2003). Inhibition of herpes simplex virus 1 gene expression by designer zinc-finger transcription factors. *Proceedings of the National Academy of Sciences of the USA* **100**, 1621–1626..

Pasqualini, R., Barbas, C.F. and Arap, W. (2002). Vessel manoeuvres: Zinc fingers promote angiogenesis. *Nature Medicine* **8**, 1353–1354.

Perkins, N.D. (2004). NF-kappaB: tumour promoter or suppressor? *Trends in Cell Biology* **14**, 64–69.

Perlmann, T. and Vennstrom, B. (1995). The sound of silence. *Nature 377*, 287.

Pierce, J.W., Read, M.A., Ding, H., Luscinskas, F.W. and Collins, T. (1996). Salicylates inhibit I kappa B-alpha phosphorylation, endothelial-leukocyte adhesion molecule expression, and neutrophil transmigration. *Journal of Immunology* **156**, 3961–3969.

Prives, C. and Manley, J.L. (2001). Why is p53 acetylated? *Cell* **107**, 815–818.

Rabbits, T.H. (1994). Chromosomal translocations in human cancer. *Nature 372*, 143–149.

Ranum, L.P. and Day, J.W. (2004). Pathogenic RNA repeats: an expanding role in genetic disease. *Trends in Genetics* **20**, 506–512.

Reynolds, L., Ullman, C., Moore, M., Isalan, M., West, M.J., Clapham, P. et al. (2003). Repression of the HIV-1 5′ LTR promoter and inhibition of HIV-1 replication by using engineered zinc-finger transcription factors. *Proceedings of the National Academy of Sciences of the USA* **100**, 1615–1620.

Ringrose, L. and Paro, R. (2001). Cycling silence. *Nature* **412**, 493–494.

Rosen, E.D. and Spiegelman, B.M. (2001). PPARgamma: a nuclear regulator of metabolism, differentiation, and cell growth. *Journal of Biological Chemistry* **276**, 37731–37734.

Rosin-Arbesfeld, R., Townsley, F. and Bienz, M. (2000). The APC tumour suppressor has a nuclear export function. *Nature* **406**, 1009–1012.

Rossi, J.J. (2004). A cholesterol connection in RNAi. *Nature* **432**, 155–156.

Schaffar, G., Breuer, P., Boteva, R., Behrends, C., Tzvetkov, N., Strippel, N. et al. (2004). Cellular toxicity of polyglutamine expansion proteins; mechanism of transcription factor deactivation. *Molecular Cell* **15**, 95–105.

Shachaf, C.M., Kopelman, A.M., Arvanitis, C., Karlsson, A., Beer, S., Mandl, S. et al. (2004). MYC inactivation uncovers pluripotent differentiation and tumour dormancy in hepatocellular cancer. *Nature* **431**, 1112–1117.

Sharpless, N.E. and DePinho, R.A. (2002). p53: Good Cop/Bad Cop. *Cell* **110**, 9–12.

Shaulian, E. and Karin, M. (2002). AP-1 as a regulator of cell life and death. *Nature Cell Biology* **4**, E131–E136.

Sugars, K.L. and Rubinsztein, D.C. (2003). Transcriptional abnormalities in Huntington disease. *Trends in Genetics* **19**, 233–238.

Taipale, J. and Beachy, P.A. (2001). The Hedgehog and Wnt signalling pathways in cancer. *Nature* **411**, 349–354.

Takeya, T. and Hanafusa, H. (1985). Structure and sequence of the cellular gene homologous to the RSV-*src* gene and the mechanism for generating transforming virus. *Cell* **32**, 881–890.

Tan, S., Guschin, D., Davalos, A., Lee, Y.-L., Snowden, A.W., Jouvenot, Y. et al. (2003). Zinc-finger protein-targeted gene regulation: genomewide single-gene specificity. *Proceedings of the National Academy of Sciences of the USA* **100**, 11997–12002.

Trottier, Y., Dewys, D. and Mandel, J.L. (1993). Fragile X syndrome: An expanding story. *Current Biology* **3**, 783–786.

Tufarelli, C., Stanley, J.A., Garrick, D., Sharpe, J.A., Ayyub, H., Wood, W.G. and Higgs, D.R. (2003). Transcription of antisense RNA leading to gene silencing and methylation as a novel cause of human genetic disease. *Nature Genetics* **34**, 157–165.

Valdez, B.C., Henning, D., So, R.B., Dixon, J. and Dixon, M.J. (2004). The Treacher Collins syndrome (TCOF1) gene product is involved in ribosomal

DNA gene transcription by interacting with upstream binding factor. *Proceedings of the National Academy of Sciences of the USA* **101**, 10709–10714.

Vassilev, L.T., Vu, B.T., Graves, B., Carvajal, D., Podlaski, F., Filipovic, Z. *et al.* (2004). In vivo activation of the p53 pathway by small-molecule antagonists of MDM2. *Science* **303**, 844–848.

Vogelstein, B. and Kinzler, K.W. (1992). p53 function and dysfunction. *Cell* **70**, 523–526.

Vogelstein, B., Lane, D. and Levine, A.J. (2000. Surfing the p53 network. *Nature* **408**, 307–310.

Wang, L., Fan, C., Topol, S.E., Topol, E.J. and Wang, Q. (2003). Mutation of MEF2A in an inherited disorder with features of coronary artery disease. *Science* **302**, 1578–1581.

Weinberg, R.A. (1993). Tumour suppressor genes. *Neuron* **11**, 191–196.

White, R.J. (2004). RNA polymerase III transcription and cancer. *Oncogene* **23**, 3208–3216.

Yang, A., Kaghad, M., Caput, D. and Mckeon, F. (2002). On the shoulders of giants: p63, p73 and the rise of p53. *Trends in Genetics* **18**, 90–95.

Yang, Y., Li, C.-C. H. and Weissman, A.M. (2004). Regulating the p53 system through ubiquitination. *Oncogene* **23**, 2096–2106.

Zhang, J., Kalkum, M., Yamamura, S., Chait, B.T. and Roeder, R.G. (2004). E protein silencing by the leukemogenic AML1-ETO fusion protein. *Science* **305**, 1286–1289.

Conclusions and future prospects

10

The extraordinary rate of progress in the study of gene regulation can be gauged from the fact that by 1980 no transcriptional regulatory protein or its DNA binding site in a regulated promoter had been defined. The tremendous advances since that time were evident by the time the first edition of this book was published. Thus, at that time (1990), a relatively clear picture of the action of individual factors which regulate gene expression was available. For example, in the case of gene induction by glucocorticoid hormone it was known that these agents act by activating specific receptor proteins (Section 8.2), that such activated receptors bind to specific DNA sequences upstream of the target gene (Section 7.2), displacing a nucleosome (Section 6.7) and that a region in the receptor protein then interacts with a factor bound to the TATA box to activate transcription (Section 8.3).

Clearly, therefore, the process by which a single agent can activate a specific transcription factor and thereby modulate gene expression was reasonably well understood in outline. However, in the period from 1990 to the publication of the second edition of this book in 1995, it became increasingly clear that the activation of a single transcription factor by a single agent cannot be assessed in isolation. Rather, the activity of a specific factor will depend upon its interaction with other transcription factors as well as with other proteins. Thus, for example, glucocorticoid hormone activates transcription by promoting the dissociation of the glucocorticoid receptor from the inhibitory hsp90 protein which otherwise anchors it in the cytoplasm and prevents it from moving to the nucleus and activating transcription (see Section 8.4).

A similar example of protein–protein interactions regulating transcription factor activity, but in the opposite direction, is observed in the Fos and Jun proteins which regulate gene expression in response to phorbol ester treatment (Section 9.4). Thus, as noted in Section 8.2, the Fos protein cannot bind to DNA on its own but can do so only after forming a heterodimer with the Jun protein. This heterodimer then binds to DNA and activates transcription. Hence, in this case, the interaction with another protein allows Fos to activate transcription whereas in the case of the glucocorticoid receptor the interaction with hsp90 has the opposite effect.

Such types of protein–protein interactions regulating transcription factor activation are not confined to the regulation of gene expression by short-term inducers but are also involved in the regulation of tissue-specific gene expression. Thus, as discussed in Section 8.2, the MyoD transcription factor can bind to the promoters of muscle-specific genes and activate their expression, thereby causing the differentiation of a fibroblast cell line into muscle cells. The production of differentiated skeletal muscle cells in this situation requires, however, that the cells are cultured in the absence of growth factors. In the presence of growth factor, the cells contain high levels of the inhibitory factor Id which heterodimerizes with MyoD and prevents it binding to DNA and activating transcription. The production of the skeletal muscle phenotype therefore requires not only high levels of the activating MyoD protein but also a fall in the levels of the inhibitory Id protein which is induced by growth factor removal.

It is clear, therefore, that in families of transcription factors which are capable of dimerization by means of specific motifs, such as the leucine zipper or helix-loop-helix (see Section 8.2), such dimerization with another member of the family can result in activation (Fos-Jun) or repression (MyoD-Id).

The Id example also illustrates another theme which has become of increasing importance in gene regulation, namely the importance of specific inhibition as opposed to activation of gene expression (see Section 8.3). As well as transcription factors, such as the Id factor, which function only as inhibitors, cases also exist where the same factor can also function as a direct activator or repressor depending on the circumstances. This is seen in the case of the thyroid hormone receptor which, unlike the glucocorticoid receptor, binds to its DNA target site even in the absence of hormone. As discussed in Section 8.3, this factor can directly inhibit promoter activity in the absence of thyroid hormone, whereas in the presence of thyroid hormone it undergoes a conformational change which allows it to activate gene expression.

This case of the thyroid hormone receptor illustrates two of the major themes to have emerged between the publication of the second edition of this work and the present day: namely the role of co-repressor/co-activator molecules and the influence of transcription factors on chromatin structure. Thus, as discussed in Section 8.3, the inhibitory domain of the thyroid hormone receptor acts indirectly by recruiting an inhibitory molecule known as the nuclear receptor co-repressor which actually produces the inhibitory effect. Although this factor may interact directly with the basal transcriptional complex, it has also been shown itself to recruit another molecule which has the ability to acetylate histones (see Section 6.6). As such histone acetylation is associated with a tightly packed chromatin structure incompatible with transcription (see Section 6.6), the inhibitory effect of the thyroid hormone receptor appears to be mediated, at least in part, by the ability of its co-repressor to recruit a molecule which can organize the chromatin into a non-transcribable form.

As discussed in Section 8.3, the addition of thyroid hormone results in a conformational change in the thyroid hormone receptor which results in the release of its co-repressor and allows the receptor to activate transcription. Interestingly, as with transcriptional repression by the receptor, such activation requires a co-factor and involves alterations in histone acetylation. Thus, following exposure to thyroid hormone, the thyroid hormone receptor can bind the CBP co-activator. This factor was originally identified as a co-activator which binds to the CREB transcription factor but was subsequently shown to be involved in transcriptional activation by nuclear receptors such as the glucocorticoid receptor and the thyroid hormone receptor (see Section 8.3).

As discussed in Section 8.3, CBP is able to act as a co-activator by inter-acting with components of the basal transcriptional complex to stimulate its activity. In addition, however, it also has histone deacetylase activity (see Section 6.6) indicating that by deacetylating histones it can reverse the effects of the nuclear receptor/co-repressor complex and convert the chromatin into a form where transcription can occur. Hence, the addition of thyroid hormone converts the thyroid hormone receptor from a form which can bind an inhibitory co-repressor complex with histone acetylase activity to a form where it can bind a stimulatory co-activator with histone deacetylase activity.

This example thus indicates the critical importance of co-activators and co-repressors, as well as providing a critical link between the transcription factors discussed in Chapter 8 and the chromatin structure changes discussed in Chapter 6. Indeed, it has become increasingly clear that modi-fication of histones by transcriptional regulators is a key event in the modulation of chromatin structure, leading to transcriptional activation or repression. As discussed in Section 6.6, such histone modifications include not only acetylation but also methylation, phosphorylation, ubiquitination and sumoylation with the existence of one modification at a particular amino acid, influencing positively or negatively the occur-rence of another modification at another amino acid. This has led to the idea of a "histone code" in which a particular pattern of modification on the histones in a particular region of DNA promotes or inhibits the changes necessary for transcription to occur by affecting the recruitment of chromatin remodeling complexes and DNA binding transcriptional activators or repressors.

Interestingly, a number of co-activator/co-repressor molecules appear to interact respectively with a number of different activating or inhibitory transcription factors. In the case of the CBP co-activator which is used by transcription factors activated by different signaling pathways, this has been shown to result in interactions between these pathways. Thus, both the glucocorticoid receptor which is activated by steroids and the AP1 (Fos-Jun) complex which is activated by phorbol esters require the CBP co-activator in order to mediate this stimulatory effect on gene transcrip-tion. Because CBP levels in the cell are relatively low, when both pathways are activated simultaneously, they will compete for the limited amount of CBP and neither pathway will be able to activate transcription. Hence, the requirement of both of these pathways for CBP results in a mutual inhi-bition of gene activation by each of these factors.

This example illustrates how the effect of one signaling pathway can be affected by the activation of another signaling pathway. Hence, in the absence of glucocorticoid hormone, the glucocorticoid receptor will be anchored in the cytoplasm by hsp90 and the Fos–Jun complex will be able to interact with CBP and activate gene expression in response to growth factors or phorbol esters. However, when the glucocorticoid receptor is freed from hsp90 by the presence of glucocorticoid, it can interact with CBP removing it from Fos–Jun and preventing them activating gene expression. This is seen in the case of the gene encoding the collagenase enzyme which is activated by the Fos–Jun complex in response to phor-bol esters and can produce severe tissue destruction in an inflamed area. As expected on the basis of the above model, the activation of this gene by phorbol esters is inhibited by treatment with glucocorticoid hormone, accounting for the anti-inflammatory effort of this steroid hormone. Conversely, the activation of glucocorticoid-responsive genes by hormone treatment will be inhibited by the presence of high levels of Fos and Jun induced by growth factor or phorbol ester treatment since they will compete for the limited amount of available CBP.

These examples indicate, therefore, that the effect of a particular factor on gene expression will depend on the nature of the other factors present in the cell and the state of activation of specific signaling pathways. Moreover, it is clear that our simple picture of the glucocorticoid receptor acting in isolation has to be modified to consider its inhibition by inter-action with hsp90, as well as its need to recruit other factors in order to stimulate transcription such as CBP and, as discussed in Section 6.7, the SWI-SNF complex which also reorganizes chromatin structure.

Over the last few years it has become clear that the regulation of a single gene in an intact cell is vastly more complex than could be predicted from the study of an isolated transcription factor interacting with its binding site. At first sight it may appear that recent progress in this area has merely complicated the issue. However, the process of gene regulation must not only produce all the different types of cells in the body but also ensure that each cell type is produced in the right place and at the right time during development.

It is inevitable, therefore, that gene regulation must be a highly complex process with numerous factors interacting with each other and their respective DNA binding sites to activate or inhibit the expression of specific genes in specific cell types. Moreover, such transcriptional control is significantly supplemented by post-transcriptional control processes such as RNA splicing or RNA editing which can produce multiple mRNAs encoding related proteins from a single gene, as well as by the regulation of RNA stability or RNA translation which can produce a rapid change in protein levels (see Chapter 5). Indeed, it is becoming increasingly clear that the processes which control transcriptional initiation and elongation, interact with post-transcriptional processes such as capping, splicing and polyadenylation to ensure the final mRNA is produced in an effective and coordinated manner (Sections 3.3 and 4.3).

Interestingly, the CBP factor which, as discussed above, plays a critical role in regulating gene activation in response to specific signaling pathways also appears to be involved in gene regulation during development. Thus, as discussed in Section 9.1, mutations in the CBP gene in humans result in Rubinstein-Taybi syndrome, which is characterized by abnormal development leading to mental retardation and physical abnormalities. Interestingly, this disease occurs when only one of the two copies of the CBP gene is mutated with the other being capable of producing functional CBP protein and individuals with two mutant CBP genes have never been identified. This indicates that CBP is likely to be of such importance that its complete loss by inactivation of both CBP genes is incompatible with survival. Moreover, as in the signaling processes described above, the precise level of CBP in a cell is critical for proper development and a single functional gene evidently cannot produce sufficient CBP for normal development to occur.

The central role of CBP in many different signaling pathways and in development together with the critical role of the basal transcription factor TBP in transcription by all three RNA polymerases (see Section 3.2) has often made gene regulation in the last few years appear to be "The study of CBP and TBP". Of course, despite the central role of these factors, it is far more complex than that. Nonetheless, the progress in understanding gene regulation in terms of factor–factor interactions, changes in chromatin structure, etc., which has taken place in the last few years indicates that an understanding of the ultimate problem of mammalian development in terms of differential gene expression can ultimately be achieved. Similarly, our enhanced understanding of the role of aberrant gene regulation in human cancer and other diseases (see Chapter 9) offers hope of improved therapies for such diseases based on the artificial manipulation of gene expression in human patients.

INDEX

Page numbers in **bold** type refer to illustrations or methods boxes; those in *italic* to tables. Note that prefixes are ignored in alphabetization; c-*myc*, for example, appears under the letter 'm'.